Studies of Vortex Domina

Studies of
Vortex Dominated Flows

Proceedings of the Symposium on Vortex Dominated Flows
Held July 9–11, 1985, at NASA Langley Research Center,
Hampton, Virginia

Edited by
M.Y. Hussaini and M.D. Salas

With 219 Figures

Springer-Verlag
New York Berlin Heidelberg
London Paris Tokyo

M.Y. Hussaini
M.D. Salas

Institute for Computer Applications
  in Science and Engineering (ICASE)
ICASE/NASA
NASA Langley Research Center
Hampton, Virginia 23665 U.S.A.

Library of Congress Cataloging-in-Publication Data
Symposium of Vortex Dominated Flows (1985: NASA
   Langley Research Center)
   Studies of vortex dominated flows.
   Proceedings of the symposium conducted by NASA
Langley Research Center and the Institute for
Computer Applications in Science and Engineering.
   1. Vortex-motion—Congresses.   2. Fluid dynamics—
Congresses.   I. Hussaini, M. Yousuff.   I. Salas,
M.D. III. Langley Research Center.   IV. Institute
for Computer Applications in Science and Engineering.
V. Title.
QA925.S94   1985   532'.0595   86-26058

Printed and bound by R.R. Donnelley & Sons, Harrisonburg, Virginia.
Printed in the United States of America.

9 8 7 6 5 4 3 2 1

ISBN 0-387-96430-4 Springer-Verlag New York Berlin Heidelberg
ISBN 3-540-96430-4 Springer-Verlag Berlin Heidelberg New York

Introduction

From the astrophysical scale of a swirling spiral galaxy, through the geophysical scale of a hurricane, down to the subatomic scale of elementary particles, vortical motion and vortex dynamics have played a profound role in our understanding of the physical world. Kuchemann referred to vortex dynamics as "the sinews and muscles of fluid motion." In order to update our understanding of vortex dominated flows, NASA Langley Research Center and the Institute for Computer Applications in Science and Engineering (ICASE) conducted a workshop during July 9–11, 1985. The subject was broadly divided into five overlapping topics— vortex dynamics, vortex breakdown, massive separation, vortex shedding from sharp leading edges and conically separated flows. Some of the experts in each of these areas were invited to provide an overview of the subject. This volume is the proceedings of the workshop and contains the latest, theoretical, numerical, and experimental work in the above-mentioned areas.

Leibovich, Widnall, Moore and Sirovich discussed topics on the fundamentals of vortex dynamics, while Keller and Hafez treated the problem of vortex breakdown phenomena; the contributions of Smith, Davis and LeBalleur were in the area of massive separation and inviscid-viscous interactions, while those of Cheng, Hoeijmakers and Murman dealt with sharp-leading-edge vortex flows; and Fiddes and Marconi represented the category of conical separated flows.

The opening article of this volume by Leibovich deals with the principal features of weakly nonlinear bending waves (in the form of solitons) on infinitely long, initially straight vortex filaments. Such studies, while of interest in their own right, also provide possible insight into the highly nonlinear vortex breakdown phenomena. In this article, in outline form, Leibovich treats the global bifurcation of axially symmetric, steady, inviscid vortex flows, and suggests the connection with Benjamin's vortex breakdown theory.

Widnall gives a brief review of the linear stability theory for concentrated vortex structures. She distinguishes between three types of instability—two-dimensional, three-dimensional long-wave and three-dimensional short-wave instability. Then she analyzes the three-dimensional instability of a single simple vortex shedding from a cylinder and the Foppl vortices modelling the flow behind the cylinder. Such instability mechanisms are proposed as sources of three-dimensionality in separated and turbulent flows.

Sirovich and Lim provide a historical background for Karman vortex street, and they interpret it afresh in the light of a recent experiment on the flow behind a circular cylinder. This experiment relates the flow structure to that of a low order nonlinear dynamical system. They give a fairly complete treatment to the initial value problem for the linear evolution of an initial perturbation to the Karman vortex trail. They find "encouraging" similarities between their theoretical results and some experiments for the rotation number (the ratio of the second frequency to the shedding frequency) behavior with Reynolds number. Such a similarity is also found for the phenomenon of wave propagation along the trail.

The relevance of vortex sheets to separated flows and the origin of turbulent shear flows is very well established. In their article, de Bernardinis and Moore discuss the ring-vortex representation of an axi-symmetric vortex sheet for practical purposes. This is an extension of Van der Vooren's procedure for two-dimensional vortex sheets. This representation loses accuracy if the vortex sheet intersects the axis of symmetry. This loss of accuracy is discussed in the case of an instantaneously spherical vortex sheet.

The work of Keller et al. extends Benjamin's variational principle for axi-symmetric flows to include free-surface flows. They apply it to a Rankine vortex, and show a loss-free transition between two vortex states, and discuss its relevance to a certain type of vortex breakdown.

Hafez and Salas present results (based on two entirely different numerical methods) which show that the pertinent equations governing a steady axi-symmetric inviscid flow with swirl yield solutions with closed streamlines. They also present solutions to the steady axi-symmetric Navier–Stokes equations, but the Reynolds numbers calculated are too low to allow any conclusions on whether the inviscid solutions obtained are limiting solutions of the viscous problem.

The high-Reynolds-number, large-scale separated flows have been one of the central problems of fluid dynamics. The theory is far from complete, and the full-scale computations are not absolutely convincing. Smith presents a succinct review of the available theory for two-dimensional incompressible flows with massive separation and discusses their properties and numerical solutions. He adopts the view that since small-scale separated flows can be described completely within the framework of triple deck theory, they can be used to get an insight into the physical mechanisms involved in massive separations. The work of Rothmayer is such an attempt. The paper of LeBalleur is a fairly comprehensive review of the status of inviscid-viscous interaction procedures for the computation of massively separated flows.

Hoeijmakers reviewed the prevalent computational methods for the simulation of aerodynamic flow configurations involving a leading-edge vortex. These methods are categorized into rigid-vortex methods, fitted-vortex methods and captured-vortex methods. The rigid-vortex methods consist of classical potential flow techniques (such as vortex-lattice methods) which incorporate empirical concepts to account for vortical interactions without resolving the vortical flow details. Fitted-vortex methods fix vortex sheets in potential flow models to allow for vortical interactions and include discrete vortex methods and nonlinear vortex lattice methods. Captured-vortex methods are based on Euler equations or Navier–Stokes equations. A very useful comparative study of these methods is given. Cheng et al. employ two types of fitted-vortex methods to simulate the vortex dynamics of a leading edge flap, while Murman and Powell study the leading edge vortex shed from a delta wing. The latter solve the Euler equations by Jameson's finite-volume technique on a rather fine grid. They obtain good agreement with the measured pitot pressures.

Fiddes and Marconi both consider separated flows about cones at incidence. Fiddes focuses on modelling the vortical flow by line vortex models and vortex sheet models. A major interest of this work is in understanding the development of asymmetry in the flow field. Marconi solves the Euler equations by Moretti's lambda-scheme, and fits both the bow shock and the cross-flow shock. This work is one of the most systematic studies of the conically separated flows on record.

MYH, MDS

Contents

Section IV Vortex Shedding From Sharp Leading Edges

Section V Conically Separated Flows

Contributors

de Bernardinis, B.
Instituto di Idraulica
University of Genoa, Italy

Cheng, H.K.
Department of Aerospace
  Engineering
University of Southern California
Los Angeles, CA 90089, USA

Davis, R.T.
Department of Aerospace
  Engineering and Applied
  Mechanics
University of Cincinnati
Cincinnati, OH 45221, USA

Edwards, R.H.
Department of Aerospace
  Engineering
University of Southern California
Los Angeles, CA 90089, USA

Egli, W.
Brown Boveri Research Centre
CH 5405 Badeu, Switzerland

Exley, J.
Department of Mathematics
Imperial College
London SW7, UK

Fiddes, S.P.
Aerodynamics Department
Royal Aircraft Establishment
Farnborough
Hants GU14 6TD, UK

Hafez, M.M.
Department of Mechanical
  Engineering
University of California at Davis
Davis, CA 95616, USA

Hoeijmakers, H.W.M.
National Aerospace Lab NLR
Amsterdam–1017
The Netherlands

Jia, Z.X.
Department of Aerospace
  Engineering
University of Southern California
Los Angeles, CA 90089, USA

Keller, J.J.
Brown Boveri Research Centre
CH 5405 Badeu, Switzerland

LeBalleur, J.C.
Office National d'Etudes et de
  Recherches Aerospatiales
  (ONERA)
92320 Chatillon, France

Leibovich, Sidney
Sibley School of Mechanical and
    Aerospace Engineering
Cornell University
Ithaca, New York 14853, USA

Lim, C.
Division of Applied Mathematics
Brown University
Providence, RI 02912, USA

Marconi, F.
Grumman Aerospace
M/S A08-35
Research and Development
Bethpage, NY 11714, USA

Moore, D.W.
Department of Applied Mathematics
Imperial College
London, SW7, UK

Murman, Earll M.
Computational Fluid Dynamics
    Laboratory, Department of
    Aeronautics and Astronautics
Massachusetts Institute
    of Technology
Cambridge, MA 02139, USA

Powell, Kenneth G.
Computational Fluid Dynamics
    Laboratory, Department of
    Aeronautics and Astronautics
Massachusetts Institute
    of Technology
Cambridge, MA 02139, USA

Rothmayer, A.P.
Department of Aerospace
    Engineering
Iowa State University,
Ames, Iowa 50010 USA

Salas, M.D.
NASA Langley Research Center
Hampton, VA 23665, USA

Sirovich, L.
Division of Applied Mathematics
Brown University
Providence, RI 02912, USA

Smith, F.T.
Department of Mathematics
University College London
London WC1E 6BT, U.K.

Widnall, Sheila E.
Department of Aeronautics and
    Astronautics
Massachusetts Institute
    of Technology
Cambridge, MA 02139, USA

Section I

Vortex Dynamics

Waves and Bifurcations in Vortex Filaments

SIDNEY LEIBOVICH

ABSTRACT

Weakly nonlinear bending waves on vortex filaments are briefly described. These take the form of solitons governed by the cubically nonlinear Schrodinger equation. In addition, conditions leading to the global bifurcation of inviscid, axially-symmetric vortices are outlined.

1. Introduction

Concentrated vortices, such as those commonly found in flows of aerodynamical interest, allow waves to propagate along their cores much like water waves on a running stream. In addition, vortices often are generated which, due to instabilities, cannot persist without change of form. In this paper, we consider inviscid phenomena that may happen in infinitely long, initially straight vortex filaments having vortex cores with prescribed structures. These idealized vortices are assumed to exist in an unbounded incompressible fluid.

This work is motivated in part by the phenomenon of vortex breakdown (see Leibovich (1984)). These coherent, highly nonlinear, disruptions of nearly columnnar vortex flows (that is, flows with cylindrical streamsurfaces) do not yet have a satisfactory theoretical description. At their best, existing theories for waves on vortices provide a description of weakly nonlinear motions, and therefore cannot be expected to amount to a fully acceptable theory for vortex breakdown. Despite this, it has been argued (Faler and Leibovich, 1977; Leibovich, 1983, 1984; Maxworthy et al., 1983; Escudier et al., 1982) that wave theories have provided useful insights concerning vortex breakdown events. Those wave theories which have been heretofore invoked in relation to this phenomenon admit only axially symmetric disturbances, even though it is known that vortex breakdown does not have this symmetry. In particular, "bending waves", those which cause the vortex centerline to move radially, are known to exist in all forms of vortex breakdown (Leibovich, 1984). The first part of this paper concerns weakly nonlinear bending waves: these, of course, are of interest beyond their possible connections to vortex breakdown, as the recent work of Hopfinger et al. (1984) and Maxworthy et al. (1985) attest.

A second topic addressed here concerns the bifurcation of a class of fully nonlinear vortex flows. We ask when multiple solutions can occur for steady, axially symmetric vortex flows with specified boundary data. The method used is global in the sense that the existence of multiple solutions, and the location of the bifurcation criterion are established without use of local expansions.

The treatment of both topics of this paper will be in outline form only; more detailed presentations of the subject matter will be given elsewhere.

2.   Bending waves and spiral solitons

We deal with the Euler equations and use cylindrical $(r,0,z)$ coordinates. Any columnar flow with velocity vector

$$\underset{\sim}{U}(r) = (0,V(r),W(r)) \tag{1}$$

as a possible inviscid motion. Perturbations to this flow are considered with velocity vectors

$$\underset{\sim}{v} = U(r) + \varepsilon\underset{\sim}{u}(r,\theta,z,t;\varepsilon);$$

$$\underset{\sim}{u} = (u,v,w) = \underset{\sim}{u}_0(r,\theta,z,t,\tau,Z) + \varepsilon\underset{\sim}{u}_1 + \varepsilon^2\underset{\sim}{u}_2 + \cdots , \tag{2}$$

where ε is a small amplitude parameter and dots stand for higher order terms which will not be considered. The arguments τ and Z are slow time and space variables to be specified; they are included in anticipation of the need for a multiple scaling analysis in the accounting for nonlinearities.

2.1 Linear analysis

The assumed expansion is substituted into the Euler equations and terms higher order than those linear in ε are neglected, leaving a problem to be solved for $\underset{\sim}{u}_0$. All disturbance quantities are assumed to be in normal mode form:

$$\underset{\sim}{u}_0 = \underset{\sim}{u}_0(r) \exp i(kz + m\theta - \omega t) . \tag{3}$$

and we let the radial component of $\underset{\sim}{u}_0$ be f. The modal amplitude f(r) satisfies a second order differential equation first given by Howard and Gupta (1962). This equation may be compactly written as follows

$$D(SD_*f) - (1 + \frac{a}{\gamma} + \frac{b}{\gamma^2})f = 0 \tag{4}$$

where

$$\gamma = \underset{\sim}{k}\cdot\underset{\sim}{U} - \omega; \quad \underset{\sim}{k} = mr^{-1}\underset{\sim}{e}_\theta + k\underset{\sim}{e}_z; \quad S = 1/|\underset{\sim}{k}|^2;$$

$$a = r^2 \underset{\sim}{V} \cdot [r^{-2}(\underset{\sim}{k} \times \underset{\sim}{\zeta})/|\underset{\sim}{k}|^2]; \quad b = -2(\underset{\sim}{k} \cdot \underset{\sim}{\Omega})(\underset{\sim}{k} \cdot \underset{\sim}{\zeta})/|\underset{\sim}{k}|^2; \quad \underset{\sim}{\Omega} = \underset{\sim}{e}_z V r^{-1}$$

Here $\underset{\sim}{e}_\theta$ and $\underset{\sim}{e}_z$ are unit vectors and $\underset{\sim}{\zeta}$ is the unperturbed vorticity vector.

The function $f(r)$ must vanish as r tends to infinity, and, for $|m| = 1$, single-valuedness requires $Df = 0$ at $r = 0$. We consider here only the bending modes $|m| = 1$.

The parameter k is regarded as given, so that the problem for f is an eigenvalue problem with the frequency ω as eigenvalue. For general velocity fields $\underset{\sim}{U}$ and arbitrary k, the eigenvalue problem must be solved numerically. For long waves ($k \ll 1$) numerical computation reveals at least two branches of the dispersion relation, $\omega = \omega(k)$. Both have zero frequency at $k = 0$; one has zero frequency and phase speed at $k = 0$, and we call this the "slow branch". The other(s) (generally more than one) branch(es) of the dispersion relation has nonzero frequency and infinite phase speed at $k = 0$; such a branch we call a "fast branch".

Long waves are of considerable interest, since many phenomena arising in vortex filaments occur on length scales large compared to the radius of the vortex core. It is therefore significant that the eigenfunction and dispersion relation on the slow branch in the limit $k \to 0$ can be found by a singular perturbation analysis of the Howard-Gupta equation for <u>arbitrary</u> vortex core structure. The analysis will be described elsewhere; here we cite only the final results. The composite expansion of the eigenfunction valid to $O(k^2)$ is

$$f(r) = \Omega(r) + m|k|W(r)$$

$$- k^2 \left\{ \Omega r^2/2 + \int_0^r [x^3 \Omega^2(x)]^{-1} dx \int_0^x [y^3 \Omega^2(y) + 2y^2 W W'(y)] dy \right\}$$

$$- m\omega - \frac{\Gamma_0}{2\pi r^2} + \frac{\Gamma_0}{4\pi} k^2 [\ln(|k|r/2) + \gamma_e + \frac{1}{2}]$$

$$- |k| \frac{\Gamma_0}{2\pi} \frac{d}{dr} K_1(|k|r) \tag{5}$$

where K_1 is the modified Bessel function of the second kind, and the dispersion relation to this order is

$$\omega = -mk^2 \frac{\Gamma_0}{4\pi} (\beta + \ln(2/|k|)) \tag{6}$$

where

$$\beta = \int_0^1 \frac{\Gamma^2}{\Gamma_0^2} \frac{dr}{r} + \int_1^\infty \frac{\Gamma^2 - \Gamma_0^2}{\Gamma_0^2} \frac{dr}{r} - \frac{8\pi^2}{\Gamma_0^2} \int_0^\infty r W^2 dr - \gamma_e. \tag{7}$$

Here γ_e is Euler's constant, and

$$\Gamma_0 = 2\pi \lim_{r \to \infty} r V(r) \tag{8}$$

which is assumed nonzero.

This agrees with the long wave dispersion relation developed by Moore and Saffman (1972), although their derivation of it is ad hoc. They consider only the rigid rotational motion of an ideal helical vortex filament, using the Biot-Savart formula with a cutoff established by comparison with a steady calculation of Widnall et al. (1971). The connection between such a problem and the desired dispersion relation is obscure.

Numerical computations of the primary fast mode were given by Leibovich and Ma (1983) for the particular vortex

$$W(r) = 0, \quad V(r) = [1 - \exp(-r^2)]/r . \tag{9}$$

There are many fast branches of the dispersion relation for this example, all of which merge as k tends to zero, and all have zero group velocity there. Fast branches, such as these, also turn out to be amenable to analysis for general velocity fields in the limit of vanishing wavenumber. The details have been worked out by S. N. Brown, and we will report the details in a joint paper. It is, however, the slow branch which is of immediate interest, since Maxworthy et al. (1984) report the discovery of slow branch solitons in experiments on vortex filaments. We note that their attempts to compare with the soliton results of Leibovich and Ma (1983) were not generally successful. This presumably is so because the Leibovich and Ma soliton is on the (primary) fast branch.

We turn briefly now to consider solitons centered on the slow branch.

2.2 Solitons

To extend the analysis to higher order, modulation of the wave over long spatial and temporal scales must be allowed. This may be done by permitting the amplitude A in equation (1) to depend on a slow space variable X and a slow time variable τ where, to balance the nonlinearities, we take

$$X = \epsilon(z - c_g t), \quad \tau = \epsilon^2 t ,$$

and c_g is the group velocity,

$$c_g = d\omega/dk .$$

Thus the "carrier wave" of frequency ω and wavenumber k is modulated by the wave envelope $A(X,\tau)$, and the motion of the wave packet is determined by the evolution of A, which is controlled by the nonlinear Schrodinger equation

$$i\partial A/\partial\tau + \mu\partial^2 A/\partial X^2 + \nu A|A|^2 = 0 \tag{10}$$

where

$$\mu = d^2\omega/dk^2 ,$$

and the constant ν is found from an orthogonality condition. Soliton solutions to this equation exist provided $\mu\nu > 0$, and in this case the velocity vector has the form

$$\underset{\sim}{v} = U(r) + \underset{\sim}{\hat{u}}_0(r) \frac{\epsilon}{\lambda} \sqrt{2\mu\nu} \; \text{sech}\{\frac{\epsilon}{\lambda} [z-(c_g+\epsilon c_s)t]\} \; \exp[i(k'z+m\theta-\omega't)] \tag{11}$$

where k' is a wavenumber, modified by nonlinear effects,

$$k' = k + \epsilon c_s/2\mu , \tag{12}$$

and ω is a nonlinearly modified frequency,

$$\omega' = \omega + \epsilon^2(\frac{c_s c_g}{2\mu} + \frac{c_s^2}{4\mu} - \frac{\mu}{\lambda^2}) . \tag{13}$$

In these expressions, c_s and λ are free parameters specifying the soliton (beyond the free parameters k and ϵ of the linear carrier wave).

The values of μ and ν depend upon the wavenumber of the carrier wave. As an example, Mr. Yunas Patel has computed μ and ν for me for the particular vortex (9) for a range of wavenumbers k along the slow branch. This is a repetition of the calculations of Leibovich and Ma (1983), except that they explored the (primary) fast branch of the dispersion relation. Leibovich and Ma found both a low and a high wavenumber cut-off for solitons, in that $\mu\nu > 0$ only in an interval (k_L,k_H) where $k_L \approx .65$ and $k_H \approx 1.0$. On the slow branch, the behaviour is different, in the sense that solitons are permitted down to $k = 0$. Here, in fact, there appear to be two soliton windows, one in the (approximate) wavenumber range $(0,0.4)$ and the other in $k > 1$. Solitons are excluded in the range $(0.4,1.)$.

Note that the lower branch of permitted solitons, extending as it does to $k = 0$, provides a more plausible match between solitons computed from the Euler equations, as here, and the Hasimoto soliton (Hasimoto, 1972, computed from an approximation to the Biot-Savant formula) than the calculations of Liebovich and Ma — restricted to the fast branch — were able to provide.

3. Bifurcation of Axisymmetric Vortices

The Euler equations for steady, incompressible, axially-symmetric flow may be reduced (by integration of the Helmholtz vorticity equation) to the following equation for the Stokes streamfunction, ψ (Squire, 1956; Benjamin, 1962):

$$D^2\psi = r^2 H'(\psi) - FF'(\psi) \tag{14}$$

where

$$D^2\psi \equiv \psi_{rr} - \frac{1}{r} \psi_r + \psi_{zz}$$

and $H(\psi)$, the total head or Bernoulli function

$$H(\psi) = \frac{1}{\rho} p + \frac{1}{2} \underset{\sim}{v} \cdot \underset{\sim}{v} , \tag{15}$$

is a function of ψ alone (Bernoulli) and F is (apart from a factor of 2π) the circulation about the symmetry axis, so

$$F(\psi) = rv(r) \tag{16}$$

where v is the azimuthal velocity component. That F is a function of ψ alone is a consequence of conservation of angular momentum in an inviscid fluid.

As (14) embodies the (integrated) Euler equations under the stated restrictions, differing problems arise from the specifications of the heretofore arbitrary pair of functions H and F. For example, if at some plane $z = z_1$, there should be no radial component of velocity, and if the axial (w) and azimuthal (v) velocity components are specified on this plane to be W(r), V(r) with W $\neq$ 0, then H'(ψ) and F are determined. The way this may be done is explained by Benjamin (1962) (see also Leibovich, 1985).

Now we suppose that, at $z = z_1$, the functional form of the swirl V(r) is fixed, but the level is adjustable, so that

$$F(\psi) = \lambda \, f(\psi) \tag{17}$$

where f is a fixed function, and λ is an adjustable constant. Equation (14) for ψ may be written (Leibovich, 1985)

$$D^2\psi = r^2 A(\psi) - \lambda^2 B(\psi, r^2) \tag{18}$$

where

$$A(\psi) \equiv \frac{1}{R(\psi)} \frac{dW}{dR} (R(\psi)) \tag{19}$$

and

$$B(\psi, r^2) = (R^2(\psi) - r^2) \, ff'(\psi)/R^2(\psi) \tag{20}$$

In these equations $R(\psi)$ is the radius at at which the streamfunction attains the value ψ at the specifying plane $z = z_1$.

Suppose the boundaries of a region D in the (r,z) plane are designated ∂D, that two boundaries of the fluid is are cylinders at r = a and r = b > a, then one solution of (18) holding for __all__ λ is the columnar flow

$$\psi \equiv \Psi(r)$$

Are there any other solutions having the same f(ψ) and the same normal velocity at the boundaries r = a, r = b, $z = z_1$, and $z = z_2$?

Let

$$u = \psi - \Psi(r), \tag{21}$$

then, if there are other solutions, there is a nontrivial solution to the problem

$$D^2 u + \Omega(u, r, \lambda) = 0 \text{ in } D,$$
$$u = 0 \text{ on } \partial D, \tag{22a}$$

where

$$\Omega \equiv \lambda^2 \left[B(\Psi + u, \; r^2) - B(\Psi, r^2) \right]$$

$$-r^2 \left[A(\Psi + u) - A(\Psi) \right] \equiv + \lambda^2 \; \Delta B - r^2 \Delta A. \tag{22b}$$

The last line defines the functions ΔA and ΔB.

The linearized problem associated with (22) is

$$D^2 v + \left. \frac{\partial \Omega}{\partial u} \right|_{u = 0} v = 0 \text{ in } D \tag{23a}$$

$$v = 0 \text{ on } \partial D$$

and, in terms of the data at $z = z_1$ used to construct $H(\psi)$ and $F(\psi)$,

$$\left. \frac{\partial \Omega}{\partial u} \right|_{u = 0} = \frac{\Phi}{W^2} - \frac{r^2 d}{W \, dr} \frac{1}{r} \frac{dW}{dr} \tag{23b}$$

where Φ is Rayleigh's function

$$\Phi = r^{-3} \frac{d}{dr} (r^2 v^2) \tag{24}$$

which, according to our scheme, we write as

$$\Phi = \lambda^2 \Lambda(r) = \lambda^2 r^{-3} \frac{d}{dr} f^2.$$

We note a connection between the problem posed here and one posed by Benjamin (1962): if v were assumed to be independent of z, then equation (23a) is identical to that satisfied by Benjamin's "test function", which establishes whether a given columnar flow is sub or super-critical according to his classification scheme (supercritical flows allowing waves with positive phase speeds only, subcritical flows allowing waves with both positive and negative phase speeds).

Let μ_1, be the smallest eigenvalue of the linearized problem (23), which we write as

$$D^2 v + \left[\mu \, p(r) - r^2 q(r) \right] v = 0, \tag{25}$$

$$v = 0 \text{ on } \partial D,$$

with

$$p(r) \equiv \Lambda(r)/W^2(r) \tag{26}$$

$$r^2 q(r) \equiv \frac{r}{W} \frac{d}{dr} \left(\frac{1}{r} \frac{dW}{dr} \right). \tag{27}$$

The eigenfunction associated with μ_1 is called the "principal eigenfunction" and has no zeros in the interior of D.

"Sensible" problems have $p(r) > 0$, since a negative value for $\Lambda(r)$ implies centrifugal instability (Synge, 1933), and we assume $p(r)$ is positive. We can give a variational characterization of μ associated with a variational principle underlying this problem: equation (14) is the Euler equation associated with the problem of finding functions ψ rendering the functional

$$\int_D \frac{1}{2} \left[(\frac{1}{r} \psi_r)^2 + (\frac{1}{r} \psi_z)^2 + 2H - \frac{1}{r^2} F^2 \right] r \, dr \, dz$$

stationary (physically, this is the volume integral of $p/\rho + u^2 + w^2$). The integral over area D

$$\int_D \frac{v}{r} D^2 v \, d\sigma = \int_D \frac{v}{r} \left[\nabla^2 v - \frac{1}{r} \frac{\partial v}{\partial r} \right] d\sigma$$

$$= \int_D \left[\nabla \cdot (\frac{1}{r} v \nabla v) - \frac{1}{r} |\nabla v|^2 \right] d\sigma$$

$$= - \int_D \frac{1}{r} |\nabla v|^2 \, d\sigma \tag{28}$$

where ∇, ∇^2, and $d\sigma$ are the two-dimensional gradient and Laplacian and area element in the (r,z)-plane, and the divergence theorem and boundary condition $v = 0$ have been used. If we multiply (25) by v/r, integrate over D and use (28), we find

$$\mu = \int_D \frac{1}{r} \left[|\nabla v|^2 + r^2 q \, v^2 \right] d\sigma / \int_D (pv^2/r) d\sigma. \tag{29}$$

The functional on the right in (29) has no upper bound, and therefore

$$\mu_1 = \min \int_D \frac{1}{r} \left[|\nabla v|^2 + r^2 q v^2 \right] d\sigma / \int_D pv^2/r \, d\sigma \tag{30}$$

where the minimum is taken over differentiable functions v vanishing on ∂D. A further physical restriction is applicable: since $|\frac{1}{r} \nabla v|$ is the perturbation fluid speed, it is reasonable to require this to be finite at $r = 0$.

We now assume that $\mu_1 > 0$: this is immediate if $q > 0$, but this condition is not necessary. As λ^2 is increased from zero, a bifurcation is possible when $\lambda^2 = \mu_1$, and in the neighborhood of this point a second branch with norm vanishing as $\lambda^2 \to \mu_1$ may occur. The existence and direction (i.e. do nontrivial solutions for v exist for $\lambda^2 < \mu_1$ or for $\lambda^2 > \mu_1$?) for specific flows may in general be determined by a local analysis. On the other hand, some global features may be obtained if the functions A and B (or their negatives) are convex as functions of ψ. We outline this kind of analysis only.

We define

$$H_1(u,r) = \Delta A - qu, \quad H_2(u,r) = \Delta B - pu ;$$

and assume that the functions A and B satisfy convexity conditions of the form

$$C(i) \quad H_1(u,r) > C_1(r)u^2, \quad H_2(u,r) < - C_2(r) u^2$$

or

$$C(ii) \quad H_1(u,r) < -C_3(r)u^2, \quad H_2(u,r) > C_4(r)u^2$$

where the $C_j(r)$, $j = 1,2,3,4$ are nonnegative functions and the inequalities hold for all $r \leq a$ and for all u. Furthermore, we assume that $H_{1,2} = O(u^2)$ as $u \to 0$.

Note that the convexity conditions C(i) are satisfied if $\partial^2 A/\partial\psi^2 > 0$, $\partial^2 B/\partial\psi^2 < 0$, while the convexity conditions C(ii) are satisfied if these inequalities are reversed.

By adapting an argument given by Sattinger (1972), it can be shown (Leibovich, 1985) that, when $\lambda^2 > \mu_1$, there is at least one positive solution to (22) under condition C(i) and that this solution is positive and, under condition C(ii), there is at least one solution that is negative. When $\lambda^2 < \mu_1$, we are unable to assert existence or nonexistence of solutions to (22). On the other hand, we can show that, if a solution exists for $\lambda^2 < \mu_1$ under conditions C(i), then it is negative in the interior of D; similarly, under C(ii), any solution is positive in the interior.

Thus multiple solutions exist for flows satisfying C(i) or C(ii) when $\lambda^2 > \mu_1$, and the schematic bifurcation diagrams of Figure 1 summarize the findings. Solid lines indicate nontrivial solutions which are known to exist. The dotted curves indicate the character (positive or negative) of any solutions which might exist in $\lambda^2 < \mu_1$. There may, of course, be more solutions than have been shown here to exist.

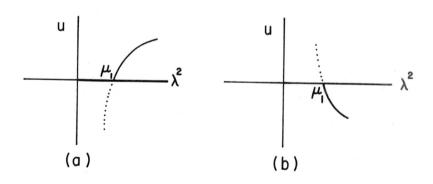

Figure 1. Bifurcation diagram. (a) Case (i), (b) Case (ii).

Positive solutions u correspond to flows with streamlines displaced towards the vortex centerline, corresponding to an acceleration of the flow near the axis and a deceleration near the outer wall r=a, relative to the base flow. Similarly, negative solutions correspond to flows with streamlines displaced away from the centerline, leading to flow deceleration near the axis, relative to the base flow, and an acceleration near the outer wall.

The results presented are global in character; they do not depend on local constructions in the neighborhood of the bifurcation point $\lambda^2 = \mu_1$. This aspect of generality and simplicity is achieved by the formulation of conditions (i) and (ii), which restrict the class of flows for which this approach yields results. If these conditions are not satisfied, branching may (and likely will) still occur in the neighborhood of $\lambda^2 = \mu_1$ (if $\mu_1 > 0$). The present approach concludes nothing about such flows, but a local analysis would determine the issue.

To attempt to apply these results to a given flow, we must compute μ_1 and then check the convexity conditions. The first step, in nearly all interesting cases, must be carried out numerically, but is in any event required for any bifurcation study. The second step may not be straightforward. To show that there are flows likely to be of interest that satisfy either (i) or (ii), we give an example of each.

Example 1. (Condition (ii) satisfied).

The profiles

$$W(r) = e^{-\alpha r^2}$$

$$V(r) = \frac{1}{\alpha r} \left\{ \frac{1 - e^{-\alpha r^2}(1 + \alpha r^2)}{2} \right\}^{1/2}$$

lead to

$$R(\psi) = \left[\frac{1}{\alpha} \ln(1 - 2\alpha\psi)^{-1} \right]^{1/2}$$

$$f(\psi) = \frac{1}{\alpha} \left\{ 1 + (1 - 2\alpha\psi)[-1 + \ln(1 - 2\alpha\psi)] \right\}$$

$$A(\psi) = -2\alpha(1 - 2\alpha\psi)$$

$$B(\psi, r^2) = \frac{1}{\alpha} \ln(1 - 2\alpha\psi)^{-1} - r^2.$$

Notice that, for this flow, $p(r)$ and $q(r)$ are both positive. The eigenvalue problem for μ_1, which we do not solve here, is

$$D^2 v + [\mu p(r) - r^2 q(r)]v = D \qquad \text{in } D$$

$$v = 0 \text{ on } \partial D$$

with

$$q(r) = 4\alpha^2, \qquad p(r) = 2e^{\alpha r^2}$$

and, since both p and q are positive, μ_1 is also positive.

Example 2. (Condition (i) satisfied).

Let the tube radius be $r = a = 2$ and let $\psi = 1$ at $r = a$; this establishes a normalization. The flow with velocity profiles

$$W(r) = \left(1+\sqrt{1-r^2/4}\right)^{-1}$$

$$V(r) = \frac{2}{r}\left[1-\sqrt{1-r^2/4}\right]\left\{(5-\sqrt{1-r^2/4})/16\right\}^{1/2}$$

satisfies conditions (i). Here

$$R(\psi) = 2\sqrt{\psi-\psi^2/4}$$

$$f(\psi) = 2\psi\sqrt{1-\psi/6}$$

$$A(\psi) = -(2-\psi)^{-3}$$

$$B(\psi,r^2) = 4(\psi-\psi^2/4) - r^2.$$

Again, both $p(r)$ and $q(r)$ are positive, leading to a positive value for μ_1.

4.   Concluding Remarks

We have discussed here weakly nonlinear nonaxisymmetric "bending waves" on vortex filaments, as well as global bifurcations of axially-symmetric swirling flows in circular tubes. The soliton "bending waves" presumably can be related to the Hasimoto (1972) soliton, which is derived from the so-called local induction approximation (LIA) to the Biot-Savart formula. Since the solitons in this paper are treated from the Euler equations and allow for the structure of the vortex core, in contrast to LIA, it is hoped that the connection between the two soliton theories will permit the identification of the arbitrary constants which appear in the LIA. This topic will be addressed further in a joint paper with S. N. Brown and Y. Patel that is presently in preparation.

The problem of bifurcation of axially-symmetric flows, treated in section 3, has connections with existing work by Benjamin (1962) on the problem of vortex breakdown. In particular, the eigenvalue equation here which determines critical points of the nonlinear partial differential equation governing steady motions is, when suitably specialized, closely related to Benjamin's (1962) "test" equation, but the multiple solutions found here are not related to Benjamin's "conjugate states" of flow. Further discussion of this point is given in Leibovich (1985).

Acknowledgements
This work was supported in part by NSF Grant MEA 83-06713 and in part by ONR 3RO IV. I learned of the methods described in Sattinger's book from lectures given by Professor Lars Wahlbin, to whom I am grateful.

References

Courant, R. and Hilbert, D. 1953. Method of Mathematical Physics, vol. 1,
Interscience, New York, 561pp., Chapter VI.

Benjamin, T. B. 1962. Theory of the vortex breakdown phenomenon, J. Fluid Mech.,
14, 593-629.

Escudier, M. P., Burnstein, J., and Maxworthy, T. 1982. The dynamics of
concentrated vortices, Proc. Roy. Soc., A382, 335-360.

Faler, J. H. and Leibovich, S. 1977. Disrupted states of vortex flow and vortex
breakdown, Phys. Fluids, 20, 1385-1400.

Hasimoto, H. 1972. A soliton on a vortex filament, J. Fluid Mech., 51, 477-485.

Hopfinger, E. J., Browand, F. K., and Gagne, Y. 1982. Turbulence and waves in a
rotating tank. J. Fluid Mech., 125, 505-534.

Howard, L. N. and Gupta, A. S. 1962. On the hydrodynamic and hydromagnetic
stability of swirling flows, J. Fluid Mech., 14, 463-476.

Leibovich, S. 1983. Vortex stability and breakdown, in Aerodynamics of Vortical
Type Flows in Three Dimensions, ed. A. D. Young, AGARD CP 342, 23.1-23.22.

Leibovich, S. 1984. Vortex stability and breakdown: Survey and extension, AIAA
J., 1984, 1192-1206.

Leibovich, S. 1985. A note on global bifurcation of columnar vortices, Sibley
School of Mechanical and Aerospace Engineering Report FDA 85-16, Cornell University,
Ithaca, New York.

Leibovich, S. and Ma, H-Y. 1983. Soliton propagation on vortex cores and the
Hasimoto soliton, Phys. Fluids, 26, 3173-3179.

Maxworthy, T., Mory, M. and Hopfinger, E.J. 1983. Waves on vortex cores and their
relation to vortex breakdown, in Aerodynamics of Vortical Type Flows in Three
Dimensions, ed. A.D. Young, AGARD CP 342, 29.1-29.13.

Maxworthy, T., Hopfinger, E. J. and Redekopp, L. G. 1985. Wave motions on vortex
cores, J. Fluid Mech., 151, 141-165.

Moore, D. W. and Saffman, P. G. 1972. The motion of a vortex filament with axial
flow. Phil. Trans. Roy. Soc. A272, 403-429.

Sattinger, D. H. 1972. Topics in Stability and Bifurcation Theory, Lecture Notes
in Mathematics No. 309, Springer-Verlag, New York, 190pp.

Squire, H. B. 1956. Rotating Fluids. Chapter in Surveys in Mechanics, ed. G. K.
Batchelor and R. M. Davies, Cambridge Univ. Press.

Synge, J. L. 1933. The stability of heterogeneous liquids, Trans. Roy. Soc. Can.,
27, 1-18.

Widnall, S. E., Bliss, D., and Zalay, A. 1971. Theoretical and experimental study
of the stability of a vortex pair, in Aircraft Wake Turbulence and Its Detection,
ed. J. H. Olsen et al., Plenum Press.

Sattinger, D. H.   1972.   Topics in Stability and Bifurcation Theory, Lecture Notes
in Mathematics No. 309, Springer-Verlag, New York, 190pp.

Squire, H. B.   1956.   Rotating Fluids.   Chapter in Surveys in Mechanics, ed. G. K.
Batchelor and R. M. Davies, Cambridge Univ. Press.

Synge, J. L.   1933.   The stability of heterogeneous liquids, Trans. Roy. Soc. Can.,
27, 1-18.

Widnall, S. E., Bliss, D., and Zalay, A.   1971.   Theoretical and experimental study
of the stability of a vortex pair, in Aircraft Wake Turbulence and Its Detection,
ed. J. H. Olsen et al., Plenum Press.

Review of Three-Dimensional Vortex Dynamics: Implications for the Computation of Separated and Turbulent Flows

Sheila Widnall

Abstract

Linear instability of concentrated vortex structures is examined as a source of
three-dimensionality in separated and turbulent flows. A general review of vortex
instability to long-wave and short-wave perturbations is given with special
emphasis on the role of vortex structure in determining the most unstable wave
lengths and their amplification rates. Stability calculations are presented for
two vortex configurations as models for separated flows: a quasi-stationary vortex
shed from a cylinder and the stationary Foppl vortex pair. The single vortex is
found to be most unstable to three-dimensional perturbations while the Foppl
vortices are found to be most unstable to two-dimensional perturbations. Some
implications for numerical calculations using vortex filaments are discussed.

1. INTRODUCTION

The study of three-dimensional vortex dynamics plays an important role in fluid
mechanics. Beyond the intrinsic interest in the properties of vortex flows --their
structure, motion and stability to three-dimensional disturbances-- vortex
dynamics, and in particular the three-dimensional instabilities of vortices can
contribute to an understanding of the origin of three-dimensionality in turbulent
flows. Vortex filaments are also used as a basis for the calculation of flows with
distributed vorticity. In this case the properties of the flow follow from the
dynamics of the collection of vortex filaments. (See Leonard, 1985, for a review
of vortex methods.)

For example the instability of shear-layer vortices to three-dimensional
perturbations gives considerable insight into the development of
three-dimensionality of the shear layer flow after the formation of concentrated
shear-layer vortices. The implications of the linear stability calculation of
Pierrehumbert and Widnall have been born out by direct Navier-Stokes calculations

such as those of Riley and Metcalf, (1985) and observed in experiments such as those of Breidenthal (1982).

Numerical calculations of aerodynamic configurations which shed vortex sheets are also affected by the three-dimensional instability of vortex filaments. (We will not discuss the instability of two-dimensional vortex sheets to Kelvin-Helmholtz instability which has played havoc with studies of vortex sheets in two dimensions; we do suggest that the additional instabilities lurking in three-dimensions are likely to be as important and have received very little attention.) Calculations of helicopter wakes, delta-wing vortex flows in three-dimensions, and jets which use vortex filaments to construct the flow field and to follow its development in time will be effect by the inherent instability of the vortex flows to both long wave and short wave instabilities.

Vortex dynamics also plays an important role in the calculation of separated flows at high Reynolds numbers. Fundamental to the calculation of such flows is the behavior of the vorticity in the separated wake. Most observations indicate that this vorticity rolls up into concentrated vortex structures. Numerical models for separated flows often take advantage of the vortex nature of these flows by using vortex elements as building blocks for flow calculations. Many of these models are two-dimensional, using point vortices to represent flow separation from airfoils and bluff bodies such as flat plates and cylinders.

It has often been found that two-dimensional point-vortex calculations of separated flows do not accurately predict drag even when great care has been taken to incorporate Reynolds number effects into the boundary-layer model. One explanation for this discrepancy is the neglect of three-dimensional effects. Even if the vortices are shed from the body in a two-dimensional manner, three-dimensional vortex instabilities may distort the filament and affect the average pressure on the body. Graham (1984) obtained improved agreement between experiment and vortex calculations by including three-dimensional distortions of the filaments of arbitrary amplitude.

Several questions arise concerning vortex models for separated flow. These concern the adequacy of two-dimensional models and the behavior of discrete vorticity as a representation of continuous vorticity. To construct models we need to know the source of three-dimensionality in separated flows. Can it be related to the stability properties of the concentrated vortex structures? Does the discrete vortex system have the same stability characteristics as the continuous system that it represents?

To gain some insight into these issues and into more general issues of how the instability characteristics of vortices play a role in more complex flows, we will review the characteristics of vortex sintability for a variety of configurations which can serve at simple models for flows in which concentrated vorticity plays a dominant role. We will also discuss the stability of vortex systems in some simple models for separated flows.

2. REVIEW OF VORTEX STABILITY

The stability of vortex configurations to small perturbations has been extensively studied. Primary attention has been devoted to the study of the instability of self-preserving configurations. These are configuration of vortex points or filaments in either two or three dimensions whose self-induced motion translates or rotates the vortex configuration without change of form. This is not a common property of vortex configurations, most of which distort continually in time. The question of the stability of configuration which distort continually in time to small disturbances is more difficult and perhaps somewhat academic.

One can distinguish three types of instability: two-dimensional; three-dimensional long-wave; and three-dimensional short-wave. Two-dimensional instability is the k=0 limit of long-wave instability of a two-dimensional configuration but it is useful to consider this case separately since so many configurations were investigated initially for two-dimensional motions. Examples of two-dimensional stability investigations include: the row of point vortices; the Karman vortex street, Karman (1911,1912) and Karman and Rubach (1912); the ring of N point vortices and its extensions, Havelock (1931). In studies of two-dimensional instability, the only property of the vortex that enters is its strength and position; details of the vortex core size or its distribution of vorticity do not affect the stability to two-dimensional motions.

2.2 Long-Wave Stability of Vortex Configurations

Long-wave stability of vortices have been studied for the vortex pair Crow (1971), the ring Widnall and Sullivan (1972), the helix Widnall (1972), the Karman vortex street and a single infinite row of point vortices, Saffman and Robinson (1982). In a general three-dimensional flow these are the simplest configurations that are self-preserving. Except for the row, they are also stable to uniform (k=0) displacement modes.

In the analysis of long-wave instability the vortex is treated as a filament and the only property of the vortex core that enters is its energy. The self-induced motion of a vortex filament is governed by the core size and a shape constant dependent upon the vorticity distribution. Therefore the response of a vortex filament to deformation can be described by its strength, shape and a single parameter called effective core size (see Widnall et al. 1971). Effective core size is defined so that for a core of constant vorticity the effective core radius is the actual core radius. Therefore all long-wave instability problems can be described in terms of vortex configuration, wave length and effective core size.

The results of long-wave stability investigations show that: the vortex pair of equal and opposite strength is unstable to symmetric perturbation and unstable to anti-symmetric perturbations, Crow (1971); the pair of equal strength, which circles at constant spacing, is stable, Jimenez (1975); the helix has two long-wave modes of instability, Widnall (1972); and the vortex ring is stable to long-wave disturbances, Widnall and Sullivan (1972). The Karman vortex street is unstable to three-dimensional perturbation while the infinite row of point vortices is unstable to both two- and three-dimensional perturbations, Saffman and Robinson (1982).

Since vortex filaments move with the local velocity field, the background flow seen by a vortex filament is to lowest order a strain or stagnation-point flow. Three-dimensional instability of vortex filaments can be understood as a balance between the destabilizing effects of the local strain field at the vortex location and the self-induced velocity due to local curvature, which for a slender filament is equivalent to a rotation of the disturbance with angluar velocity Ω. (See for example Widnall, Bliss and Tsai, 1974) Instability occurs whenever the destabilizing effects of strain overcome the stabilizing effects of self-induced rotation. An additional effect, important for long-wave instability, is the velocity induced on the vortex filament by the displacement perturbations of the distant portions of the configuration, Q_1; these velocities determine what wave length will be most unstable for a given configuration and core size.

As an example, consider the long-wave instability of the vortex pair first studied by Crow (1971). The instability configuration is shown in Fig. 1. The vortex pair is assumed to undergo a symmetric pattern of displacements and self and mutual influence velocities are calculated. (The mode shown is unstable; the antisymmetric mode is stable.)

Since the vortex as a whole moves with the velocity at its origin from the other vortex, the perturbation field seen by the displaced filament is that of a strain or corner flow. If the filament has displacement $X_o e^{ikz}$ and $Y_o e^{ikz}$, the velocities at the filament position in this strain field will be

$$Q_0 = -\Gamma/2\pi b^2 \ (X_o j + Y_o i) \ e^{izk} \tag{1}$$

where i and j are unit vectors.

In addition, perturbations of the distant filament produce a velocity at the filament position of

$$Q_1 = \Gamma/2\pi b^2 \ (V_x \ X_1 \ j + U_y \ Y_1 \ i) \ e^{ikz} \tag{2}$$

Expressions for U_y and V_x were presented by Crow (1971) as

$$U_{1y} = -X_o \ \Gamma/2 \ \pi \ b^2 \ (kb) \ K_1(kb)$$

$$V_{1x} = -Y_o \ \Gamma/2 \ \pi \ b^2 \ ((kb)^2 \ K_o(kb) + kb \ K_1(kb)) \tag{3}$$

where K_o and K_1 are modified Bessel functions of the second kind.

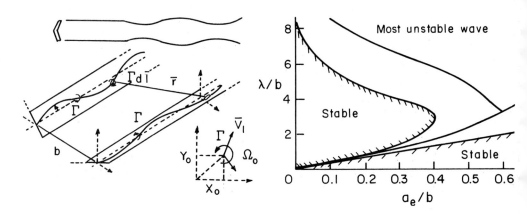

Fig. 1 Geometry of Vortex-Pair Instability   Fig. 2 Stability Diagram, Vortex Pair

The self-induced motion of a sinusoidally perturbed filament to long-wave perturbations is a rotation about its axis in a direction opposite to the direction of circulation of magnitude (Widnall et al. 1971)

$$\Omega_o = 1/2 * k^2 * (-\log(a_{eff} \ k) + .366) \tag{4}$$

where a_{eff} is the effective core radius of the filament. Stability is investigated by assuming

$$X_1 = -X_0; \ Y_1 = Y_0 \text{ and}$$

$$X_o = x_o \; e^{\alpha t} \; ; \; Y_o = y_o \; e^{\alpha t}. \tag{5}$$

Therefore,

$$dX_o/dt = \alpha x_o = U_y \; y_o - \Omega \; y_o$$

and

$$dY_o/dt = \alpha y_o = V_x \; x_o + \Omega \; x_o \tag{6}$$

where U_y and V_x are the sum of U_{1y}, V_{1x} and the velocities due to the local strain field. Solving for α we obtain

$$\alpha^2 = (U_y - \Omega_o) * (V_x + \Omega_o) \tag{7}$$

Numerical calculations give the result shown in Fig 2; The most unstable wave depends upon the effective vortex core size; for $a_{eff} = .1$, $\lambda = 8b$ and the amplification rate is $\alpha = .8 \; \Gamma/2\pi b^2$.

One defect of the formula for self-induced motion of a vortex filament is that it predicts $\Omega_o = 0$ for $ka = 1.44$ --since it holds only for long waves it is not valid at $ka = O(1)$. Thus, for reasons that will become clear after our discussion of short-wave instability, long-wave stability calculations will always pick up a spurious root, with amplification rate equal to the local strain, at $ka = 1.44$ where the stabilizing effects of rotation vanish. Such spurious short-wave instabilities were found by Crow (1971) and Widnall (1971,1972) and others in various studies. Their meaning is now understood in terms of the actual short-wave instability of vortex filaments. They do predict that instability can occur if the dispersion relation for waves on a vortex has $\Omega = 0$ for some value of ka; the amplification rate of short-wave instabilities is accurately predicted; but the instability occurs at the wrong wave number, $ka = 1.44$ related to the spurious root.

2.2 Short-Wave Instabilities of Vortices

Short-wave instabilities of vortex rings were observed experimentally Krutzsch (1939) before theory was able to explain their occurrence. The mechanism for short-wave instability turns out to be remarkably simple and the results indicate that every vortex with distributed vorticity, except the straight, isolated vortex in an infinite fluid, is linearly unstable to short-wave disturbances with an amplification rate equal to the local strain. This has implications for the origin of three-dimensionality in vortex-dominated separated flows. The first identification of the mechanism of short-wave instability was the study of Widnall, Bliss and Tsai (1974) who identified the connection between the short-wave modes of vortex oscillation that had no self-induced rotation and the short-wave instabilities of vortices. Short-wave instabilities of straight vortex filaments were considered by Moore and Saffman (1975) and Tsai and Widnall (1976); the short-wave instability of the vortex ring was analyzed by Widnall and Tsai (1977).

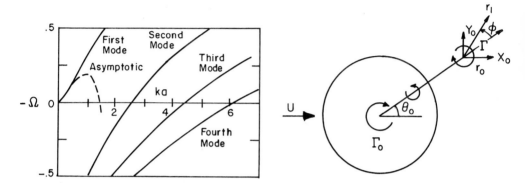

Fig. 3 Dispersion Relation, Bending Waves   Fig. 4 Vortex Near a Circular Cylinder

The details of the solution for short-wave modes depends upon core size and vorticity distribution and must be calculated numerically in most cases. For the simple core of constant vorticity the frequencies of the modes of oscillation are shown in Fig. 3 along with the rotation rate predicted by (4). As can be seen, the mode for which $\Omega_o=0$ for the lowest value of ka is the second bending mode, in which the center of the core moves in opposition to the outer boundary. Since such a mode cannot rotate itself out of the destabilizing effects of the local strain field a vortex placed in a non-uniform field would be locally unstable to strain at a wave number such that $\Omega=0$; for cores of constant vorticity this occurs at ka=2.4. For vortices with more continuous distributions, instability occurs at ka=3.8 Widnall et al. (1974). Saffman (1978) calculated ka for a family of vortex profiles representing the core of a vortex rolled up to form a sheet and obtained $1.65<ka_1<2.51$ with a_1 taken as the radius of maximum swirl velocity; $a_1<a$. These predictions are in good agreement with observations of the instability of vortex rings when the core vorticity has been measured or estimated from roll-up models.

Short-wave instability of concentrated vortices in rotational flows has also been studied. Pierrehumbert and Widnall (1982) identified two modes of instability of vortices in a shear layer. The long-wave instability is related to vortex pairing. The short-wave instability is a bending-wave instability; however it occurs at the first bending mode, possibly because of enhanced strain from the surrounding vorticity. Maximum amplification was found at spanwise wave numbers of k=2 to 3 based on vortex separation; however a broad range of shorter waves had comparable amplification. Short-wave instabilities of vortices surrounded by potential flow occur at bending modes where $\Omega_o=0$.

3. APPLICATION TO SEPARATED FLOWS

To gain some insight into the source of three-dimensionality in separated flows we analyze the three-dimensional instability of simple vortex configurations that model some aspects of separating flow. We consider the stability of a single vortex separating from a cylinder as well as that of the Foppl vortices behind the cylinder. (This was also presented in Widnall, 1985)

3.1 Stability of a Shed vortex Near a Cylinder

During the process of flow separation behind a cylinder, strong vortices are shed. Although the process is unsteady, it is of some interest to examine the three-dimensional stability of the shed vortex considering the shedding process to be locally quasi-steady relative to the occurrence of three-dimensionality.

We consider the vortex configuration shown in Fig. 4. As an approximation of the quasi-steady process of vortex shedding, we are interested in vortices that are stationary at positions in the cylinder wake. We therefore analyze conditions under which a single vortex will be stationary at a point in the wake of a cylinder, r_o, θ_o. We find that except for points along the y axis ($\theta_o=\pi/2$) we cannot find a strength Γ for which both the r and θ velocities will be zero at the vortex position. We therefore choose Γ with the condition $u_\theta=0$ and assume that the vortex is at least quasi-stationary in our references frame due to radial velocities induced by the shed vortex sheet. We include the effects of an arbitrary circulation Γ_o about the cylinder. The configuration is sketched in Fig. 4. The radius of the cylinder is taken as 1; the velocity of the free stream as 2π (consistent with normal scaling of vortex instability).

The condition that $u_\theta=0$ at the vortex position gives the following relationship between Γ, Γ_o and the vortex position r_o, θ_o.

$$\Gamma=(2 \pi U \sin\theta_o (1+1/r_o^2) + \Gamma_o/r_o) (r_o-1/r_o) \tag{8}$$

where Γ_o is the circulation about the cylinder.

The stability of this vortex configuration to three-dimensional perturbations is examined by assuming that the filament undergoes a sinusoidal perturbation of magnitude

$$r_o = X_o i + Y_o j \tag{9}$$

Velocities are induced at the filament as a result of this perturbation from several different sources. First, perturbation velocities are seen at the vortex position as a result of its change of position. These are obtained by derivatives of the mean flow field of the cylinder U_o, V_o.

$$Q_o = (U_{ox} X_o + U_{oy} Y_o)i + (V_{ox} X_o + V_{oy} Y_o) \tag{10}$$

The second interaction arises as a result of the sinusoidal perturbation of the vortex filament in the presence of an impervious cylinder. For $k=0$ it reduces to the effect of the displacement of the image vortex initially located at $r = 1/r_o$ as it is constrained to move in opposition to the free vortex. We represent this effect as

$$Q_1 = (U_{1x} X_0 + U_{1y} Y_o)i + (V_{1x} X_o + V_{1y} Y_o) \tag{11}$$

where the velocities U_{1x}, U_{1y}, V_{1x} and V_{1y} are functions of the perturbation wave number k; they are obtained by the procedure outlined in the Appendix using the properties of Bessel functions. The final contribution is the self-induced rotation Ω_o given by (4).

Assuming a time dependence of $e^{\alpha t}$ we obtain the following form for the motion of the vortex

$$dX_o/dt = \alpha x_o = U_x x_o + U_y y_o - \Omega_o y_o$$

and

$$dY_o/dt = \alpha y_o = V_x x_o + V_y y_o + \Omega_o x_o \tag{12}$$

where U_x, V_y etc. is the sum of U_{ox} and U_{1x} etc.

The eigenvalue α is given by

$$\alpha = (U_x + V_y)/2 + 1/2 \ ((U_x + V_y)^2 - 4(U_x V_y - U_y V_x + \Omega_o(V_x - U_y) + \Omega_o^2))^{1/2} \tag{13}$$

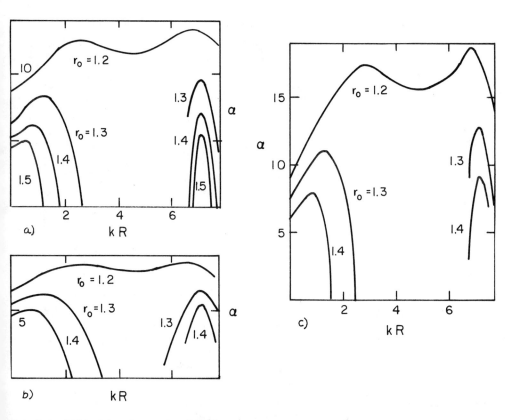

Fig. 5 Amplification Rates for 3-D instability of a Vortex Near A circular
Cylinder; θ=45. a)Γ=0; b)Γ negative c)Γ positive

Fig. 5 shows the results of numerical calculations for three-dimensional
instability of the vortex filament in the wake of a cylinder for various values of
r_0 and Γ_0. The vortex position was assumed to be at θ=45. Calculations were
done for other angles; the results and conclusions were essentially the same. The
magnitude of Γ_0 was chosen as $2(2)^{1/2}\pi U$, the value for which the stagnation
point on a cylinder in a free stream would be located at 45; both positive and
negatives values were taken. A vortex core size of a=.2 was used for all of the
calculations. For r_0=1.2 the core touches the cylinder so the results should be
viewed with caution.

These results show that the most unstable mode of vortex instability for the vortex
separating from a circular cylinder is three-dimensional. Thus in the formation of
the wake, three-dimensional motions should be found. A short-wave instability near

k=7.2 is also shown corresponding to the spurious zero at ka=1.44. Actually the short-wave instability would occur at a value of ka dependent upon the vorticity distribution; however the indication of instability and the amplification rate are correctly predicted.

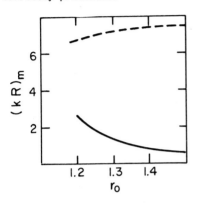 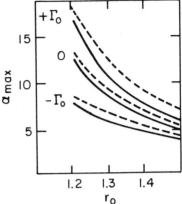

Fig. 6 Amplification Rate & Most Unstable wave number for the 3-D Instability of a Vortex Near a Cylinder

Increasing Γ_0 increases the amplification rate but makes no noticeable change in the most unstable wave number, as shown in Fig. 6. A cylinder shedding alternately signed vortices will have a positive value of Γ_0 when the vortex is shed from the upper-half of the cylinder. The wave number and amplification rate decrease as the vortex moves away from the cylinder boundary. Such three-dimensional instabilities have been observed on vortices shed from circular cylinders Wei and Smith (1985). The relevance of this simple model to experimental observations is only suggested since actual vortex shedding is a process of considerable complexity involving boundary layers and separated shear layers.

3.2 Stability of The Foppl Vortices

The previous analysis can be extended to investigate the stability of the Foppl vortices to three-dimensional perturbations. For the Foppl configuration the strength and position of the vortices are such that the velocity is zero at each vortex (see Milne-Thomson 1960). Foppl vortices are more of a curiosity than a model for separating flows since they are known to be unstable to two-dimensional disturbances; however, the extension to three-dimensional perturbations has not been worked out.

All of the expressions needed to calculate the velocity field for the response of the Foppl vortices to three-dimensional perturbations are available from the previous analysis. The various influences which must be included are:

1. The effect of displacement in the mean field consisting of cylinder nd vortex pair.

2. The effects of sinusoidal perturbations of the filament in the presence of an impervious cylinder.

3. The effects of sinusoidal perturbations of the companion vortex in the presence of an impervious cylinder.

4. The self-induced rotation of the filament.

The form of the equation for the amplification rate α is identical to 12 with suitable modifications of U_x, U_y etc. Both symmetric ($X_1 = X_o$, $Y_1 = -Y_o$) and antisymmetric ($X_1 = -X_o$, $Y_1 = Y_o$) modes are considered, as sketched in Fig. 7.

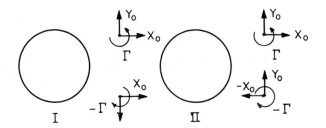

Fig. 7 Symmetric (I) and Antisymmetric (II) Modes, Foppl-Vortex Instability

The symmetric mode is stable to two-dimensional perturbations while the antisymmetric mode is unstable. The two-dimensional case was worked out by Smith (1973), who corrected some earlier errors of Foppl. Calculations were done for the three-dimensional instability for each of these modes. The results for the antisymmetric mode are shown in Fig. 8 which shows the amplification rate as a function of wave number for various vortex positions r_o. For this mode the two-dimensional perturbation is most unstable; three-dimensional perturbations have lower amplification rates becoming completely stable for some kR for $r_o > 1.2$. The value of α for $k=0$ agrees with that found by Smith (1973) allowing for a difference of 2π in the scaling. The stability diagram for this mode is shown in Fig. 9. Both of these figures show the short-wave instability predicted for $ka = 1.44$. In actuality short-wave instability would occur at a value ka predicted by an analysis of the actual vortex dispersion relation.

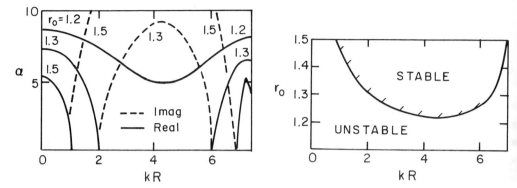

Fig. 8, 9 Amplification Rates and Stability Diagram, Antisymmetric Mode

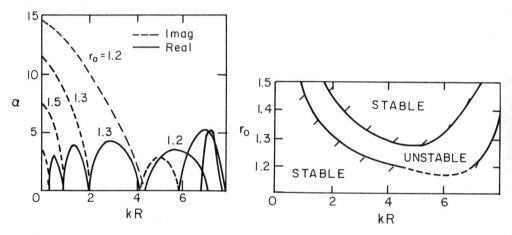

Fig. 10, 11 Amplification Rate and Stability Diagram, Symmetric Mode

The amplification rate vs kR for the symmetric mode is shown in Fig. 10 and the stability diagram in Fig. 11. This mode is stable for k=0 but becomes unstable for a range of three-dimensional waves. Short-wave instability near ka=1.44 is also predicted. For vortex positions close to the cylinder the long- and short-wave instabilities merge.

These results are related to the results for the stability of an isolated vortex pair as previously discussed. The symmetric mode which is stable for the Foppl vortices is unstable for the vortex pair for k>0. Therefore as the vortex moves away from the cylinder the three-dimensional instability that develops is a vortex-pair instability in the cylinder wake; the antisymmetric mode is unstable for the Foppl vortices at k=0 and stable for the vortex pair. Therefore, three-dimensional effects stabilize this mode as the distance from the cylinder increases.

4. Implications for Numerical Calculations

Calculation of flows containing concentrated vortex structures are affected by vortex instability. In a two-dimensional model, neglect of three-dimensionality can lead to disagreements with experimental results. In a fully three-dimensional model, the calculation must deal with vortex instability either by using sufficient resolution to follow the details or by using turbulence or instability modelling to capture the effects of vortex stability on the overall flow. Another issue concerns the discretization of a continuous flow in vortex filaments. The possibility exists that this process will introduce spurious instabilities into the numerical simulation that are not representative of the behavior of the continuous system.

5. APPENDIX: VELOCITY FIELD INDUCED BY THE SINUSOIDAL PERTURBATION OF A VORTEX FILAMENT IN THE PRESENCE OF AN IMPERVIOUS CIRCULAR CYLINDER

We consider the vortex configuration shown in Fig 4. The vortex is located at the position r_o, θ_o. The vortex undergoes sinusoidal perturbations $r(z)=r_o \sin kz$; and $\theta(z)=\theta_o \sin kz$. In a local cylindrical coordinate system at the point r_o, θ_o, the velocity potential for this perturbation is

$$\Phi = k\,\theta_o\,\cos(\phi)K_1(kr_1) - k\,r_o\,\sin(\phi)\,K_1(kr_1)= \theta_o\,\Phi_1 - r_o\,\Phi_2 \tag{1}$$

where r_1 and ϕ are local cylindrical coordinates at the vortex location. The symbols Φ_1 and Φ_2 have been introduced to represent the response to θ_o and r_o displacements. Φ, Φ_1 and Φ_2 satisfy Laplaces equation.

This velocity potential can be re-expressed in the cylindrical coordinate system centered in the impervious cylinder, r,θ by using the following property of Bessel functions, Erdelyi et al. (1953, eq. 7.15.35).

$$\cos(\phi)\,K_1(kr_1)= -\sum_{n=-\infty}^{n=\infty} K_{n+1}(kr_o)\,I_n(kr)\,\cos(n(\theta-\theta_o)) \tag{2}$$

and

$$\sin(\phi)\,K_1(kr_1)= \sum_{n=-\infty}^{n=\infty} K_{n+1}(kr_o)\,I_n(kr)\,\sin(n(\theta-\theta_o)) \tag{3}$$

which converges for $r < r_o$. These expressions were verified numerically by summing to $n = 21$.

When this expression is applied to the form of the velocity potential in (1) we obtain

$$\Phi_1 = -\sum_{n=0}^{n=\infty} a_n^1 \, I_n(kr) \, \cos(n(\theta-\theta_o))$$

$$\Phi_2 = -\sum_{n=0}^{n=\infty} a_n^2 \, I_n(kr) \, \sin(n(\theta-\theta_o)) \qquad (4)$$

where

$$a_n^1 = K_{n+1}(kr_o) + K_{n-1}(kr_o)$$

and

$$a_n^2 = -K_{n+1}(kr_o) + K_{n-1}(kr_o) \qquad (5)$$

This is the velocity potential at the cylinder due to the perturbations of the external vortex. However, the boundary condition on the cylinder requires that the induced radial velocity be canceled by radial velocities induced by a potential of the form

$$\Phi_i = \sum_{n=0}^{n=\infty} b_n^1 \, K_n(kr) \, \cos(n(\theta-\theta_o)) + b_n^2 \, K_n(kr) \, \sin(n(\theta-\theta_o))$$

(6)

so that the boundary condition $u_r = 0$ at $r = 1$ requires that

$$b_n^1 = a_n^1 \, I_n'(k)/K_n'(k) \quad \text{and} \quad b_n^2 = a_n^2 \, I_n'(k)/K_n'(k)$$

(7)

The expression for the induced potential can be written

$$\Phi_i = \theta_o \, \Phi_{i1} - r_o \, \Phi_{i2} \qquad (8)$$

The radial and tangential velocities for the influence of the cylinder can be obtained from this expression for any r, θ point in the flow field.

$$u_r = \theta_o \, \partial\Phi_{i1}/\partial r - r_o \, \partial\Phi_{i2}/\partial r$$

$$u_\theta = \theta_o \, 1/r \, \partial\Phi_{i1}/\partial\theta - r_o \, \partial\Phi_{i2}/\partial\theta \qquad (9)$$

The x and y velocities due to the perturbations x_o, y_o can be obtained from

u_r and u_θ through the relations

$$\theta_o = y_o \cos \theta_o - x_o \sin \theta_o$$

$$r_o = x_o \cos \theta_o + y_o \sin \theta_o$$

$$u = u_r \cos \theta - u_\theta \sin \theta$$

$$v = u_r \sin \theta + u_\theta \cos \theta \tag{10}$$

The influence coefficients are defined as

$$U_x = \partial u / \partial x_o ; U_y = \partial u / \partial y_o ; V_x = \partial v / \partial x_o ; V_y = \partial v / \partial y_o \tag{11}$$

They are found by combining the relations given above.

5. REFERENCES

BREIDENTHAL, R.E. (1978) A Chemically Reacting Turbulent Shear Flow. PhD thesis Cal. Tech.

CROW, S.C. (1971), "Stability Theory for a Pair of Trailing Vorticies", AIAA JOURNAL, 8, pp. 2172.

ERDELYI, A. et al (1953), Higher Transcendental Functions, McGraw-Hill Book Co., Inc. New York,

GRAHAM, J.M.R. (1984), Int. Workshop on Numerical Simulation of Bluff Body Aerodynamics, Meersburg, Germany, May,

HAVELOCK, T.H. (1931) "The Stability of Motion of Rectilinear Vortices in Ring Formation" Phil. Mag. S.7. Vol 11 no.70 Feb.

JIMENEZ, J. (1975), "Stability of a Pair of Co-Rotating Vortices", Phys. of Fluids, 18, no. 11, Nov.,

KARMAN, T. von (1911) Gottinger Nachrichten, Math. Phys. Kl. 509-517

KARMAN, T, von (1912) Gottinger Nachrichten, Math. Phy. Kl. 547-556

KARMAN, T. von and RUBACH, H.L. (1912) pphys. Zeitschr. 13 49-59

KRUTZSCH, C.H. (1939), Ann. Phys., 35, pp. 497.

LEONARD, A. (1985) "Computing Three-Dimensional Flows With Vortex Elements", Annual Reviews of Fluid Mechanics 17 pp.523-59

MILNE-THOMPSON, L.M. (1960), Theoretical Hydrodynamics, 4th Edition, MacMillan, London,

MOORE, D.W. and SAFFMAN, P.G. (1975) The Instability of a Straight Vortex in a Strain Field Proc. Roy. Soc. London Series A 346 p.413

PIERREHUMBERT, R.T. and WIDNALL, S.E. (1982) , "The Two-and Three-Dimensional Instabilities of a Spatially Periodic Shear Layer", J. Fluid Mech., 114, pp. 59-82.

RILEY, J.J. and METCALF, R.W. "Direct Numerical Simulation of Chemically Reacting Flows" AAAS Annual Meeting, Los Angeles Ca. May 1985

SAFFMAN, P.G. (1978), "The Number of Waves on Unstable Vortex Rings", J. Fluid Mech., 84, part 4, pp. 625-639.

SAFFMAN, P.G. and ROBINSON, A.C. (1982), "Three-Dimensional Stability of Vortex Arrays", Applied Mathematics, Calif. Instit. of Technology, Pasadena, CA,

SMITH, A.C. (1973), "On the Stability of Foppl's Vortices", J. Appl. Mech., 70, pp. 610.

TSAI, C-Y and WIDNALL, S.E. (1976), "The Stability of Short Waves on a Straight Vortex Filament in a Weak Externally Imposed Strain Field", J. Fluid Mech. 73, part 4, pp. 721-733.

WEI, T. and SMITH, C.R. (1985), "Secondary Vortices in the Wake of Circular Cylinders", Dept. of Mech. Engineering, Lehigh University, PA, to be published.

WIDNALL, S.E., BLISS, D. and ZALAY, A. (1971), "Theoretical and Experimental Study of the Stability of a Vortex Pair", Proceedings of the BSRL/AFOSR Aircraft Wake Turbulence Symposium, Sept. 1970, Seattle, Washington, Plenum Press.

WIDNALL, S.E., BLISS, D. and TSAI, C-Y. (1974), "The Instability of Short Waves on a Vortex Ring", J. Fluid Mech. 66, part 1, pp. 35-47.

WIDNALL, S.E. (1975), "The Structure and Dynamiics of Vortex Filaments", Reprinted from Annual Review of Fluid Mechanics, 7,

WIDNALL, S.E. (1972), "The Stability of a Helical Vortex Filament", J. Fluid Mech., 54, part 4, pp. 641-663.

WIDNALL, S.E. and SULLIVAN, J.P. (1972), "On the Stability of Vortex Rings", Proc. Roy. Soc. London, A332, pp. 335-353.

WIDNALL, S. E. and TSAI, C.Y. (1977), "The Instability of the Thin Ring of Constant Vorticity", PhTrans. Roy. Soc. of London, 287, pp.273-305

WIDNALL, S.E. (1985) "Three-dimensional Instability of Vortices in Separated flows" Symposium on Separated Flow around Marine Structures, Norwegian Institute of Technology, Trondheim Norway, June 26-28

A Ring-Vortex Representation of an Axi-Symmetric Vortex Sheet

B. DE BERNARDINIS AND D. W. MOORE

Abstract

We give a representation of an axi-symmetric vortex sheet by a set of
vortex rings with precisely determined self-induced velocities. The
representation loses accuracy at points near the axis of symmetry.

§1. INTRODUCTION

The point vortex method was introduced by Rosenhead (1), who took the
intuitive step of replacing a vortex sheet in two dimensions by an
array of infinite straight line or 'point' vortices. A point vortex
experiences no self-induced velocity, so the velocity of any member of
the array is a sum the velocities induced by the other vortices.

For an axi-symmetric vortex sheet, this intuitive approach runs into
difficulty. The axi-symmetric analogue of an infinite straight-line
vortex is a circular vortex ring of zero core radius. However such
a vortex ring has an infinite self-induced velocity. This difficulty
has been recognised and overcome (2) by the attribution of a core radius
to the ring vortices, the choice being arbitrary; however the final
results were not sensitive to the choice.

A more satisfactory approach is to follow Van der Vooren (3) and recog-
nise the singular nature of the integral which gives the velocity of any
point of the vortex sheet. In two dimensions a systematic discretization of the singular integral leads to both a formal justification of

Rosenhead's point-vortex method and a correction term which improves its accuracy.

Van der Vooren proceeded by means of a cancellation function which rendered the integrand analytic. We carry out an analogous calculation for the axi-symmetric case in §2 of this paper.

The integrand, or influence function, is much more complicated in the axi-symmetric case and we have not been able to remove the singularity entirely; however it has been weakened sufficiently to enable a standard integration formula to be applied.

To leading order, the result - given in equation (2.27) below - is the same as in two dimensions and the velocity of any elemental ring is the sum of contributions from the other rings in the array. Next in the ordering is a term which can be thought of as the self-induced velocity of a vortex ring with a numerically imposed core radius. However, equally important in practice is a term which reduces to Van der Vooren's correction as the radius of the vortex sheet becomes infinite. Unfortunately, the radius dependent terms become large as the radius gets small and the correction formula fails. Thus our integration formula loses accuracy if the vortex sheet intersects the axis of symmetry.

This defect is emphasised in §3 in which our integration formula is tested for the case of an instantaneously spherical vortex sheet, for which exact results are available.

§2. CANCELLATION OF THE SINGULARITY IN THE INFLUENCE FUNCTION

We use cylindrical polar coordinates (x,y,φ) with $0x$ the axis of symmetry. The vortex sheet is prescribed by the parametric equation

$$\underline{x}(\xi) = (x(\xi),y(\xi)) \qquad\qquad (2.1)$$

of its trace in the plane $\varphi = 0$. The parameter ξ is chosen so that the point $\underline{x}(\xi)$ moves with the average of the velocities on either side of the sheet. This choice of the parameter ξ means that the circulation $d\Gamma$ of the element $(\xi,\xi+d\xi)$ will be unchanging in a homogeneous fluid of zero viscosity, so that the density $V(\xi)$ defined by

$$V(\xi) = \frac{d\Gamma}{d\xi} \qquad (2.2)$$

is a known function, set by the initial conditions. We choose (+) and (-) to denote quantities evaluated on the left or right of the sheet when described in the ξ increasing direction.

Suppose that the velocity at the point $\underline{x}_0$, with Lagrange parameter $\xi=0$, due to a vortex ring of circulation $d\Gamma$ at $\underline{x}$ is $d\Gamma \ \underline{G}(\underline{x}_0,\underline{x})$: Let $\hat{n}$ be the unit normal to the vortex sheet at $\underline{x}_-$ and define, for $n \geqslant 0$

$$\underline{x}_+(n) = \underline{x}_0 + \hat{n} n$$
$$\underline{x}_-(n) = \underline{x}_0 - \hat{n} n \qquad (2.3)$$

so that the points $\underline{x}_+$, $\underline{x}_-$ are displaced normally off the sheet by a small distance n from the point $\underline{x}_0$ where the velocity is required. Then

$$\dot{\underline{x}}_0 = \frac{1}{2} \lim_{n\to 0_+} \int V(\xi)(\underline{G}(\underline{x}_+,\underline{x}(\xi)) + \underline{G}(\underline{x}_-,\underline{x}(\xi))d\xi \quad , \qquad (2.4)$$

where a time derivative at a fixed value of the Langrangian parameter is denoted by a dot. The problem is thus to evaluate the right-hand side of equation (2.4) numerically.

We could in principle find a good approximation to $\dot{\underline{x}}_0$ by choosing a small fixed n and evaluating the integral numerically. The trouble is that the resulting integrand would be large and rapidly varying for $\xi \sim 0$, so that a large number of integration points would be needed. Moreover, we would need such a locally refined mesh for each point x_0. We prefer to partly cancel the singularity of $\underline{G}$ before attempting numerical integration. Thus we must determine the singularity in $\underline{G}$ at $\underline{x}=\underline{x}_0$ and use this information to construct a cancellation function.

The function $\underline{G}$ can be constructed from the stream function for a vortex ring given in (4) and it follows that $\underline{x} \to \underline{x}_0$,

$$2\pi \ \underline{G}(\underline{x}_0,\underline{x}) = \hat{\underline{i}}(- \frac{\bar{y}}{s^2} + \frac{1}{2y} (\ln \frac{8y}{s} - 1 + \frac{\bar{y}^2}{s^2})) + \hat{\underline{j}}(\frac{\bar{x}}{s^2} - \frac{\bar{x}\bar{y}}{2ys^2})$$

$$+ O(\frac{s \ \log s}{y^2}). \qquad (2.5)$$

Here $\bar{x} = x_0-x$, $\bar{y} = y_0-y$ and $s^2 = \bar{x}^2 + \bar{y}^2$. To leading order, the velocity field is $\underline{H}_1^*(\underline{x}_0,\underline{x}_1)$ where

$$2\pi \ \underline{H}_1^* = (-\hat{\underline{i}}\bar{y} + \hat{\underline{j}}\bar{x})/s^2 \qquad (2.6)$$

and is thus the velocity of an infinite line vortex tangential to the ring at $(x,y,0)$. The higher terms $\underline{H}_2^*(\underline{x}_0,\underline{x})$, where

$$2\pi\ \underline{H}_2^*(\underline{x}_0,\underline{x})\ =\ (\hat{\underline{i}}(\ln\frac{8y}{s}\ -\ 1\ +\ \frac{\bar{y}^2}{s^2})\ +\ \hat{\underline{j}}(-\ \frac{\overline{xy}}{s^2}))/2y, \qquad (2.7)$$

are only logarithmically infinite, but vary rapidly as $\underline{x} \to \underline{x}_0$. We order our expansion by adopting the intuitive rule that terms which are only logarithmically infinite should be counted as $O(1)$; thus we must retain $\bar{y}^2/s^2$, etc., in (2.7). The higher terms in (2.5) also vary rapidly when $\underline{x} \to \underline{x}_0$, but the variations themselves are $o(1)$ and their effect wi be neglected here.

We can already see that the expansion (2.5) will fail as $y \to 0$ for fixe (x_0,y_0) and, as stressed in §1, this will lead to degradation of our integration formula at points near the axis of symmetry.

We need to expose the variation of $\underline{G}$ with the integration variable ξ fo $\xi \to 0$ and thus we expand the vector $\underline{x}(\xi)$ as

$$\underline{x}(\xi)\ =\ \underline{x}_0\ +\ \xi\ \frac{\partial\underline{x}}{\partial\xi}\ +\ \frac{1}{2}\ \xi^2\ \frac{\partial^2\underline{x}}{\partial\xi^2}\ +\ \dots \qquad (2.8)$$

so that, after some algebra, we find

$$2\pi\ \underline{H}_1^*(\underline{x}_+,\underline{x})\ =\ (\xi^2f^2+n^2+\xi^3g+\dots)^{-1}\ \left\{\hat{\underline{i}}(\xi b-n\cos\vartheta\ +\ \frac{1}{2}\ \xi^2\ \frac{\partial b}{\partial\xi}\ +\ \dots)\right.$$
$$\left.+\ \hat{\underline{j}}(-\xi a-n\sin\vartheta\ -\ \frac{1}{2}\ \xi^2\ \frac{\partial a}{\partial\xi}\ +\ \dots)\right\} \qquad (2.9)$$

and

$$2\pi\ \underline{H}_2^*(\underline{x}_+,\underline{x})\ =\ \frac{1}{2(y_0+\xi b+\dots)}\ \left\{\hat{\underline{i}}\left(\ln\left[\frac{8(y_0+\xi b+\dots)}{(\xi^2f^2+n^2+\dots)^{\frac{1}{2}}}\right]-\ 1+\ \frac{(\xi b-n\cos\vartheta+\dots)}{\xi^2f^2+n^2+\dots}\right.\right.$$
$$\left.\left.-\ \hat{\underline{j}}\ \left(\frac{(\xi b-n\cos\vartheta+\dots)\ (\xi a+n\sin\vartheta+\dots)}{\xi^2f^2+n^2+\dots}\right)\right)\right\} \qquad (2.10)$$

Here $a = (\frac{\partial x}{\partial\xi})_0$, $b = (\frac{\partial y}{\partial\xi})_0$, $f^2 = a^2 + b^2$, $g = a(\frac{\partial a}{\partial\xi})_0 + b(\frac{\partial b}{\partial\xi})_0$, and ϑ i is the inclination of the tangent to the sheet at $\underline{x}_0$, so that

$$\tan\vartheta\ =\ b/a \qquad (2.11)$$

Similar expressions for $\underline{H}_s^*(\underline{x}_-,\underline{x})$ can be constructed by replacing n by − in (2.9) and (2.10). The rapid variation of the integrand in the fundamental equation (2.4) for small, fixed, n is made clear by these formulae.

We construct a cancellation function $\underline{H}^{(+)} = \underline{H}_1^{(+)} + \underline{H}_2^{(+)}$ for the plus side by keeping only the leading terms in (2.9) and (2.10) and so, explicitly,

$$2\pi\, \underline{H}_1^{(+)}(n,\xi) = (\xi^2 f^2 + n^2)^{-1} \left\{ \hat{\underline{i}}(\xi b - n\cos\vartheta) + \hat{\underline{j}}(-\xi a - n\sin\vartheta) \right\} . \qquad (2.12)$$

and

$$2\pi\, \underline{H}_2^{(+)}(n,\xi) = \frac{1}{2y_0} \left\{ \hat{\underline{i}}\left(\ln\left[\frac{8y_0}{(\xi^2 f^2 + n^2)^{\frac{1}{2}}} \right] - 1 + \frac{(\xi b - n\cos\vartheta)^2}{\xi^2 f^2 + n^2} \right) \right.$$

$$\left. - \hat{\underline{j}}\left(\frac{(\xi b - n\cos\vartheta)(\xi a + n\sin\vartheta)}{\xi^2 f^2 + n^2} \right) \right\} . \qquad (2.13)$$

We now write (2.4) in the form

$$\dot{\underline{x}}_0 = \frac{1}{2}(\underline{I}_+ + \underline{J}_+ + \underline{I}_- + \underline{J}_-) + \int\limits_{|\xi|>L} V(\xi)\, \underline{G}(\underline{x}_0, \underline{x}(\xi))d\xi \qquad (2.14)$$

where L is arbitrary and where

$$\underline{I}_+ = \lim_{n\to 0+} \int_{-L}^{L} (V(\xi)\underline{G}(\underline{x}_+, \underline{x}(\xi)) - V(0)\underline{H}^{(+)}(n,\xi))d\xi \qquad (2.15)$$

and

$$\underline{J}_+ = \lim_{n\to 0+} V(0) \int_{-L}^{L} \underline{H}^{(+)}(n,\xi)d\xi \qquad (2.16)$$

with similar definitions of I_- and J_-. The last term in equation (2.14) represents contributions from distant parts of the sheet and its evaluation by the rectangle rule is immediate, since the integrand is smooth and bounded. Note that we cannot proceed in this way if $\underline{x}_0$ is near an end of the sheet.

Now if any function $f(\xi,n)$ satisfies the conditions

P(i), $f(\xi,n)$ is bounded for all ξ and n;

P(ii), $f(\xi,n)$ is a continuous function of ξ for $n \neq 0$;

P(iii), $f(\xi,0)$ is a continuous function of ξ in any interval excluding $\xi = 0$;

then it can be proved that

$$\lim_{n\to 0} \int_{\xi_0}^{\xi_1} f(\xi,n)d\xi = \int_{\xi_0}^{\xi_1} f(\xi,0)d\xi . \qquad (2.17)$$

for $\xi_1 > 0$ and $\xi_0 < 0$. By construction, the integrands of $\underline{I}_+$ and $\underline{J}_-$

satisfy conditions P(i), P(ii), and P(iii) and we can thus let n → 0 to obtain

$$\frac{1}{2} (\underline{I}_+ + \underline{I}_-) = \int_{-L}^{L} (V(\xi)\underline{G}(\underline{x}_0, \underline{x}(\xi)) - V(0)\underline{H}(0, \xi))d\xi \quad ; \tag{2.18}$$

$\underline{H}$ is the common value of $\underline{H}^{(+)}$, $\underline{H}^{(-)}$ when n = 0 and $\xi \neq 0$.

Before tackling the manipulations which lead eventually to our integration formula we note that we can evaluate the jump in velocity across the sheet, $\Delta\underline{U}$ say. This is given by

$$\Delta U = \lim_{n\to 0+} \int_{-L}^{L} V(\xi)[\underline{G}(\underline{x}_+, \underline{x}(\xi)) - \underline{G}(\underline{x}_-, \underline{x}(\xi))]d\xi ,$$

and the integral can be written, using (2.15) and (2.16) as

$$\Delta\underline{U} = \underline{I}_+ - \underline{I}_- + \underline{J}_+ - \underline{J}_- \quad .$$

We can let n → 0+ in the first two terms to find they cancel, so that

$$\Delta\underline{U} = \lim_{n\to 0+} V(0) \int_{-L}^{L} (\underline{H}^{(+)}(n,\xi) - \underline{H}^{(-)}(n,\xi)) \, d\xi \quad .$$

Use of properties P(i), P(ii) and P(iii) enable us to cancel the contributions from $\underline{H}_2^{(+)}$ and $\underline{H}_2^{(-)}$ to leave us with

$$\Delta\underline{U} = \lim_{n\to 0+} \frac{V(0)}{2\pi} \int_{-L}^{L} \frac{(-2n\cos\vartheta \, \hat{\underline{i}} - 2n\sin\vartheta \, \hat{\underline{j}})}{\xi^2 n^2 + f^2} \,)d\xi$$

We can evaluate the integral in closed form and then let n → 0 to find that

$$\Delta\underline{U} = -\hat{\underline{t}} \, \frac{V(0)}{f}$$

where $\hat{\underline{t}}$ is the unit tangent to the sheet in the ξ increasing direction. Now if $\gamma(0)$ is the vortex sheet strength

$$V(0) = \gamma(0) \left(\frac{\partial s}{\partial \xi} \right)_0 = \gamma(\xi)f$$

where s is arc distance along the sheet in the ξ increasing direction. Thus

$$\Delta\underline{U} = -\hat{\underline{t}} \, \gamma(0) \quad .$$

This equation is just the definition of γ and thus we have a check on our analysis.

The integrand in equation (2.18) is bounded and can be evaluated by rectangle integration. To introduce rectangle integration we subdivide the interval $(-L,L)$ into $2N+1$ equal sub-intervals of length $\Delta\xi$, whose centres are at the points $\xi_p = p\Delta\xi$ $(-N \leqslant p \leqslant N)$. The rectangle rule leads to the estimate

$$\tfrac{1}{2}(\underline{I}_+ + \underline{I}_-) \approx \Delta\xi \sum_{p=-N}^{p=N}{}' V(\xi_p)\underline{G}(\underline{x}_0,\underline{x}(\xi_p)) - \Delta\xi \sum_{p=-N}^{p=N}{}' V(0)\underline{H}(0,\xi_p) +$$

$$+ \Delta\xi \lim_{\xi\to 0} (V(\xi))\underline{G}(\underline{x}_0,\underline{x}(\xi)) - V(0)\underline{H}(0,\xi))$$

$$(2.19)$$

The prime attached to the summation sign means $p = 0$ is omitted, so that the first term in equation (2.19) is the contribution from elementary vortex rings at points $\underline{x}(\xi_p)$ $(p \neq 0)$ distinct from the point $\underline{x}_0$, where the velocity is being calculated. The second term can be calculated explicitly. Inspection of the cancellation function $\underline{H}$ reveals that the component $\underline{H}_1$ is an odd function of p and makes no contribution to the sum and that the contribution from $\underline{H}_2$ reduces to

$$-\Delta\xi \, V(0) \sum_{p=-N}^{p=N}{}' \underline{H}_2 = \frac{\Delta\xi}{2\pi y_0} V(0) \left\{ \underline{\hat{i}}(N \, \ln(\frac{8y_0}{f\Delta\xi}) - \ln(N!) + N(-1 + \frac{b^2}{f^2})) + \underline{\hat{j}}(-N \, \frac{ab}{f^2}) \right\}$$

$$(2.20)$$

The last term in equation (2.19) can be evaluated using the asymptotic results (2.9) and (2.10) and the definitions (2.12) and (2.13). It emerges that only $\underline{H}_1^{(+)}$ and $\underline{H}_1^{(-)}$ contribute and that this contribution is $\underline{P}$ where

$$\underline{P} = \frac{\Delta\xi}{2\pi} \left\{ \underline{\hat{i}}(\frac{\frac{\partial V}{\partial\xi}b}{f^2} + \frac{V}{2f^4}(b'(a^2-b^2)-2a'ab)) + \underline{\hat{j}}(-\frac{\frac{\partial V}{\partial\xi}a}{f^2}) + \right.$$

$$\left. + \frac{V}{2f^4}(a'(a^2-b^2)+2b'ab)) \right\} \; . \qquad (2.21)$$

In this equation $a' = \frac{\partial^2 x}{\partial\xi^2}$ and $b' = \frac{\partial^2 y}{\partial\xi^2}$.

It remains to evaluate the integrals $\underline{J}_+$ and $\underline{J}_-$ defined by equation (2.17). We cannot let $n \to 0+$ before evaluation, because not all the integrands are bounded. However, all the integrals are elementary and can be evaluated for finite n and the limit $n \to 0$ taken _after_ evaluation. Thus we find

$$\tfrac{1}{2}(\underline{J}_+ + \underline{J}_-) = \frac{L}{2\pi y_0} V(0) \left\{ \underline{\hat{i}}(\ln(\frac{8y_0}{Lf}) + \frac{b^2}{f^2}) - \underline{\hat{j}}(\frac{ab}{f^2}) \right\} \; . \qquad (2.22)$$

We can now collect the results (2.20), (2.21) and (2.22) and substitute into (2.14) to find that

$$\dot{\underline{x}}_0 = \Delta\xi \sum_p' V(\xi_p)G(\underline{x}_0,\underline{x}(\xi_p)) + \underline{P} + \frac{\Delta\xi_0\, V(0)}{2\pi y_0}\left\{\tfrac{1}{2}\frac{b^2}{f^2}\hat{\underline{i}} \quad -\tfrac{1}{2}\frac{ab}{f^2}\hat{\underline{j}}\right\}$$
$$+ \frac{\Delta\xi\, V(0)}{2\pi y_0}\, A(N)\hat{\underline{i}} \tag{2.23}$$

where

$$A(N) = -N\,\ln\left(\frac{8y_0}{f\Delta\xi}\right) + \ln(N!) + N + (N+\tfrac{1}{2})\,\ln(8y_0/f\Delta\xi(N+\tfrac{1}{2})) \tag{2.24}$$

and where in deriving (2.24) and (2.25) the relation

$$L = (N+\tfrac{1}{2})\Delta\xi \tag{2.25}$$

has been used to eliminate L. The final step is use of Stirling's formula to show that as $N \to \infty$

$$A(N) = \tfrac{1}{2}\,\ln(16\pi y/f\Delta\xi) - \tfrac{1}{2} + O(1/N)\quad, \tag{2.26}$$

so that our integration formula is

$$\dot{\underline{x}}_0 = \Delta\xi \sum_p' V(\xi_p)\underline{G}(\underline{x}_0,\underline{x}(\xi_p)) + \frac{\Delta\xi}{2\pi}\left\{\hat{\underline{i}}\left(\frac{\frac{\partial V}{\partial\xi}b}{f^2} + \frac{V(0)}{2f^4}(b'(a^2-b^2) -\right.\right.$$
$$\left.\left. - (2ab'ab)\right) + \hat{\underline{j}}\left(-\frac{\frac{\partial V}{\partial\xi}a}{f^2} + \frac{V(0)}{2f^4}(a'(a^2-b^2)+2b'ab)\right)\right\} +$$
$$+ \frac{\Delta\xi\, V(0)}{4\pi y_0}\left\{\hat{\underline{i}}(\ln(16\pi y_0/f\Delta\xi)-1 + \frac{b^2}{f^2})-\hat{\underline{j}}\frac{ab}{f^2}\right\}. \tag{2.27}$$

Equation (2.27) is the main result of this paper. We remark first tha if only the summed terms are retained, the velocity at ξ is the sum of the velocities due to all other elementary rings. This is the direct analogue of Rosenhead's point vortex method. Next, we note that if we let $y_0 \to \infty$ we recover Van der Vooren's correction for vortex sheets in two dimensions. The new result is contained in the last term in (2.27). The contribution

$$\frac{\Delta\xi\, V(0)}{4\pi y_0}(\ln(16\pi y_0/f\Delta\xi)-1)\hat{\underline{i}}$$

can be thought of as arising from a numerically determined core size. The remaining contributions, which involve the inclination of the vorte sheet to the axis of symmetry, seem to lack any intuitive significance.

§3. A TEST OF THE INTEGRATION FORMULA

We have made no attempt to establish a rigorous estimate of the accuracy of our integration formula (2.27). Instead we have tested it in a case for which the solution is known exactly.

Suppose the centre of a thin spherical shell is moving along the x axis with velocity U, the fluid inside being at rest relative to the shell. If the shell is instantaneously dissolved, the resulting vortex sheet has strength $\frac{3}{2}$ U sinϑ, where ϑ is the angular displacement from the symmetry axis of a point on the sheet. This result is a consequence of the potential flow solution for the exterior motion. The exterior flow at the instant of dissolution at a point of the sheet in the axial plane $\varphi = 0$ is

$$U(\cos^2\vartheta - \tfrac{1}{2}\sin^2\vartheta)\underline{\hat{i}} + \frac{3}{2} U \cos\vartheta\sin\vartheta \; \underline{\hat{j}}$$

while the interior velocity is just $U\underline{\hat{i}}$. Thus the mean velocity at a point of the spherical sheet at angular displacement ϑ is $\underline{\nu}_t$, where

$$\underline{\nu}_t = U(\cos^2\vartheta + \tfrac{1}{4}\sin^2\vartheta)\underline{i} + \frac{3}{4} U \sin\vartheta\cos\vartheta \; \underline{j} \qquad .$$

In Figure 1a we display $\log_{10}|e_x|$ and in Figure 1b $\log_{10}|e_y|$ where the absolute error vector $\underline{e}$ is defined by

$$\underline{e} = \underline{\nu}_N - \underline{\nu}_t \qquad .$$

The estimate $\underline{\nu}_N$ is obtained from equation (2.27) using N = 50,100,200 equally spaced points. Five point formulae were used for the derivatives occurring in the correction terms and the complete elliptic integrals needed for the computation of the influence function $\underline{G}$ were obtained from the fourth degree polynomial fits given in (5).

The most striking feature of the results is the great loss of accuracy near $\vartheta = 0°$ and $180°$ where the sheet intersects the axis of symmetry. Moreover, increasing N from 50 to 200 yields only a small improvement in the accuracy in the polar regions. On the other hand, the formula (2.27) is quite accurate away from the points $\vartheta = 0°$ and $\vartheta = 180°$. In the central regions, doubling N reduces the error by roughly a factor of eight. We believe this due to applying the rectangle rule to a function whose derivatives vanish at the ends of the range of integration, as is the case when the Lagrangian variable ξ is chosen to be just the angle ϑ.

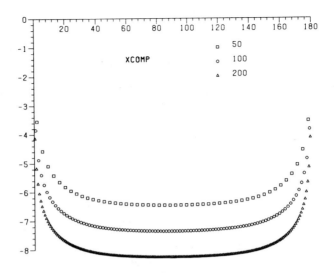

Figure 1a

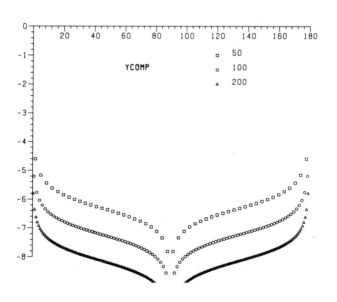

Figure 1b

When N = 50 the first term in (2.27) gives between 86% and 96% of the
current velocity, the precise fraction depending on both the position
on the sphere and the component examined.   Soh (6) has proposed the

ring sum with the self-induced velocity omitted as an integration formula.   This is formally correct in the limit $N \to \infty$, as is clear from (2.27), but the errors can be significant in practice.

Nor is the logarithmic term dominant in practice, which casts doubt on the accuracy of the correction term given in (7).

ACKNOWLEDGEMENT   We are grateful to Professor H. Reuter for explaining to us the convergence properties used in §2.   We are indebted to Professor G.R. Baker for helpful discussion.

REFERENCES

(1) Rosenhead, L. (1931).   Proc. Roy. Soc. A <u>34</u>, p.170.
(2) Acton, E. (1980).   J.F.M. <u>98</u>, p.1.
(3) Van der Vooren, A.I. (1980).   Proc. Roy. Soc. A<u>373</u>, 67.
(4) Lamb, H. (1932).   Hydrodynamics.   Cambridge University Press.
(5) Abramowitz, M. and Stegun, I.A. (1964).   Handbook of Mathematical Functions.   National Bureau of Standards.
(6) Soh, W.K. (1984).   Computation Techniques and Applications: CTAC-83. Elsevier.
(7) Edwards, A.V.J. and Morfey, C.L. (1981).   Computing & Fluids, <u>9</u>, 205.

Comparison of Experiment with the Dynamics of the von Karman Vortex Trail

L. Sirovich and C. Lim

1. Introduction

von Karman[1,2,3] introduced the double rowed assembly of staggered vortices, Fig. 1, as a model of the vortex trail observed behind bluff bodies and in particular behind the circular cylinder. He found this array to be unstable for all values of the spacing ratio

$$(1) \qquad k = h/\ell$$

except when

$$(2) \qquad k = k_0 = \frac{1}{\pi} \sinh^{-1} 1 \approx .281$$

in which case the array is (linearly) neutrally stable. A number of extensions of the von Karman treatment appeared in the 1930's[4-8]. For our purposes the most noteworthy achievement of this period was the proof by Kochin[8,9] that the von Karman model is unstable even for the distinguished value, (2). The growth found by Kochin is due to cubic non-linearities and is therefore slow. In fact, a direct calculation shows that hundreds of vortex pairs would be shed by a body before a significant divergence appears[10]. Domm[11] later demonstrated growth due to quadratic terms, but his mathematically elegant treatment can be reduced to the linear theory of a case for which the spacing ratio (1) does not satisfy (2), and his result becomes trivial[12]. These results are based on periodic initial disturbances. More recently by considering aperiodic disturbances we have been able to show that within linear theory itself arbitrarily small initial disturbances can be constructed which become arbitrarily large in a finite time.[12] A somewhat more abstract result due to Coddington[13] is that the l_2-norm of an aperiodic disturbance in general grows at least linearly in time.

Any unstable flow is generally regarded as experimentally unrealizable except possibly with great care. It is therefore ironic that the von Karman model, which is surrounded by unstable neighboring states, and itself is unstable, should so closely approximate the experimental situation. However into this picture one must factor the departures between the actual flow and that generated by the model:

(1) Physically we have a *boundary value problem* in which only a *half infinite trail* appears. In view of *long range forces*, $O(1/r)$, this can produce effects significantly different than for the fully infinite trail in the von Karman model. (For an attempt at considering the semi-infinite trail see Weihs[14].)

(2) *Viscous effects* have been neglected. A simplified picture is that the flow is basically inviscid with each vortex being eaten away from within by viscosity. A simple criterion for the point at which viscosity matters is

$$x/d \; \alpha \; Re$$

(d the cylinder diameter) gotten by elementary dimensional arguments. We therefore expect a significant inviscid trail behind a body.

(3) The *point vortex* idealization may be questioned. Certainly the way in which they are formed (see the photographs in van Dyke[15]) indicates that a distributed vorticity model would be more faithful to the physics. However the work by Meiron, Saffman & Schatzman[15] shows that the qualitative features of the von Karman model are *unaltered* by adopting finite core vortices. (This is not true in three dimensions where the zero core limit produces divergent velocities[17,18].)

(4) A number of experimental investigations do not fully confirm the von Karman model. For example variations of roughly 25% in $k \simeq .281$ are found and the ratio of h/d is found to have dynamic behavior[19-21]. Reviews of this work have been given by Chen[22] and Wille[23].

2. Review of Experiments by Sreenivasen[24]

Sreenivasen recently carefully measured velocity fluctuations behind a circular cylinder and subjected the data to a high resolution spectral analysis[24]. From this there emerged a provocative picture of the seemingly regular pattern of vortex shedding usually described. (See Refs. 25-28 for earlier evidence of irregularities.)

Fig. 2 contains Sreenivasen's results in the form of power spectra for three cases. The first at Re $\simeq$ 36, just beyond the onset of vortex shedding, shows one frequency (and the *harmonics* which arise from non-linearity). This represents limit cycle behavior. The next case at Re $\simeq$ 58 shows two frequencies and their combination tones. This represents two torus motion. The third case, for Re $\simeq$ 66 shows a *spectrum broadly placed around the shedding frequency and its harmonics*. Sreenivasen calls this the chaotic regime. If the Reynolds number is incremented further the motion against becomes orderly. In fact windows of order and chaotic motion then appear over a range of Reynolds' numbers.[24] Fig. 3 displays the rotation number. This is the ratio of second frequency to shedding frequency and provides a characterization of the low energy spectrum under study.

3. Review of the von Karman Theory[29]

For the vortex trail configured in Fig. 1, the complex potential is

$$(3) \qquad w = \frac{-i\kappa}{2\pi} \ln \sin \frac{\pi}{\ell} (z - z_1) + \frac{i\kappa}{2\pi} \ln \sin \frac{\pi}{\ell} (z - z_2)$$

where z_1 and z_2 are typical positions of vortices in upper and lower rows. The entire system moves to the left with speed

$$(4) \qquad U_s = \frac{\kappa}{2\ell} \tanh \pi k = \frac{\kappa}{\ell 2^{3/2}}$$

where the numerical value corresponds to the distinguished spacing ratio, (2). A nominal case to treat is that of the four-group perturbation, indicated by arrows in Fig. 1. In the linearized approximation, for the spacing ratio (2), this leads to the oscillation frequency

$$(5) \qquad \Omega_0 = \frac{\kappa}{8\ell^2} hz.$$

Observe that the above perturbation can be viewed as four rows of vortices and in analogy with the two row case, where the potential can be written in terms of two typically placed vortices, we can now write the potential in terms of four typically placed vortices. Thus the dynamics can be described in terms of four complex ordinary differential equations describing these. This in summary is the basis of the treatment by Kochin.[8,9] Actually after all the invariance properties are used only three real first order equations result.

This is an appropriate place to mention the contribution of Domm[11]. If the positions of typical vortices in the four-group case are denoted by Z_i, $i = 1, ..., 4$ as in Fig. 1, then an invariant is

$$(6) \qquad C = \frac{Z_1 - Z_2 + Z_3 - Z_4}{2}.$$

Observe that

$$(7) \qquad C = \left(\frac{X_1 + X_3}{2} - \frac{X_2 + X_4}{2} \right) + i \left(\frac{Y_1 + Y_3}{2} - \frac{Y_2 + Y_4}{2} \right)$$

which is the average relative positions of upper and lower typical vortices. Kochin[8,9] only considers initial data for which

$$(8) \qquad C = C_0 = \frac{l}{2} + ih$$

corresponding to the distinguished value, $k = k_0$. Domm[11] requires

$$(9) \qquad \Delta C = C - C_0 \neq 0$$

to show instability. The role of this condition is characterized by the following[12]:

Theorem. *If $\Delta C \neq 0$ then the problem may be transformed to a new problem for which the spacing ratio does not have the distinguished value but for which $\Delta C = 0$.*

But if $k \neq k_0$ then the configuration is linearly unstable. Thus although Domm's demonstration is elegant the result is trivial.

4. Flow Behind a Cylinder[10,30]

We now relate the von Karman model to flow past a cylinder with uniform upstream flow U_0. The velocity of the vortex trail, (4), relative to the cylinder is

$$(10) \qquad U_0 - U_s = U_0 - \kappa/(\ell 2^{3/2})$$

and the shedding frequency Ω is therefore given by

$$(11) \qquad \frac{\Omega_0}{\Omega} = \frac{U_s/U_0}{2^{3/2}(1 - U_s/U_0)}$$

where we have normalized Ω by the four group modal frequency (5). As noted earlier the frequency ratio is referred to as the rotation number.

Equation (11) contains U_s which in turn depends on κ and ℓ all of which are a priori unknown. von Karman[1,2,3], through the use of a momentum argument relates the drag coefficient to U_s,

$$(12) \qquad C_D \simeq \frac{\ell}{d} \left\{ 1.588 \frac{U_s}{U_0} - 0.628 \left[\frac{U_s}{U_0}\right]^2 \right\},$$

where d is the cylinder diameter. If we also introduce the Strouhal number

$$(13) \qquad S = \Omega d/U_0$$

then (12) may be written as[10]

$$(14) \qquad SC_D = \left(1 - \frac{U_s}{U_0}\right) \left[1.588 \frac{U_s}{U_0} - 0.628 \left[\frac{U_s}{U_0}\right]^2\right].$$

The left hand side is now regarded as a function of the Reynolds number Re and hence (14) determines U_s/U_0 as a function of Re. We use the formula given by Roshko[31]

$$(15) \qquad S \simeq 0.212 - 4.5/R; \quad 50 < R < 150$$

for Strouhal number, and the drag data of Tritton[26] (given in graphical form) for C_D. The dependence of U_s/U_0 on Re is substituted in (11) and the result of this is plotted in Fig. 3 as the heavy continuous line.[10] We emphasize the fact that (11) has been normalized by (5), the four group modal frequency.

5. Initial Value Problem

We will now consider the von Karman model in somewhat more detail than is usual. In fact we will see that the linear evolution of an initial disturbance to von Karman trail can be given a complete treatment.

In equilibrium the vortices are located at

$$(16) \qquad Z_m^+ = m\boldsymbol{\ell} + U_s t + \frac{ih}{2}, \qquad Z_m^- = (m + \tfrac{1}{2})\boldsymbol{\ell} + U_s t - \frac{ih}{2}$$

with respect to a fluid at rest as $y \uparrow \pm\infty$. U_s represents the earlier stated translational velocity of the street, (4). If we denote the perturbed positions of the vortices by

$$(17) \qquad z_m^\pm = x_m^\pm + iy_m^\pm$$

then

$$(18) \qquad \frac{d}{dt}\,\overline{z}_m^\pm = \mp \frac{i\kappa}{2\pi} \sum_k{}' \frac{z_m^\pm - z_k^\pm}{(m-k)^2 \boldsymbol{\ell}^2} \pm \sum_k{}' \frac{z_m^\pm - z_k^\mp}{((m - k \mp \tfrac{1}{2})\boldsymbol{\ell} \pm ih)^2}$$

where a bar denotes the complex conjugate.

Symmetries: (A) If $z_m = (z_m^+, z_m^-)$ and $Tz_m = (z_{m+1}^+, z_{m+1}^-)$ then $T^n z_m$, for all integer n is also a solution. (B) It also follows that

$$(19) \qquad \hat{z}_m = \hat{T}z_m = (z_{-m}^-(-\kappa),\ z_{-m}^+(-\kappa))$$

is a solution.

The first symmetry implies we introduce the *generating functions*

$$(20) \qquad \begin{cases} (X^+, Y^+) = \sum_m (x_m^+, y_m^+)\exp[im\phi] \\[2mm] (X^-, Y^-) = \sum_m (x_m^-, y_m^-)\exp[i(m + \tfrac{1}{2})\phi] \end{cases}$$

and the second symmetry that we introduce

$$(21) \qquad \begin{cases} \alpha = (X^+ + X^-)/2, & \beta = (Y^+ - Y^-)/2 \\[2mm] \mu = (X^+ - X^-)/2, & v = (Y^+ + Y^-)/2 \end{cases} \quad .$$

This leads to

(22) $\quad \dfrac{d}{d\tau}\begin{bmatrix}\alpha\\ \beta\end{bmatrix} = \begin{bmatrix} ib & a-c \\ a+c & ib \end{bmatrix}\begin{bmatrix}\alpha\\ \beta\end{bmatrix}$

where time has been normalized by

(23) $\quad \tau = \kappa t/(2\pi l^2)$.

The elements in the above matrix are given by

(24) $\quad \begin{cases} a = [\phi(2\pi-\phi)-\pi^2]/2, \quad b = [\sqrt{2}\,\pi\phi\sinh\dfrac{h}{l}(\pi-\phi) + \pi^2\sinh\dfrac{h}{l}\phi]/2 \\[2mm] c = [\pi^2\cosh\dfrac{h}{l}\phi - \pi\phi\sqrt{2}\,\cosh\dfrac{h}{l}(\pi-\phi)]/2 \end{cases}$

which are closely related to the functions given in Lamb[29] in his presentation of the von Karman analysis.

If we adopt the vector notation

(25) $\quad \boldsymbol{\alpha} = (\alpha,\beta), \quad \mathbf{v} = (\nu,\mu)$

and write the above differential equation a

(26) $\quad \dot{\boldsymbol{\alpha}} = \mathbf{F}\boldsymbol{\alpha}$

with $\mathbf{F}$ the matrix appearing in (22) then

(27) $\quad \dot{\mathbf{v}} = \bar{\mathbf{F}}\mathbf{v}$.

Thus the solution to the problem is reduced to evaluating the solution operator of the system. This can be accomplished directly and we find

(28) $\quad \exp[\mathbf{F}\tau] = \begin{bmatrix} \cos K\tau & \dfrac{a-c}{K}\sin K\tau \\[3mm] \dfrac{a+c}{K}\sin K\tau & \cos K\tau \end{bmatrix}\exp(ib\tau); \quad K = (c^2-a^2)^{1/2}.$

An alternate version is the modal decomposition,

(29) $\quad \exp[i\mathbf{F}\tau] = \dfrac{1}{2}\exp(i\omega^+\tau)\begin{bmatrix}1\\ -i\,\rho\end{bmatrix}[1,i/\rho] + \dfrac{1}{2}\exp(i\omega^-\tau)\begin{bmatrix}1\\ i\,\rho\end{bmatrix}[1,-i/\rho]$

where

$$(30) \qquad \omega^{\pm} = b \pm K, \qquad \rho = \sqrt{(c+a)(c-a)}$$

To relate this to the general initial value problem for the evolution of the perturbation to the vortex trail it is convenient to introduce the 'Green's problems';

$$(31) \qquad \begin{cases} \mathbf{g}_m = (g_m^+, g_m^-): \ \mathbf{g}_m(\tau = 0) = (1,0)\delta_{m0} \\ \mathbf{G}_m = (G_m^+, G_m^-): \ \mathbf{G}_m(\tau = 0) = (i,0)\delta_{m0} \end{cases}$$

These are discrete analogs to δ-function initial data and give the evolution of an initial displacement in x and y directions respectively of the $m = 0$ vortex in the upper row. Once $\mathbf{g}_m$ and $\mathbf{G}_m$ are known

$$(32) \qquad \begin{cases} \hat{\mathbf{g}}_m = \hat{T}\mathbf{g}_m \\ \hat{\mathbf{G}}_m = \hat{T}\mathbf{G}_m \end{cases}$$

are such that

$$(33) \qquad \begin{cases} \hat{\mathbf{g}}_m(\tau = 0) = (0,1)\delta_{m0} \\ \hat{\mathbf{G}}_m(\tau = 0) = (0,i)\delta_{m0}. \end{cases}$$

Next for a sequence of real numbers $\{a_m\}$ we define the convolution produce

$$(34) \qquad (z*a)_m = (\Sigma_n z_{m-n}^+ a_n, \ \Sigma_n z_{m-n}^- a_n).$$

Then if the general initial disturbance is expressed by

$$(35) \qquad z_m^{\pm}(\tau = 0) = a_m^{\pm} \pm ib_m^{\pm}$$

its evolution is given by

$$(36) \qquad z_m = (\mathbf{g}*a^+)_m + (\mathbf{G}*b^+)_m + (\hat{\mathbf{g}}*a^-)_m + (\hat{\mathbf{G}}*b^-)_m.$$

Each of the Green's function problems translates to simple data for the ordinary differential equation problem, e.g., corresponding to $\mathbf{G}_m$ we have

(37) $\alpha(\tau = 0) = (0, \tfrac{1}{2}), \qquad \mathbf{v} = (\tfrac{1}{2}, 0).$

The actual construction of a solution now follows from elementary linear operations. Thus once α and $\mathbf{v}$ are determined, then e.g.,

(38) $X^+ = \alpha + \mu$

and once the generating functions are determined the actual solution follows from inversion, e.g.,

(39)
$$x_m^+ = \frac{1}{2\pi} \int_0^{2\pi} X^+ \exp(-im\phi)d\phi$$
$$= \frac{1}{2\pi} \int_0^{2\pi} (\alpha + \mu)\exp(-im\phi)d\phi.$$

6. The Periodic Case

A particularly simple case is that of periodic initial data. In brief if the initial data is n-periodic, e.g.,

(40) $z_m(0) \propto \exp\left[im\left(\frac{2\pi}{n}\right)\right]$

then the corresponding generating functions are proportional to delta functions as are α and y, viz.,

(41) $\alpha, \mathbf{v} \propto \delta\left[\phi - \frac{2\pi}{n}\right] = \delta(\phi - \phi_n).$

If this is substituted for example into (39), we obtain the solution as linear combinations of the four terms

(42) $\exp[\pm i\omega^\pm(\phi_n) - im\phi_n]$

where $\omega^\pm(\phi)$ is given by (30). The roots $\omega^\pm$ are plotted in the Figure 4.[10] From the condition of realness we need only consider $0 \leqslant \phi \leqslant \pi$. To make contact with our earlier deliberations if $n = 2$, $\phi_2 = \pi$ the initial disturbance is 2 periodic and this gives the four-group perturbation already discussed. In fact as can be seen $\omega^\pm(\phi_2) = 1$ in Figure 4 and $\omega^\pm(\phi_2)$ has been used as the normalization of the plot.

In our preliminary estimates for rotation number we used the four group frequency. In Fig. 3 we have also plotted the rotation number versus Reynolds number for other periodic groups besides the four-group case, as well as the experimental points of Sreenivasen.[24] As can be seen better agreement is obtained with other modal oscillations.

7. Wave Propagation on the Vortex Trail

By considering the original von Karman model we find in the modal oscillations a mechanism by which there is reasonable agreement with experiment for the rotation number. But as forms such as (28) indicate this also implies the presence of wave propagation. Sreenivasen did not look into this question, however Tritton[26] reports on measurements taken simultaneously at two different locations. From these Tritton concluded the presence of a wave traveling along the vortex trail in the negative direction with a speed

$$(43) \qquad U_p \simeq .3U_0$$

relative to the vortex trail. This is an average value and is based on the nominal value of the street velocity given by

$$(44) \qquad \hat{U}_s \simeq .8U_0.$$

According to Tritton[26] there is considerable variation in (44) and (43). We now sketch the calulation of a wave speed from the analysis given thus far.

A typical term in the solution of the initial value problem has the form

$$(45) \qquad I = \int f(\phi) \exp[i\omega(\phi)\tau - im\phi]d\phi$$

where $f(\phi)$ arises out of initial data and terms from the solution operator (29). Then a standard argument associated with Kelvin's formula[32] states that as $\tau \uparrow \infty$ the leading contribution arises from the neighborhood of the value of ϕ such that

$$\omega'(\phi) = \frac{m}{\tau}$$

where $\omega' = d\omega/d\phi$ is the group speed. Thus the possible wave speeds are gotten by taking the slopes of the curves in Fig. 4.

This argument must be modified if the coefficient $f(\phi)$ in (45) is singular. In this instance the contribution is largest when the singularity and group speed location coincide. The approach we take is to find the wave speeds coinciding with the greatest singularity. This in turn will lead to the largest amplitude wave, and it is this speed which will be compared with the experimental value (43).

For the case under study the largest singularity occurs when $\phi = 0$ (or $\phi = 2\pi$). In fact in this case

$$\rho = 0(\phi)$$

at the origin and the integral (45) doesn't exist! In order to correctly treat this case we must use the intermediate representation given by (28) rather than the modal form (29). From this we can write the integral in question as

$$(46) \qquad I = \int_0^\pi \frac{a - c}{K} \sin(K\tau)\exp[ib\tau - im\phi]d\phi$$

where $K = 0(\phi)$. Initial conditions have been neglected in (46) since their only effect in the asymptotic evaluation will be to introduce a multiplicative constant.

The major contribution to (46) comes from the neighborhood of the origin where $a - c$ tends to a constant and $K = 0(\phi)$. Therefore it will suffice if we consider

$$(47) \qquad I \propto 2i \int_0^\pi \frac{\sin K\tau}{\phi} \exp[ib\tau - im\phi]d\phi$$

$$= \int_0^\pi \frac{\exp(i\omega^+\tau) - \exp(i\omega^-\tau)}{\phi} \exp(-im\phi)d\phi$$

under the limit;

$$(48) \qquad \tau \uparrow \infty, \quad \frac{m}{\tau} = \frac{d\omega^\pm(0)}{d\phi}, \quad \text{fixed.}$$

To treat (47) we consider

$$I \propto \lim_{\epsilon \downarrow 0} \int_0^\pi \frac{\exp(i\omega^+\tau - im\phi) - \exp(i\omega^-\tau - im\phi)}{\phi^{1-\epsilon}}d\phi$$

in which case each term yields to standard methods.[32] The result of this is

$$(49) \qquad I = 0(\ln \tau).$$

The corresponding wave-speeds are

$$\frac{d\omega^{\pm}(0)}{d\tau} = b'(0) \pm K'(0) = 5.3, \ 1.9.$$

The units are vortex spaces, l, per unit time τ as given by (23). Therefore in dimensional units the speeds are

$$U_p = \frac{5.3 \ \kappa}{2\pi\mathit{l}}, \ \frac{1.9\kappa}{2\pi\mathit{l}}.$$

To evaluate κ/l we appeal to the relation which gives the velocity of the vortex relative to the upstream

$$\hat{U}_s = U_0 - \frac{\kappa}{\mathit{l}2^{3/2}} = .8U_0$$

where we have substituted Tritton's value (44). The two wave therefore become

$$U_p = .47U_0, \ .17U_0$$

which is to be compared with Tritton's value (43). In view of the loose connection between theory and experiment as well as the degree of imprecision in experiment this comparison must be regarded as encouraging.

8. Conclusions

While the arguments linking the von Karman model to the physical vortex make are still unclear, their qualitative and quantitative agreement remains striking. In addition to similar values of the spacing ratio, we now find similarities in the rotation number and in the phenomenon of wave propagation along the vortex trail.

Acknowledgment. The authors are grateful to K. R. Sreenivasen for many helpful comments, his generosity in providing them with data, and permission to display Figure 2.

56

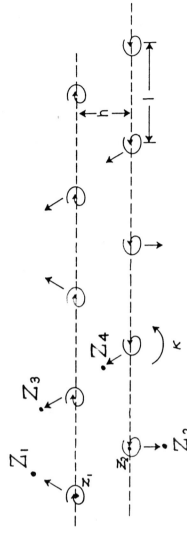

Figure 1. Schematic of a vortex trail aspect ration h/ℓ and circulation κ. Arrows indicate a 4-group perturbation (see text).

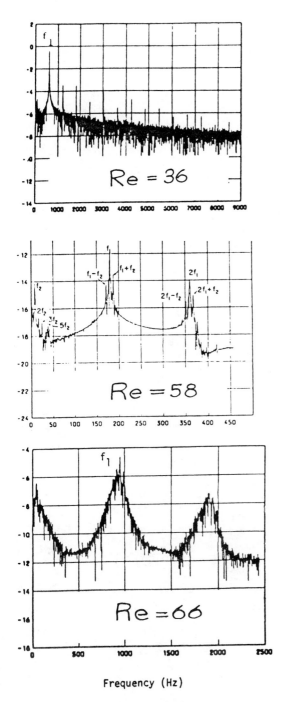

Frequency (Hz)

<u>Figure 2.</u> Normalized frequency spectrum for streamwise velocity perturbations at the three Reynolds numbers indicated. Further explanation is given in the text.

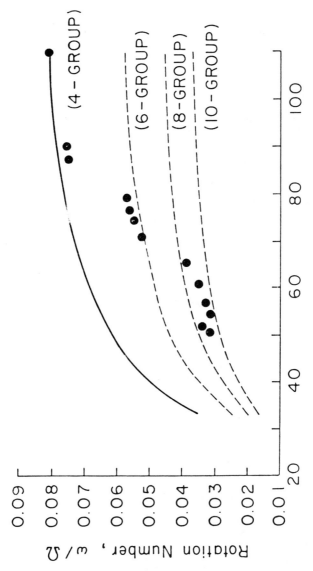

Figure 3. Plots of rotation number, Φ/Ω, versus Reynolds number, Re. Continuous curve represents theoretical prediction based on 4-group perturbation. Dashed curves are theoretical predictions based on indicated perturbations. Heavy dots represent measurements obtained by Sreenivasen.[24]

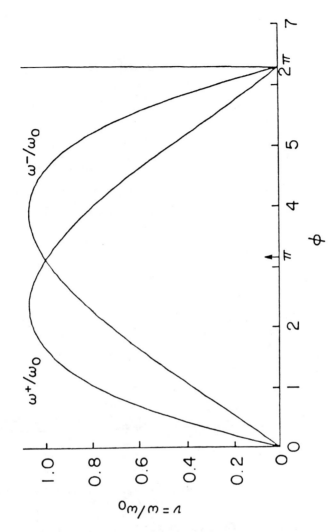

Figure 4. Dispersion relation showing possible oscillation frequencies for the vortex trail. Frequencies have been normalized with respect to the four-group frequency ω_0. Abscissa, ϕ, is the angular wave number.

60

References

1. v. Karman, Th. (1911), *Gottinger Nachrichten, Math. Phys. Kl.*, 509-519.
2. v. Karman, Th. (1911), *Gottinger Nachrichten, Math Phys. Kl.*, 547-556.
3. v. Karman, Th., and Rubach, H. (1912), *Phys. Zeitschrift* **13**, 49-59.
4. Schmieden, C. (1936), *Ing. Arch.* **7**, 215-241.
5. Durand, G. (1933), *Compt. Rend. de l'Acad. des Sci.* **196**, 382-385.
6. Dolapschiew, B. (1937), *ZAMM* **17**, 313-323.
7. Maue, A. (1940), *Z. angew. Math. u. Mech.* **20**, 129-137.
8. Kochin, N. E. (1939), *Compt. Rend.* (*Doklady*) *de l'Acad. des Sci. de l'URSS* **24**, 18-22.
9. Kochin, N. E., I. A. Kibel' and N. V. Roze (1964), Theoretical Hydrodynamics, Sec. 5.21, Interscience.
10. Sirovich, L. (1985), *Phys. Flu.* (to appear September).
11. Domm, U. (1956), *ZAMM* **36**, 367-371.
12. Lim, C. and Sirovich, L. (1985), An investigation of the von Karman vortex trail model (in preparation).
13. Coddington, E. A. (1952), *Jour. Math. Phys.* **30**, 171-199.
14. Weihs, D. (1972), *J. Fluid Mech.* **54**, 679.
15. van Dyke, M. (1982), An Album of Fluid Motions, The Parabolic Press, Stanford.
16. Meiron, D. I., Saffman, P. G. and Schatzman, J. C. (1984), *J. Fluid Mech.* **147**, 182-212.
17. Saffman, P. G. (1970), *Stud. App. Math.* **49**, 371-380.
18. Moore, D. W. and Saffman, P. G. (1972) *Phil. Trans. R. Soc.* **A272**, 403-429.
19. Schaefer, J. W. and Eskanazi (1959), *JF f.* **6**, 241.
20. Taneda, S. (1956), *J. Phys. Soc. Japan* 1 , 302.
21. Berger, E. and Wille, R. (1974), *Ann. Rev. Flu. Mech.* **4**, 313.
22. Chen, Y. N. (1972). *Trans. A.S.M.E. Jour. Eng. Ind.* **94**, 613.
23. Wille, R. (1960), *Adv. Appl. Mech.* **6**, 273-287.
24. Sreenivasen, K. R. (1985), in Frontiers in Fluid Mechanics (S. H. Davis and J. L. Lumley, eds.), 41-67, Springer-Verlag, New York.
25. Taneda, S. (1959) *J. Phys. Soc. Japan* **14**, 843.
26. Tritton, D. J. (1959), *J. Fluid Mech.* **6**, 547.
27. Tritton, D. J. (1971), *J. Fluid Mech.* **45**, 203.
28. Friehe, C. A. (1980), *JFM* **100**, 237-241.
29. Lamb, H. (1945), Hydrodynamics, Sec. 156, Dover, New York.
30. Goldstein, S. (1965), Modern Developments in Fluid Mechanics, vol. II, Sec. 247, Dover, New York.
31. Roshko, A. (1954), On the development of turbulent wakes from vortex sheets, Report of the National Advisory Committee for Aeronautics, no. 1191, Washington.
32. Sirovich, L. (1971), Techniques of Asymptotic Analysis, Springer-Verlag, New York.

Section II

Vortex Breakdown

Force- and Loss-Free Transitions Between Vortex Flow States

J. J. KELLER, W. EGLI AND J. EXLEY

SUMMARY

It is suggested that a certain kind of vortex breakdown can be interpreted as a two-stage transition. The analysis is based on an extension of a variational principle proposed by Benjamin. Complete numerical results are presented which show loss-free transitions (including their internal structure) in a Rankine vortex. Moreover, an apparent paradox is resolved which is associated with the fact that a vortex flow in a diffuser approaches the critical state.

INTRODUCTION

Obviously transitions between flow states which are both force- and loss-free represent rather special phenomena if they exist at all. Force-free transitions between flow states (e.g. hydraulic jumps, shock waves in gases etc.) generally lead to an entropy increase, and in many cases, including the two just mentioned, the only transition which is both force- and loss-free is the trivial one.

In the present context it is useful to consider the entropy increase as a necessary degree of freedom of a transition leading from a supercritical to a subcritical flow state. Wavelets which move upstream in a subcritical flow after a transition can carry information to the transition and cause it to respond in a particular way to changes in the downstream conditions. In this case the transition must be able to satisfy imposed downstream boundary conditions. Such causal arguments are well-known in the theory of shock waves.

However, with respect to the present consideration of force- and loss-free transitions it is preferable to use the logical inversion of the previous statement: For a transition between two flow states which is both force- and loss-free the downstream flow state cannot be subcritical.

A well-known example of a loss-free transition to a supercritical flow state is Benjamin's solution (ref. 1) for a gravity current in a closed channel flow. In this case, a loss-free transition leads to a supercritical flow with a free surface (see figures 1b and 1c). The Froude number of this intermediate flow state is $\sqrt{2}$. Finally a hydraulic jump or a more general dissipative transition leads from the intermediate flow state to a subcritical downstream flow state.

The situation is as follows: We consider a channel which ends in a large pool filled with water up to a level which is initially above the upper channel wall. Assuming now that the water level of the pool drops to a position just below the upper channel wall, as illustrated by figure 1a, a wave moves into the channel and produces a flow into the pool. However, to simplify the problem we assume that the undisturbed flow in the channel is uniform and is just set to cancel the wavespeed. In this case the wave appears stationary with respect to the channel exit. Considering now the flow downstream of this wave shows that the second flow state is subcritical so long as the level of the pool remains

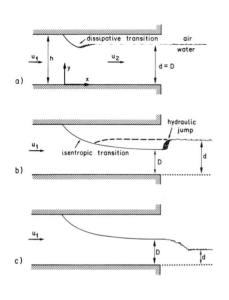

Fig. 1: a) Single (dissipative) transition to a subcritical channel
flow.

b) Depth d for which two solutions are possible: A single
dissipative transition or a
two-stage transition consisting of the isentropic transition and a hydraulic jump.

c) Small depth d for which the
first transition is always
isentropic.

above a critical height (∼ 65 % of the
channel height above the lower channel
wall). For lower levels of the pool the
downstream flow would become supercritical and the existence of a single dissipative transition to the second state has
to be rejected on physical grounds.

In this case the initially single dissipative wave splits into a unique loss-free
wave, leading to a second flow state which is supercritical, and a hydraulic jump
which leads to the subcritical downstream flow state. The loss-free wave corresponds to Benjamin's gravity current. This interpretation of events has been
confirmed experimentally and is generally accepted.

Based upon a great deal of experimental evidence and the analogy with the
hydraulic problem mentioned we postulate the existance of two-stage transitions
in certain vortex flows. As illustrated in figure 2 three flow regimes may occur
which are connected by two fundamentally different transitions: an initial loss-
free transition from the supercritical upstream to an intermediate flow state
which is again supercritical, followed by a dissipative jump to the final downstream state. The first transition diverts the flow around the nose of a zone of
essentially stagnant fluid. However, to avoid a longer discussion of stability
properties we may simply assume that the fluid in the stagnant zone is replaced
by a fluid of much lower density, e.g. an air-filled bubble surrounded by water
flow. The two-stage transition (and in particular the criterion for its occurrence) is uniquely defined by the first (loss-free) transition. Hence, no detailed considerations need be given to the second transition.

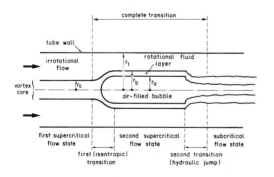

Fig. 2: Schematic diagram of the
transitions of a vortex
with a core of rotational
flow in a tube.

To tackle this problem we first
consider the extension of Benjamin's variational principle (ref.
2) for axisymmetric vortex flows
to cases where free surfaces appear.

It has been shown by Benjamin that the variational principle

$$\delta S = 0, \quad S = 2\pi \int_0^a \{\tfrac{1}{2} \psi_y^2 + H(\psi) - \frac{I(\psi)}{2y}\} \, dy, \tag{1}$$

where $\quad I = \tfrac{1}{2} C^2, \quad y = \tfrac{1}{2} r^2,$

r denotes the radial coordinate, H is the total head and C the angular momentum, yields solutions $\psi(y)$, bounded by $y = 0$ and $y = a$, which represent stream functions of possible cylindrical vortex flows, hereafter called flow states (see figure 3). As we consider loss-free flows both the total head H and the function I remain constant on stream surfaces. For this reason H and I depend on the stream function ψ only. S is simply the sum of pressure and momentum flux expressed in the stream function ψ and integrated over a flow cross-section. It is easy to show that Euler's equation, which corresponds to this variational principle, is the equation for loss-free cylindrical flow.

EXTENSION OF BENJAMIN'S VARIATIONAL PRINCIPLE

As a natural extension of this principle we consider the case where either the upper or lower boundary of the solution is permitted to depart from that of the upstream state. To be specific we shall consider the case where the lower boundary can be displaced from the corresponding upstream boundary, as would be the case, for example, where the downstream flow surrounds a cavity (see figure 4). In a second step the variational principle will be extended further to axisymmetric but non-cylindrical flows with or without free surfaces. As we will see, this extended variational principle provides a basis to investigate the internal structure of the transitions. We consider a stream function ψ_A of an upstream flow occupying an annulus between $y = y_c$ and $y = a$ and the stream function ψ of a flow between $y = y_0 \geq y_c$ and $y = a$. The special case of a simple vortex tube corresponds to $y_c = 0$.

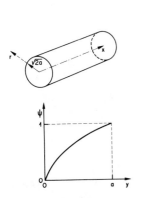

Fig. 3: Coordinates used to define the variational principle.

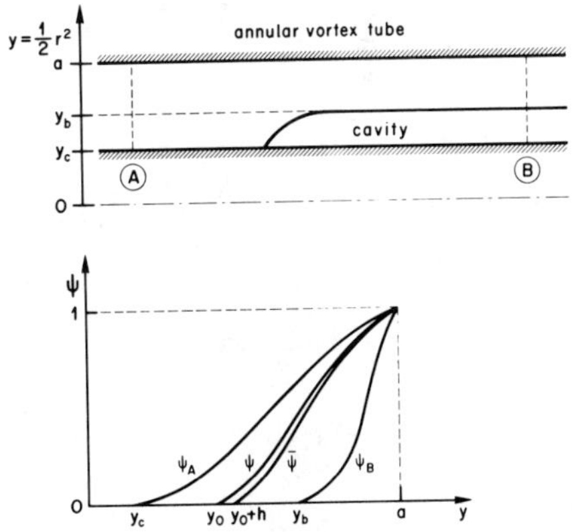

Fig. 4: Illustration related to the extended variational principle.

If we let:

$$F(y, \psi, \psi') = 2\pi \{\tfrac{1}{2} \psi'^2 + H(\psi) - \frac{I(\psi)}{2y}\}, \tag{2}$$

where primes denote differentiation with respect to y, then we have the upstream flow force given by

$$S_A = \int_{y_c}^{a} F(y, \psi_A, \psi_A') \, dy. \tag{3}$$

To find the possible downstream flow states $\psi_B(y)$ with free lower endpoint $(y_b, 0)$ we study the variational principle

$$\delta S = 0, \quad S = \int_{y_0}^{a} F(y, \psi, \psi') \, dy \tag{4}$$

where S is now varied both with respect to y_0 and ψ. We consider a stream function $\bar{\psi}$ displaced slightly from ψ (see figure 4) and intersecting the y-axis at $y = y_0 + h$.

Let: $\bar{\psi} = \psi(y) + \varepsilon\eta(y)$, $\eta(a) = 0$, $\eta(y_0 + h) \neq 0$. \hfill (5)

Following the section on variable endpoints in Bolza's "Calculus of Variations", for example, we find after some calculation

$$\Delta S = \bar{S} - S = \int_{y_0+h}^{a} F(y, \bar{\psi}, \bar{\psi}') \, dy - \int_{y_0}^{a} F(y, \psi, \psi') \, dy \tag{6}$$

$$= -h \left[\{F(y_0, \psi(y_0), \psi'(y_0))\} - \psi'(y_0) F_{\psi'}(y_0, \psi(y_0), \psi'(y_0)) \right]$$

$$- \varepsilon \int_{y_0}^{a} \eta(y) \left\{ \frac{d}{dy} F_{\psi'} - F_{\psi} \right\} dy + 0(h^2)$$

By the usual argument $\delta S = 0$ now yields Euler's equation

$$\frac{1}{2\pi} \left\{ \frac{d}{dy} F_{\psi'} - F_{\psi} \right\} = \psi_{yy} - H'(\psi) + \frac{I'(\psi)}{2y} = 0 \tag{7}$$

and the auxiliary equation

$$\frac{1}{2\pi} \left[F(y_0, \psi(y_0), \psi'(y_0)) - \psi'(y_0) F_{\psi'}(y_0, \psi(y_0), \psi'(y_0)) \right] \tag{8}$$

$$= H(0) - \frac{I(0)}{2y_0} - \frac{1}{2} \psi'^2(y_0) = 0.$$

Euler's equation may be recognized as the equation of motion derived by Long (ref. 3) and Squire (ref. 4) for cylindric vortex flows and the auxiliary equation corresponds to Bernoulli's theorem applied to the free surface of the bubble. To calculate the flow-force difference S_B-S_A between the downstream state ψ_B and the upstream state ψ_A we note that the variational principle for free endpoints can be interpreted as a two-stage procedure. First the variational principle is applied for fixed endpoints y_0, which leads to a one-parametric set of extremals $\tilde{\psi}$ $(y; y_o)$ (say). Subsequent variation with respect to the endpoint y_o finally yields the downstream state

$$\psi_B(y) = \tilde{\psi}(y, y_b).$$ (9)

If the one-parametric set of extremals $\tilde{\psi}$ is introduced in the expression (6) for ΔS the term containing Euler's equation vanishes identically, whilst the term containing the auxiliary equation generally remains.

Hence, we have

$$\Delta S = - h \; [F(y_o, \; \tilde{\psi}(y_o; \; y_o), \; \tilde{\psi}_y(y_o; \; y_o))$$ (10)

$$- \tilde{\psi}_y \; (y_o; \; y_o) \; F_{\tilde{\psi}_y} (y_o) \; \tilde{\psi}(y_o; \; y_o), \; \tilde{\psi}_y(y_o; \; y_o))].$$

Integrating this expression from the upstream lower boundary y_c to the downstream lower boundary y_b and resubstituting for the sum F of pressure and momentum flux yields

$$S_B - S_A = 2\pi \int_{y_c}^{y_b} \{ \frac{1}{2} \tilde{\psi}_y^2(y_o; \; y_o) - H(0) + \frac{I(0)}{2y_o} \} \; dy_o \; .$$ (11)

This expression should correspond to equation (4.15) in Benjamin's paper (see ref. 2) on vortex breakdown. However, Benjamin's flow-force expression for free surfaces does not contain the second and third terms in the present integrand. These are the contributions due to the variation of the free boundary. Using Weierstrass' theorem Benjamin argued that for fixed endpoints the flow-force difference S_B-S_A is strictly positive and tried to extend his result to free endpoints. We now see that his argument does not hold in the case of free endpoints.

We return now to the general discussion:

Combining Euler's equation and the auxiliary equation with the condition for vanishing flow-force difference yields the discrete solution for the second flow state. The condition for vanishing flow force $S_B-S_A = 0$ now has a rather plausible meaning. It simply says that the integral over Bernoulli's equation along the lower boundary vanishes.

A priori it is not at all clear that a flow state _does_ exist which is different from the first flow state and for which mass flow, flow force and entropy are the same as for the first flow state. However, it turns out that in general there is exactly one flow state of this kind - similarly as in the case of the hydraulic problem discussed previously.

To investigate the internal structure of this isentropic force-free transition we introduce an axial variable x and consider the local flow force

$$S(x) = 2\pi \int_{y_c(x)}^{a} \{ \frac{1}{2} \psi_y^2 + \frac{1}{4y} \psi_x^2 + H(\psi) - \frac{I(\psi)}{2y} \} \; dy.$$ (12)

To avoid the complication of the variation with free endpoints we interchange variables

$$\psi(x, y) \; \rightarrow \; y(x, \psi)$$ (13)

which yields

$$S = 2\pi \int_0^1 \left\{ \frac{1}{2y_\psi} + \frac{y_x^2}{4yy_\psi} + y_\psi \left[H(\psi) - \frac{1}{2y} I(\psi) \right] \right\} d\psi. \tag{14}$$

Integrating this expression between upstream ($x = x_a$) and downstream ($x = x_b$) states we obtain

$$\tau = 2\pi \int_{x_a}^{x_b} \int_0^1 F(y_\psi, y_x, y; x, \psi) \, d\psi \, dx \tag{15}$$

where $F = \dfrac{1}{2y_\psi} + \dfrac{y_x^2}{4yy_\psi} + y_\psi \left[H(\psi) - \dfrac{1}{2y} I(\psi) \right].$

Now we postulate the variational principle for fixed endpoints, $\Delta\tau = 0$, which yields

$$y_{\psi\psi} \cdot (y + \frac{1}{2} y_x^2) + \frac{1}{2} y_\psi^2 y_{xx} - y_\psi y_x y_{x\psi} + yy_\psi^3 H'(\psi) - \frac{1}{2} y_\psi^3 I'(\psi) = 0 \tag{16}$$

as Euler's equation and

$$\frac{1}{2y_\psi^2} + \frac{y_x^2}{4yy_\psi} = 0 \tag{17}$$

as the auxiliary equation.

Interchanging now variables again

$$y(x, \psi) \rightarrow \psi(x, y)$$

we see that Euler's equation is none other than the general equation for loss-free axisymmetric flows derived by Long (ref. 3) and Squire (ref. 4) and that the auxiliary equation again represents Bernoulli's equation which now includes the radial velocity contribution.

ANALYSIS AND NUMERICAL METHODS

To compute the complete flow field various methods have been applied success-fully. The best way to proceed is to apply a Jacobi relaxation scheme in the x-ψ-plane and to transform the solution afterwards back into the x-y-plane. Prior to this last step, however, the criterion for the occurrence of the isentropic transition has to be found.

This may be done as follows: First we prescribe the azimuthal and axial velocity profiles to define the upstream flow state, whereby a relative scaling factor Ω (the swirl number, say) is kept open. The reason for keeping the swirl number open is that the force- and loss-free transition occurs for a discrete condition only. If the shapes of azimuthal and axial velocity profiles are prescribed it occurs at one particular swirl number Ω only. Secondly we determine the genera-ting functions

$$H(\psi; \Omega), \quad I(\psi; \Omega)$$

which now depend on ψ and the scaling parameter Ω. Then, to obtain the second flow state, we guess a value for Ω and solve the x-independent version of Euler's equation for the boundary conditions

$$y(1) = a, \quad y_\psi \to \infty \quad \text{as} \quad \psi \to 0. \tag{18}$$

The second condition agrees with the auxiliary condition for simple vortex flows. After this the flow forces which correspond to the upstream and downstream flow states are calculated and the value of the swirl number Ω is correct iteriatively (using Newton's method) until the flow-force difference is equal to zero within the required accuracy.

For special vortex flows the main part of the calculation can be carried out analytically.

Among these special flows are all vortex flows with thin cores for which a boundary-layer type of analysis can be used. Further cases are the potential vortex within concentric cylinders and all vortex flows for which the generating functions H and I are the same as those for a Rankine vortex. As an example we consider the force-free transition in a Rankine vortex. In this case the axial velocity of the upstream flow is uniform. A vortex core of radius r_c in solid body rotation is embedded in a potential vortex whereby the azimuthal velocity components are matched at r_c (see figure 2). The transition diverts the rotational fluid coming from the upstream core around the bubble nose and a layer of rotational fluid between the radii r_a and r_b is formed in the downstream domain, which separates the air-filled bubble from the irrotational flow. According to the auxiliary equation the fluid is at rest everywhere on the bubble surface. This peculiar result confirms the views of Legendre (see ref. 5) who arrived at the same conclusion making use of Neumann's theorem. To calculate the downstream flow state we represent the stream function as the sum of the stream function ψ_A, which defines the upstream flow, and a departure $r\phi$,

$$\psi_B = \psi_A + r\phi. \tag{19}$$

The scaling parameter k between azimuthal and axial velocity profiles is now chosen to be twice the angular velocity of the upstream core divided by the upstream axial velocity

$$k = \frac{2\omega}{w_A}. \tag{20}$$

In both flow regimes, i.e. in the rotational layer between the airfilled bubble and the potential vortex as well as in the potential-flow regime, Euler's equation assumes the form of Bessel's equation:

$$\left[\frac{d^2}{dr^2} + \frac{1}{r}\frac{d}{dr} - \frac{1}{r^2}\right] \phi = \begin{cases} -k^2\phi & \text{if } r_a \le r < r_b \\ 0 & \text{if } r_b \le r < r_t \end{cases}. \tag{21}$$

Accounting now for the fact that the auxiliary equation requires that the fluid at the bubble surface be at rest Euler's equation yields a one-parametric set of solutions with the scaling parameter k as the set parameter. The condition for equal flow forces finally yields a unique value for k and, correspondingly, a unique second flow state.

The procedure described leads to the equations

$$\zeta^4 = [\eta^2 - \xi^2] \cdot [\eta^2 - 3\xi^2] + 2\xi^4 \ln\left(\frac{\eta^2}{\xi^2}\right), \tag{22}$$

$$\alpha J_1(\eta) + \beta Y_1(\eta) = -\frac{1}{\pi} \cdot \frac{\eta^2 - \xi^2}{\eta\xi}, \tag{23}$$

$$\alpha \; J_o(\eta) + \beta \; Y_o(\eta) = \frac{2}{\pi\zeta} \cdot \frac{\eta^2 - \xi^2}{\xi^2/R^2 - \eta^2},$$ (24)

where

$$\alpha = Y_1(\zeta) - \frac{1}{2} \; \zeta \; Y_o(\zeta), \quad \beta = \frac{1}{2} \; \zeta \; J_o(\zeta) - J_1(\zeta)$$

and

$$R = r_c/r_t, \quad \xi = kr_c, \quad \eta = kr_b, \quad \zeta = kr_a.$$ (25)

Introducing the defintion (20) in the expression (25) for ξ shows that ξ represents the swirl number of the upstream flow state.

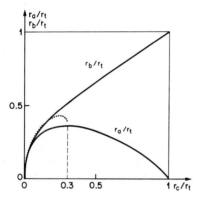

Fig. 5: Dimensionless bubble radii r_a/r_t and r_b/r_t versus dimensionless upstream core radius r_c/r_t calculated according to Eqs. (22)-(25) (solid curves) and according to an approximation for thin cores (dotted curve).

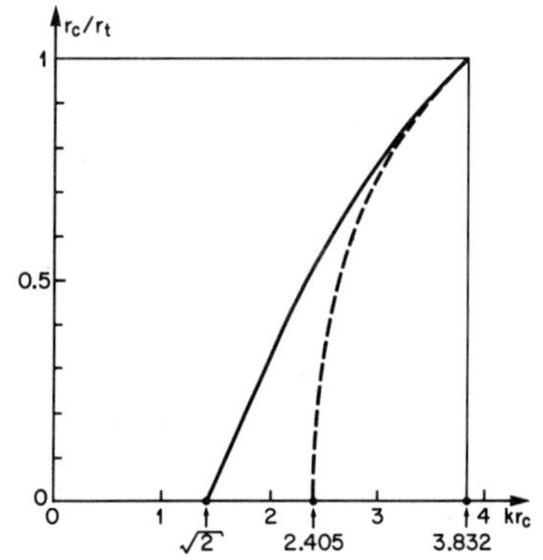

Fig. 6: Dimensionless upstream core radius r_c/r_t versus swirl number $\xi = kr_c$ of the upstream flow for the condition of a loss-free transition of a Rankine vortex (solid curve) and for a critical upstream Rankine vortex (dashed curve).

Equations (23) and (24) are obtained from matching conditions between the two flow regimes whereas equation (22) represents the condition for equal flow forces for upstream and downstream states. These three equations represent a complete set to determine the swirl number ξ and the radii r_a and r_b. Solving equations (22) - (24) numerically we obtain r_a and r_b normalized with respect to the tube radius r_t as functions of the upstream core radius (see figure 5). The maximum bubble size is reached when the upstream core radius is about 30% of the tube radius. The bubble disappears completely in the limit $r_c \rightarrow r_t$, i.e. in the limit of the solid-body vortex. The dotted curve shows the bubble radius ($r_a \sim r_b$) obtained from a boundary-layer type of analysis for thin vortex cores. On the other hand we obtain a relation between the normalized upstream core radius and the swirl number kr_c (see figure 6). The broken line illustrates the relationship between the normalized core radius and the critical swirl number. As it must be, the transition always occurs at supercritical swirl numbers. Considering the second variation of the flow force it can be shown that the second flow state is also supercritical. However, in the limit of a solid-body vortex, i.e. when

$r_c \rightarrow r_t$, both first and second flow states become just critical.

Having calculated the scaling parameter k we have obtained a criterion for the oc-currance of the loss-free transition. We can now determine the sets of generating functions H and I which correspond to a stationary transition. Solving Euler's equation with interchanged variables with the help of a Jacobi relaxation method it is now possible to compute the complete flow field. The following procedure was found to be suitable: we choose a grid with an axial extent which is large compared to the length of the transition. To start the iteration we take the up-stream flow state as an initial function in the first half of the grid and the downstream solution in the second half. As a criterion to stop the iteration pro-cedure the transformed solution has been inserted in Euler's equation (i.e. the equation derived by Long and Squire) and, using normalized stream functions, the computation was stopped when Euler's equation was satisfied within a relative ac-curacy of 10^{-6}. An obvious but interesting detail is that the transition runs away in the course of the iteration if the value for k is not compatible with the size of the upstream vortex core. Hence we could also determine k by requiring that the transition should remain stationary in the course of the iteration.

The example shown in figure 7 corresponds to the swirl-number value $kr_c = 2$. This is approximately the value at which the maximum bubble size is reached. The dashed line represents the boundary between rotational and irrotational flow.

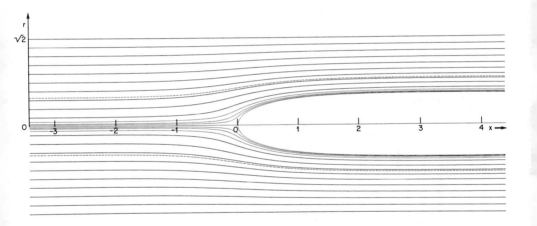

<u>Fig. 7</u>: Streamline map of the transition of a Rankine vortex for $\xi = 2$, $r_t = \sqrt{2}$ in the interval $-4 \leq x \leq 4$. The dashed line represents the boundary be-tween the core and the outer irrotational flow.

Figure 8 shows an enlargement of the nose region of the bubble shown in fi-gure 7.

The flow field shown in figure 9 corresponds to the swirl number value $kr_c = 3$. It should be pointed out that in this case the transition is longer than in the previous example.

Figure 10 shows an enlargement of the nose region of the bubble shown in fi-gure 9.

To summarize the results it can be said that the length of the transition in-creases monotonically with the upstream core size and therefore with the swirl number. For general vortex flows it seems to be the case that the smaller the

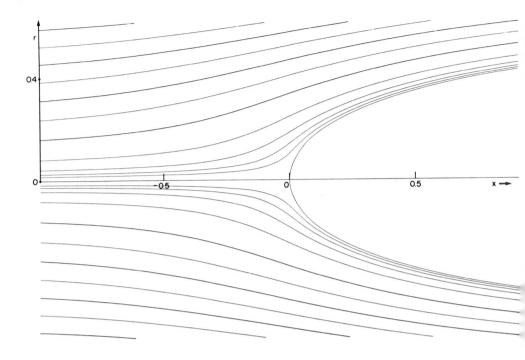

Fig. 8: Local enlargement around the point (0,0) of the streamline map shown in figure 7.

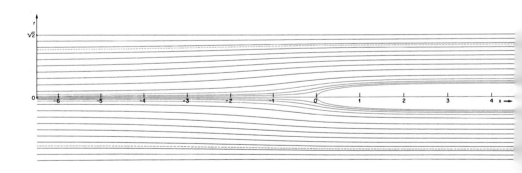

Fig. 9: Legend as for figure 7, except $\xi = 3$, $-6 \leq x \leq 4$.

departure of the value k at which the isentropic transition occurs from the value for critical flow the longer the transition will be. In the limit of a solid-body vortex for which both upstream and downstream flow states become just critical the transition becomes <u>infinitely long</u>. This limiting result is again obvious. As Euler's equation is linear in this case, any linear combination of upstream and downstream flow states represents an admissible flow.

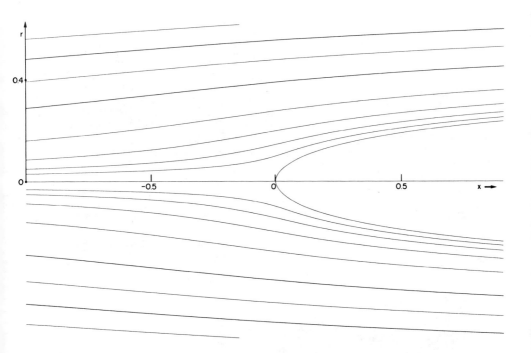

Fig. 10: Local enlargment around the point (0,0) of the streamline map shown in figure 9.

In the Appendix of a paper by Benjamin and Barnard (see ref. 6) Fraenkel showed that stationary bubbles do not exist if the upstream flow can be represented as a solid-body vortex. Obviously Fraenkel's proof agrees with the present (limiting) results.

CONCLUDING REMARKS AND INTERPRETATION

The "vortex breakdown phenomenon" which we have described is not usually observed in vortex-tube experiments. Typical experimental observations show vortex break-down structures similar to those shown in figure 11a for small Reynolds numbers and 11b for moderately large Reynolds numbers (i.e. the Reynolds number Re_r based upon the mean axial velocity component and the tube radius). The structure des-cribed here is not normally observed because it is masked mainly by viscous diffusion in the shear layers at the bubble surface at low Reynolds numbers and mainly by shear-layer instabilities (i.e. a rollup of the shear layer into helical vortex arms) at moderately large and large Reynolds numbers. Hence, the situation is much like in the case of a wake behind a bluff body which does not appear as a "shadow" either in any Reynolds-number regime.

However, if the masked zone of stagnation in a vortex breakdown bubble appearing in a water flow is filled up with air using a tublet inserted from the down-stream side, then the masked character of the phenomenon can be revealed (see figure 12). In this case the bubble is closed on the downstream side by the second transition which represents a hydraulic jump.

A further interesting property of downstream flows, which can be represented as hollow-core vortices, appears when vortex tubes with varying cross-sectional

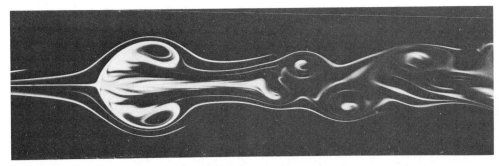

(a) $Re_r = 575$

(b) $Re_r = 1650$

Fig. 11: Examples of vortex breakdown in a cylindrical tube. (Flow from left to right). Courtesy of Dr. M.P. Escudier.

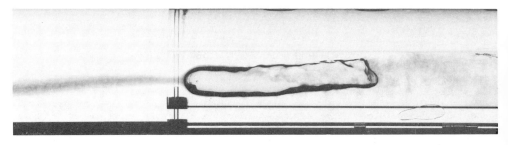

Fig. 12: An air-filled vortex-breakdown bubble. (Flow from left to right).

area are considered. Decreasing the radius of the vortex tube always brings the flow condition closer to critical for both super- and subcritical hollow-core vortices, whilst increasing the tube radius leads away from the critical state (see ref. 7, Eq. (34)). Benjamin (see ref. 2) showed that the opposite is true for simple vortex flows without free surfaces. For this reason the force-free transition discussed here alters the relatinship between flow condition and sign of cross-sectional area change. In a diffuser, for example, the flow condition is brought closer to critical until the transition occurs (which produces a hollow-core vortex) after which the flow condition is led away from the critical state.

The diffuser paradox discussed by Batchelor (see ref. 8), for example, who noted

that the behaviour of a simple vortex in a diffuser inevitably becomes cata-
strophic after a sufficient increase of the cross-sectional area, can thus be
resolved.

The (qualitative) physical interpretation of a vortex-breakdown phenomenon presented here is quite simple. The appearance of the phenomenon is based on a trade-
off between dynamic pressure contributions due to azimuthal and axial velocity
components, respectively. Hence, to provide the basis for the occurrence of a
vortex-breakdown phenomenon as discussed in this paper the presence of a real
(e.g. vortex flow in a tube or a vortex in the neighbourhood of the surface of
a wing) or virtual surface (e.g. plane of symmetry between counterrotating
vortices) is required.

REFERENCES

(1) T.B. Benjamin: Gravity currents and realted phenomena. J. Fluid Mech. $\underline{31}$,
     209 (1968).

(2) T.B. Benjamin: Theory of the vortex breakdown phenomenon. J. Fluid Mech. $\underline{14}$,
     593 (1962).

(3) R.R. Long: Steady motion around a symmetrical obstacle moving along the axis
     of a rotating fluid. J. Meteor $\underline{10}$, 197 (1953).

(4) H.B. Squire: Rotating fluids. In "Surveys in mechanics" 139-161, Ed. Batche-
     lor and Davies, Cambridge University Press (1956).

(5) R. Legendre: Remarks on axisymmetric vortex breakdown. Rech. Aérosp. $\underline{5}$, 296
     (1981).

(6) T.B. Benjamin and J.S. Barnard: A study of the motion of a cavity in rotating
     liquid. J. Fluid Mech. $\underline{19}$, 193 (1964).

(7) J.J. Keller and M.P. Escudier: Theory and observations of waves on hollow-
     core vortices. J. Fluid Mech. $\underline{99}$, 495 (1980).

(8) G.K. Batchelor: An introduction to fluid dynamics. Cambridge University
     Press (1967).

Vortex Breakdown Simulation Based on a Nonlinear Inviscid Model

M. M. HAFEZ AND M. D. SALAS

Abstract

It is shown that the inviscid equations governing steady axisymmetric flow with swirl, admit solutions with closed streamlines. Results are obtained using two different numerical algorithms. The first is based on a multigrid method for nonline· eigenvalue problems, while the second is based on a least squares formulation.

1. Introduction

The phenomena of vortex breakdown has been known for some time[1]. Numerical simulations based on steady axisymmetric Navier Stokes equations* are in good agreement with experimental results[2]. There are however, some theoretical investigations based on inviscid models[3]. In this work, it is shown that the inviscid (Euler) equations admit solutions with a vortex breakdown structure qualitatively similar to that observed experimentally. Using a stream function formulation, the conservation of mass, momentum and energy reduces to a single second order elliptic equation, in t form of a nonlinear eigenvalue problem. Two numerical algorithms are presented for solving this equation. In the first approach, the equation is discretized using central differences on a cartesian grid. The resulting algebraic system of equations (with a free swirl velocity parameter) is solved by a multigrid method using a standa relaxation procedure as a smoothing operator. The second approach is applicable, however, for any specified swirl velocity parameter and is based on a least squares formulation. The same solution can be obtained using either method. As expected, th multigrid calculation is more efficient.

In the following, the stream function equation is derived and a nonuniqueness problem (due to closed streamlines) is discussed. A model is presented where the circulation and the enthalpy are determined via the same functional behavior inside and outside the bubble. Details of the numerical algorithms are given, and the linearization and the continuation processes, as well as, the special treatment of th bifurcation point are discussed. The numerical results clearly demonstrate the existence of solutions with a closed region of recirculating flow.

---

*The steady axisymmetric assumptions are criticized in reference 1 and 3.

2. Governing Equations

An incompressible steady axisymmetric flow with swirl can be described in terms of stream function, ψ, vorticity component, ω, and circulation, k. In cylindrical coordinates (x, r, θ) the Navier Stokes equations are written in the form:

$$r(\frac{\psi_r}{r})_r + \psi_{xx} = r\omega$$

$$(u\omega)_r + (w\omega)_x + (\frac{k^2}{r^3})_x = \frac{1}{Re}[\omega_{rr} + \frac{1}{r}\omega_r - \frac{\omega}{r^2} + \omega_{xx}] \tag{2.1}$$

$$uk_r + wk_x = \frac{1}{Re}[k_{rr} - \frac{1}{r}k_r + k_{xx}]$$

where $k = rv$, $\omega = w_r - u_x$ and Re is the Reynolds number. The velocity components in the x, r, θ directions are denoted w, u, v, respectively and in terms of ψ, $w = \psi_r/r$ and $u = -\psi_x/r$. In the inviscid limit (Re = ∞), it is clear that k is constant along a streamline, it can also be shown that the total enthalpy h becomes constant along a streamline. Moreover, the vorticity component ω can be related to the gradients of the circulation k and the total enthalpy. More precisely, $\vec{q} \times (\nabla \times \vec{q}) = \nabla h$ (where $\vec{q}$ is the velocity vector) leads to,

$$r\omega = -\frac{1}{2}\frac{dk^2}{d\psi} + r^2\frac{dh}{d\psi} \tag{2.2}$$

where $h = p + \frac{1}{2}(u^2 + v^2 + w^2)$ and p is the pressure. Notice that, if the swirl velocity vanishes, ω/r becomes constant along a streamline. Also notice that the vorticity due to the circulation term does not depend on the sign of k but only on its magnitude. The functions $k(\psi)$ and $h(\psi)$ are determined, in terms of the specified $k(r)$ and $h(r)$ profiles at the upstream boundary, provided ψ is positive (i.e., outside the recirculating zone). Inside the bubble, the k and h distributions are not known. In fact, within the inviscid model a discontinuity may be allowed across a separating streamline leading to a nonuniqueness problem. One way to obtain an inviscid solution, is to use analytic continuation of the functions $k(\psi)$ and $h(\psi)$ for negative ψ. In the present work, the dependence of k and h on ψ is known analytically for positive ψ. The same functional dependence is assumed for negative ψ. Moreover, since k vanishes along a separating streamline (k = 0 on the axis), it can be assumed that k has the opposite sign inside the bubble. In solving the Navier Stokes equations, the viscous terms play an important role in the neighborhood of the separating streamline by preventing the formation of discontinuous solutions. Similarly, the artificial viscosity terms in Euler calculations (based on primitive variables) are essential in determining the flow conditions inside the bubble, but it may be argued that although the existence of such a viscosity is necessary, the solution may be independent of its form and/or its magnitude. In the above stream function formulation, there is no explicit or implicit artificial viscosity included (in fact the truncation error for central differenced approximation is of a dispersive nature) and the assumptions of analytical continuation of k and h across the separating streamline rules out any discontinuity which, in a sense, is similar to the artificial viscosity effects. It is obvious that the present inviscid model is only as good as the underlying assumptions.

Special Cases

For a rigid body motion, $v = Vr$ (where V is the swirl velocity parameter) hence, $k^2 = V^2r^4$. The stream function at the upstream boundary is given by $\psi(r, 0) = r^2/2$. Since $p_r = v^2/r$, $p = \int V^2 r\, dr = V^2\psi$ and $h = V^2\psi + \frac{1}{2}(1 + V^2r^2) = 1/2 + V^2\psi$ the vorticity distribution is obtained in terms of ψ and r as

$$r\omega = -4V^2\psi + 2r^2V^2$$

$$= 4V^2(r^2/2 - \psi) \qquad (2.3)$$

In terms of a perturbation stream function $\Psi = r^2/2 - \psi$, the governing equation becomes

$$\Psi_{xx} + r(\Psi_r/r)_r = -4V^2\Psi \qquad (2.4)$$

At the axis $\Psi = 0$. If Ψ vanishes in the far field, the only nontrivial solution is the eigensolution corresponding to an eigenvalue $\lambda = 4V^2$. In this case, an alternate variational formulation can be obtained in terms of the function of F over the domain Ω as

$$F(\Psi, V) = \iint_\Omega (\Psi_x^2/r + \Psi_r^2/r - 4V^2\Psi^2/r)\, d\Omega \qquad (2.5)$$

Bossel[5] solved the linear equation (2.4) with a prescribed downstream boundary condition via superposition of fundamental solutions (Bessel functions); this model is not satisfactory since the solution depends on an arbitrary boundary condition and the size of the domain of integration. In the present model, the swirl velocity at the upstream boundary has the same form suggested by Grabowski and Berger

$$v = Vr (2 - r^2) \qquad \text{if } r \leq 1$$

$$v = V/r \qquad \text{if } r > 1$$

the circulation distribution k is therefore:

$$k^2 = 16V^2\psi^2(1 - \psi)^2 \qquad \psi \leq 1/2$$

$$k^2 = V^2 \qquad \psi > 1/2 \qquad (2.6)$$

and the vorticity component is given by:

$$r\omega = 16V^2(1 + 2\psi^2 - 3\psi)(r^2/2 - \psi) \qquad \text{if } \psi \leq 1/2$$

and $\quad r\omega = 0 \qquad \text{if } \psi > 1/2$

In terms of Ψ, the governing equation is

$$\Psi_{xx} + r(\Psi_r/r)_r = -4V^2\alpha^2\Psi$$

where

$$\alpha^2 = 4(1 + 2\psi^2 - 3\psi) \qquad \text{if } \psi \leq 1/2 \qquad (2.7)$$

and

$$\alpha^2 = 0 \qquad \text{if } \psi > 1/2$$

Again a variational formulation can be obtained in terms of the function

$$F(\Psi, V) = \iint_{\Omega} (\psi_x^2/r + \psi_r^2/r - 4V^2G/r) \cdot d\Omega$$

where

$$G = 2[1 - 3r^2/2 + 2r \;) \; \psi^2/2 + (3 - 2r^2) \; \psi^3/3 + \psi^4/2] \qquad (2.8)$$

In the following, some numerical details of the two approaches are given.

3. Numerical Algorithm

(3.1) Multigrid Method

The governing equation for Ψ is written in the form of a nonlinear eigenvalue problem. Choosing the domain of integration as a rectangle, $(0 \leq x \leq a, 0 \leq r \leq b)$ and assuming Ψ vanishes on the boundary, discretization of equation (2.7), in conservation form, using central differences, leads to a symmetric system of algebraic equations. The trivial solution $\Psi = 0$ for any V corresponds to uniform flow (with $w = 1$) and is not interesting. To seek a solution representing a vortex breakdown, a norm of Ψ, $||\Psi||$, is forced to be different from zero, i.e., the following constraint is imposed

$$||\Psi||^2 = h^2 \sum_{ij} \psi_{ij}^2 = \text{constant} \qquad (3.1)$$

where the summation is over all points i, j and h is the mesh size (uniform grid). V is treated as a free parameter and must be obtained as part of the solution. A relaxation algorithm was developed by S. Ta'asan of ICASE to solve this problem. It consists of three steps, a local relaxation of the difference equations, a global change to satisfy the constraint, and an update of the free parameter V. For example, if a point relaxation is used, after each sweep the solution vector is scaled by a factor β to satisfy the constraint (3.1) for any prescribed constant (g) for the norm of Ψ. The factor is given by:

$$\beta^2 = g/h^2 \sum_{ij} \psi_{ij}^2 \qquad (3.2)$$

V is then updated from (3.3)

$$V^2 = -1/4 \frac{<L^h \psi^h, \psi^h>}{<\alpha^2(\psi^h)\psi^h, \psi^h>}$$

(3.3)

where L^h is the discretization of the operator $1/r\, \partial_{xx} + (\frac{1}{r}\partial_r)_r$ and $<,>$ denotes the usual inner product. This is equivalent to requiring that $F(\Psi, V) = 0$ in the corresponding linearized variational formulation. A multigrid method based on the above relaxation algorithm is used to obtain nontrivial solutions where the finest grid consists of 80 x 32 mesh points (h = 1/16). Corrections were interpolated linearly and residuals were transfered by injection.[*] At a bifurcation point, a linearized problem is solved for the perturbation C, where it is assumed that $\Psi = \Psi_0 + \varepsilon C$. Since $\Psi_0 = 0$, the equation for C is

$$C_{xx}/r + (C_r/r)_r + 4V^2\alpha_0^2(r)\, C/r = 0$$

(3.4)

Here, α_0^2 is a function of r only and C vanishes on the boundary. To obtain a nontrivial solution, C is forced to be different from zero. The solution process is identical to the one described above. With a proper choice of ε, a good initial guess for Ψ is obtained, at least for small $||\Psi||$, and a continuation process may be used for larger values of $||\Psi||$.

(3.2) Least Squares Formulation

In this approach, equation (2.7) is replaced by the following equivalent system:

$$\Psi_r/r = w'$$
$$\Psi_x/r = -u'$$
$$w'_r - u'_x = -A\Psi/r$$

(3.5)

where A stands for $4V^2\alpha^2$. A related variational formulation can easily be obtained in terms of the function F(Ψ, u', w').

$$F(\Psi, u', w') = \iint_\Omega [(\Psi_r/r - w')^2 r + (\Psi_x/r + u')^2 r$$

$$+ (w'_r - u'_x + A\Psi/r)^2] \, d\Omega$$

(3.6)

The corresponding Euler equations are:

$$\Psi_{xx}/r + (\Psi_r/r)_r - A^2/r^2\Psi = (w'_r - u'_x)(1 + A/r)$$
$$u'_{xx} - ru' = \Psi_x + w'_{rx} + (A\Psi/r)_x$$
$$w'_{rr} - rw' = -\Psi_r + u'_{xr} - (A\Psi/r)_r$$

(3.7)

[*] Equation (3.3) is properly modified for coarser meshes; the extra term is related to the difference between the injected residual and the residual based on the injected solution.

Equation (3.5) provide proper boundary conditions for u' and w'. Instead of arbitrarily discretizing equation (3.7), a variational formulation related to the discrete version of equation (3.5) automatically yields a consistant approximation of (3.7), including special operators in the neighborhood of the boundary. Details are given in ref. 7. A staggered grid is used for Ψ, u' and w'. Numerical solutions are obtained for a prescribed swirl velocity parameter V, using standard relaxation procedures, (SLOR, Zebra, ADI, ...). To avoid converging to the trivial solution $\Psi = 0$, a good initial guess is required. Since other spurious solutions may be admitted by the least squares formulation, we must check that the obtained solution satisfies equations (3.5). In this case, F must identically vanish (a local minimum is not of interest).

Initial Guess

An initial distribution of Ψ is needed to start the calculation. A vortex ring (with discontinuous vorticity distribution) can be easily represented in an analytical form as a solution of the equation.

$$\Psi_{xx}/r + (\Psi_r/r)_r = \omega_0 \tag{3.8}$$

where ω_0 is constant in $r^2 + x^2 \leqq R^2$.

Equation (3.8) has the solution

$$\Psi = \frac{\omega_0}{10} r^2(R^2 - r^2 - x^2) \tag{3.9}$$

outside the ring, Ψ is given by

$$\Psi = \frac{-B}{2} r^2(1 - R^3/(r^2 + x^2)^{3/2}) \qquad\qquad r^2 + x^2 \geqq R^2$$

where $B = 2\omega_0 R^2/15^2$.

In fact, such a solution satisfies the viscous equations provided discontinuous shear stress is allowed across a separating streamline. This can be practically realized, if the fluid inside the bubble is of different density. Equations (3.9) provide an initial guess for Ψ. At convergence, the vorticity distribution becomes continuous and it is different from that of a vortex ring.

Results

Numerical results are obtained for different values of the swirl velocity parameter V. In figure (1), $||\Psi||$ is plotted versus V^2 using both approaches, multigrid and least squares. In the first approach $||\Psi||$ is prescribed and V is a free parameter, while in the second approach, V is prescribed and $||\Psi||$ is free. The agreement between the two calculations is good. Figure (2) shows the streamlines and the vortex structures for two values of V. In figure (3) the contours of vorticity for the same two cases are shown.

Concluding Remarks

The existence of an inviscid solution with closed streamlines is demonstrated by numerically solving the steady Euler equations using a stream function/vorticity formulation. Such solutions are obtained by two different approaches, multigrid and least squares. The dependence of the solution, (for example, the size of the bubble) on the swirl velocity, may not be physically acceptable. Comparison with the viscous solution is under investigation. Of particular interest, is the limit of the viscous solution when Re becomes large and its relation to the present solution.

References

1. Leibovish, S., AIAA J., Vol. 22, No. 9, 1984, pp. 1192.

2. Grabowski, W., and Berger, S., JRM, Vol. 75, 1976, pp. 525.

3. Krause, E., AIAA Paper 83-1907, 1983.

4. Benjamin, T. B., JFM, Vol. 23, 1966, pp. 241.

5. Bossel, H. H., Ph.D. Thesis, University of Calif., Berkelely, 1967.

6. Ta'asan, S., "A Multigrid Method for Vortex Breakdown Simulation", ICASE Report to appear.

7. Hafez, M., Kuruvila, G. and Salas, M., "A Numerical Study of Vortex Breakdown", to be presented at AIAA Aerospace Sciences Meeting, Reno, 1986.

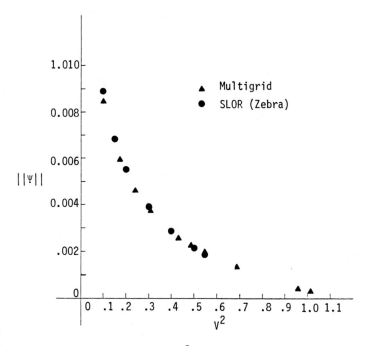

Figure 1. $||\Psi||$ vs V^2

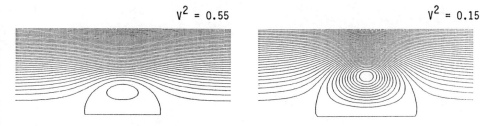

Figure 2. Streamline contours

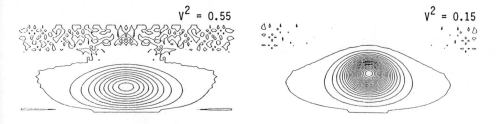

Figure 3. Vorticity contours

Section III

Massive Separation

Theory of High-Reynolds-Number Flow Past a Blunt Body

F. T. SMITH

SUMMARY

Recent computational results and theoretical developments are discussed, first for flow past a blunt body or thick airfoil, at high
Reynolds numbers. A theory based on a massive main eddy behind the
body is described and appears so far to be not inconsistent with the
computational trends. Some properties and numerical solutions of the
massive-eddy problem posed are also presented. Second, relatively
large-scale separation is reconsidered from the viewpoint of viscous-
inviscid interactive motions for smaller-sized disturbances. Throughout, viscous effects have a primary role in avoiding the arbitrariness
present in inviscid theory.

SECTION 1.   INTRODUCTION

The main concern of this contribution is the theoretical structure
of the large-scale and small-scale eddies set up in high-Reynolds-
number flow, past a blunt body or thick airfoil in a uniform
freestream, on the one hand, and past a small step, ramp or thin
trailing edge, on the other.

Depending on the Reynolds number, and with many notable exceptions,
such flows in practice often go through transition to turbulence; but
that poses difficulties that are, quite simply, beyond current understanding in any firm theoretical sense. Therefore, we turn to an
alternative strategy, namely, to try to understand the laminar steady-
flow version first. This is itself a fundamental problem, for flow
past a circular cylinder for example, and if it cannot be clarified
the hope of understanding more complex flows is correspondingly dimin-
ished. Once it is clarified, we may then consider the nonlinear stability of the laminar separated version, or rather extend the theory
to an unsteady version, subsequently, and examine the unsteady breakdown of the flow. This aspect is pursued in ref. 10, among others.
In addition, however, the limiting description for steady flow, when

the Reynolds number is large, is a very important one to have available because of its many applications. These are, for example, to separated thin-airfoil motions, including hysteresis and stall, which are often observed experimentally; to pointing to appropriate computational methods at finite Reynolds numbers; to providing comparisons or checks on numerical work; to showing the main physical mechanisms which occur, hence yielding deeper understanding; and to providing a basis for studies of unsteady flow properties. Moreover, the predictions from the limiting theory often hold good in a numerical sense at surprisingly low Reynolds numbers.

The Reynolds number Re is assumed to be large, below, and nondimensional velocities (u,v), Cartesian coordinates (x,y), and pressure p are used such that the freestream velocity is (1,0), the freestream pressure is zero, and the airfoil chord is 1. The fluid is taken to be incompressible in most of what follows. Section 2 describes previous and recent numerical and analytical studies of the large-scale separated flow past a blunt body, e.g., a circular cylinder. Then in Section 3, a proposal concerning a massive eddy-scale flow behind the body, for Re >> 1, is examined analytically and numerically. The proposal appears to be self-consistent as far as it has been taken to date. Section 4 gives an alternative approach from the viewpoint of triple-deck theory, for smaller disturbances. Further comments are made in Section 5.

SECTION 2. LARGE-SCALE SEPARATED FLOW

Let us start by considering the flow past a thick airfoil or bluff body. On an inviscid basis alone, there are many possibilities that can be considered, and indeed allowed. The following are some of the principal ones (see figure 1). The first is classical attached motion governed mainly by the potential-flow or Euler equations. A viscous boundary layer and subsequent wake are assumed, and if the boundary layer can remain attached, because of smoothed leading-edge and trailing-edge conditions for example, or due to properties of a turbulent boundary-layer model, then the large-Reynolds-number structure of the flow fits together readily and is of a fairly simple hierarchical form, including classical Prandtl boundary-layer theory. Second, if only small separations are present, for instance at a symmetric or nonsymmetric trailing edge, then again the majority of the flow takes the classical hierarchical form, while the small-scale separated motion is described by viscous-inviscid interactive theory of the triple-deck kind or similar. This type of flow is discussed more later on. Third, suppose the separation is instead of large scale, producing eddies whose sizes are comparable with the body size at least. Then one possibility is the Prandtl-Batchelor structure, with typical eddy size being comparable with the body dimensions and the vorticity in the main part of the eddy being uniform, this analytically from an appeal to viscous forces inside the eddy. A fourth candidate, and likewise a major one in the context of large-scale separated motion, is the extended Kirchhoff structure, where the flow splits into body-scale and (larger) eddy-scale parts with relatively slow flow at constant pressure inside the main eddy. A mixture of essentially the third and fourth candidates above turns out to be of most interest subsequently, in Section 3, but at the moment the central point is that according to inviscid theory all the above forms are acceptable and, once discontinuities across vortex sheets are allowed, as seems sensible physically, there are infinitely many inviscid "solutions" for large-scale separating flows.

With all this arbitrariness in purely inviscid theory, it is essential to take account of the effects of viscosity to decide matters more specifically, certainly when solid surfaces are present. Therefore, we turn first here to the Navier-Stokes equations, taking two-dimensional laminar steady motion as a starting point, and examine what happens as the Reynolds number Re increases. Accurate computations, experiments and an asymptotic theory of the extended Kirchhoff kind (figure 2) mentioned previously appear at first sight to fall encouragingly into line for medium values of Re, in the case of symmetric flow past a circular cylinder. The drag coefficient c_D for the cylinder is quite well predicted by that theory, in comparison with reliable Navier-Stokes calculations and with experimental measurements available, for Reynolds numbers as low as about 10 and up to about 300: figure 2. The theory here (reviewed in refs. 5, 6) has the Kirchhoff free-streamline form within the body scale, giving rise to an $O(1)$ drag coefficient and a parabolically widening eddy downstream at the onset of the larger eddy-scale flow. The latter scale has streamwise extent $O(Re)$, to conserve momentum in the ultimate wake far beyond, the transverse extent is $O(Re^{1/2})$, and the eddy shape is an ellipse to preserve uniform pressure at leading order. As noted before, the theory's predictions agree quite well with computations and experiments, and similarly in other flow configurations such as internal motions. But there is a drawback in the present context of external flow. The drawback (see figure 3) arises from the theoretical closure of the eddy, where the free shear layer bounding the eddy is split into two parts by inviscid means, one part going forward into the ultimate wake and the other returning upstream. To conserve vorticity [$\nabla^2\psi$ = function (ψ)] the returning part must have O(1) reversed velocities and O(1) thickness, with a jet-like profile, and so the assumption of relatively slow flow in the eddy, upstream of this, is opened to question, the more so as both the jet and the viscous free shear layer have to draw fluid from the eddy without there being any corresponding source of fluid available. No satisfactory mechanism for completing the account of the entire eddy motion has been found yet and indeed at present this extended Kirchhoff theory appears to be incomplete.

That brings us to the more recent computations of the Navier-Stokes equations at higher Reynolds numbers by Fornberg (ref. 3), and, at not quite so high Re values, in ref. 4. The computations, summarized in figure 4, show eventually an approximately linear increase of both the eddy length _and_ the width, with increasing Re. It should be mentioned that of course there are no laminar-flow experiments, as yet anyway, to compare with at these larger Reynolds numbers because in practice the flow past the cylinder is very unstable then. Nevertheless, the concept of a limiting description for Re >> 1 is still a most important one to clarify, not least because of its applications to separated airfoil motions (including, e.g., hysteresis and stall), to suggesting appropriate numerical methods, its highlighting of the physical mechanisms involved, and as a basis for unsteady-flow studies (see Section 5). The asymptotic theory (ref. 2) to which the above calculations point, and which replaces the extended Kirchhoff description, is based on a _massive_ eddy size of dimensions O(Re) by O(Re): figure 5. With such dimensions, the induced pressure and velocities inside the eddy are O(1) and so, in contrast with the earlier account, the eddy motion now affects the viscous shear layer and the viscous return wake substantially, and the earlier-noted contradiction at eddy closure and the difficulties concerning entrainment are thereby avoided.

The properties of the large-scale separated flow structure are described in the following section.

SECTION 3. LARGE-SCALE STRUCTURE AND SOLUTIONS

The limiting structure as far as the main eddy is concerned is based
on Sadovskii's (ref. 1) inviscid model supplemented by viscous
effects. The major difference from the earlier theory is that now the
recirculating eddy motion in zone II is supposed to be as strong as
the external motion in zone I and so it affects the viscous-layer
flows III substantially (Smith, ref. 2). In the inviscid regions I,
II of figure 6 we have

$$\nabla^2 \psi = 0 \text{ (in I)}, \ \nabla^2 \psi = -\bar{\Omega} \text{ (in II)} \tag{3.1}$$

for the stream function ψ in scaled terms, with $Re^{-1}\bar{\Omega}$ being the small
negative uniform vorticity. At the vortex sheet $Y = S(X)$ which is the
unknown boundary between I, II

$$\psi, \text{ p are continuous;} \tag{3.2}$$

and the end conditions, which play a significant role, require

$$\text{tangential departures at } X = 0, \text{ L}, \tag{3.3}$$

where X, Y are 0(Re)-scaled coordinates and ReL is the eddy length.
The other boundary conditions relevant impose a streamline along the
X-axis and the uniform stream in the farfield.

Generally the flow speeds q_I, q_{II} on either side of the vortex
sheet are unequal. The whole inviscid solution then depends on the
parameter C_{II}, that is, the pressure at the ends $X = 0,L$ of the
eddy where $q_{II} \to 0$; here $C_{II} = (q_{II}^2 - q_I^2 + 1)/2$ from Bernoulli's
theorem, and the constant C_{II} can take values between zero and 1/2.
The inviscid problem determines the product $\bar{\Omega}L$ for a prescribed value
of C_{II}, as discussed later in this Section.

As expected in view of earlier comments, there is still a wide range
of solutions available, on inviscid grounds. Next, however, the
properties of the three viscous layers III_1, III_2, III_3 in figure 6
narrow the choice. These layers are all governed by the boundary-

layer equations. The requirement of periodicity shown in figure 7, i.e. continuity around the viscous layers III_1 (the free shear layer or vortex sheet) and III_2 (the return wake), which border the eddy, is expected to fix the functional dependence of L and C_{II} on $\bar{\Omega}$, leaving only $\bar{\Omega}$ unknown. Also, a left-over jet with 0(1) velocities and thickness is sent back from layer III_2 into the body-scale flow (see figure 7 and below), whereas the ultimate wake, which is the third viscous layer III_3 downstream, determines the drag coefficient c_D as a function of $\bar{\Omega}$ by means of the overall momentum balance. At this stage therefore we are left with, in effect, one relation short to completely determine the motion.

The extra relation required comes from consideration of the body-scale motion, sketched in figure 8. Its main feature is that the eddy motion, turning there, again has significant vorticity (now 0(1)) but the vorticity is nonuniform, being caused by the left-over jet from the larger-scale motion. Apart from that, the body-scale structure is very like the Kirchhoff free-streamline form and produces an 0(1) drag coefficient c_D, as well as a parabolic growth of the eddy width downstream. The whole flow structure in fact is something of a mixture of the extended Kirchhoff and the Prandtl-Batchelor forms.

Some more details of the body-scale flow are the following: see also ref. 2. (a) The flow is subjected to a reduced freestream in effect, in that

$$u \to (1 - 2\, c_{II})^{1/2}, \quad p \to c_{II}$$

in the farfield, to match with the eddy-scale flow. (b) The main inviscid regions IV, V are governed by

$$\nabla^2 \psi = 0 \ (\text{in IV}), \quad \nabla^2 \psi = \frac{dU}{dy}\,(0+,\ y) \ (\text{in V}).$$

Here the latter vorticity stems from the viscous return-jet profile $U(0+, y)$ of the eddy-scale motion, layer III_2, and it sets up a significant eddy flow to be turned forward in region V. (c) The separation near the point C of figure 8 is "smooth" and described by triple-deck theory; see also Section 4, and the comparison with experiments in ref. 5. The smoothness requirement plays a part in fixing the body-scale solution. (d) Secondary separation is likely in the backward boundary layer VI_2 approaching separation but there is some question as to whether this separation is relatively small or not. (e) Because the eddy width grows like $x^{1/2}$ downstream, buffer zones are necessary to join certain aspects of the body- and eddy-scale flows. (f) The body-scale problem determines c_D as a function of $\bar{\Omega}$, since the vorticity dU/dy depends on $\bar{\Omega}$. This provides the final relation necessary to fix c_D, $\bar{\Omega}$, L and C_{II}. Also, no difficulty occurs now during the eddy-closure process, since a backward jet is an admissible feature of the present flow structure: see figure 9.

The first task, then, is to solve the inviscid problem of the massive eddy motion, the second task is to calculate the viscous-layer problems (layers III), and the third task is to solve for the body-scale flow.

Concerning the first task, for which numerical results are described subsequently, we note that ref. 1 obtained a few solutions near one end of the range (C_{II} near 1/2), while at the other end, for C_{II} small, there is a thin-eddy limit which is helpful analytically, as well as having application in Section 4 below. For small values of C_{II}, or relatively large vorticity, the equation of constant vorticity inside the eddy reduces to $\partial^2\psi/\partial Y^2 = -\bar{\Omega}$ since the eddy is relatively thin, and so, from integration, the eddy pressure $p_{II}(X)$ and the eddy shape $S(X)$ are related by

$$p_{II}(X) = C_{II} - \frac{1}{8}\bar{\Omega}^2 S^2(X).\tag{3.4}$$

But the external motion past the thin eddy yields the surface pressure, in terms of a Cauchy-Hilbert principal value,

$$p_I(X) = -\frac{1}{\pi}\fint_0^L \frac{S'(\xi)d\xi}{(X-\xi)}\tag{3.5}$$

Hence continuity of pressure leads to the nonlinear integro-differential equation

$$\frac{1}{8} \bar{\Omega}^2 S^2(X) - C_{II} = \frac{1}{\pi} \int_0^L \frac{S'(\xi)d\xi}{(X-\xi)} \qquad (3.6a)$$

for the unknown eddy shape $S(X)$ between $X = 0$, L, subject to the constaints

$$S = S' = 0 \text{ at } X = 0, L. \qquad (3.6b)$$

The above is an integrated steady form of the Benjamin-Ono equation. Further analysis using Fourier series may be possible, but computation seems most desirable for later purposes. The computations were performed by means of a Carter-like approach as follows. A guess is made for $S(X)$, then $p_{II}(X)$, $p_I(X)$ are calculated from (3.4), (3.5) respectively, and the pressure difference $(p_I - p_{II})$ is used as the basis for updating $S(X)$, and so on. The updating takes the new S as a linear combination of the old S and $(p_I - p_{II})$, at each X, in such a way that short-scale wave growth is avoided. Also, a normalization including setting $L = 2$ is applied to avoid the trivial solution $C_{II} = S(X) \equiv 0$. It is noted here that ref. 1 uses a parameter for these flows which is inappropriate as in effect it produces dual solutions, unlike with the parameter C_{II}. Solutions of the thin-eddy case (3.6a,b) are shown in figure 10 with $\hat{S}$, $\hat{p}$ denoting normalized variables and with symmetry about the eddy center.

When two characteristic properties of the full problem are plotted (figure 11), namely the maximum eddy width and a measure of the vorticity, the comparison between the thin-eddy predictions for small C_{II} and ref. 1's results near $C_{II} = 1/2$ is found to be fairly close. So the thin-eddy theory seems useful in numerical terms (see next - but - one paragraph), in addition to giving more analytical backing to the whole model and having relevance to the smaller-scale separations discussed in Section 4.

A possibility noted here is that the thin-eddy or other solutions of the massive-eddy problem may open the door to Prandtl-Batchelor solutions. No solutions of the Prandtl-Batchelor type for flow past a smooth body have been found yet, as far as we know, even on an inviscid basis, let alone allowing for viscous effects. Yet it is possible that a smooth thin body can be inserted near the start of the massive eddy without drastically disrupting the flow structure. The same conclusion holds if a thin body is placed along the x-axis even in the middle of the eddy, certainly in the thin-eddy case. Again, a small Prandtl-Batchelor eddy adjoining a smooth airfoil surface is governed by the thin-eddy analysis. So continuation would tend to suggest the existence of Prandtl-Batchelor flows having eddy sizes comparable with the airfoil dimensions, at least on inviscid grounds.

Returning to the massive-eddy problem (3.1)-(3.3) for general values of C_{II}, we have obtained preliminary solutions by a numerical procedure outlined in (a)-(f) below. This is built on the thin-eddy method since (see above) that gives a good first estimate numerically for the entire range of values of C_{II}.

(a) Specify C_{II} and guess $S(X)$, $\overline{\Omega}L$.

(b) Take the thin-eddy formula (3.5) for p_I first and iterate to include full nonlinearity, using Cauchy-Hilbert types of integrals throughout.

(c) Next, take the thin-eddy formula (3.4) for p_{II} and iterate to include full nonlinearity, using an S.O.R. scheme with the coordinate $Y/S(X)$.

(d) Update $S(X)$ by a Carter technique, in the form $S^{(n)} = S^{(n-1)} + r(p_I - p_{II})$ at each X, where r is a relaxation factor and (n) denotes the level of the sweep.

(e) Renormalize to update $\overline{\Omega}L$ by imposing the conditions (3.3).

(f) Return to (b), until convergence is achieved.

The scheme works reasonably well at lower values of the parameter C_{II} but the convergence at higher values of C_{II} has proved very slow so far. Improvements of the method and alternatives (e.g., a Veldman-Davis scheme), and extended and more accurate runs, are currently being considered and will be reported along with more de-tails in ref. 10.

Results obtained so far are presented in figure 12. It should be stressed here that these are solutions extrapolated from converged results on rather coarse grids: see also above. Nevertheless the results, for values of C_{II} up to 0.3, approach the earlier ones for the thin-eddy case, as C_{II} is decreased, and they are also tending towards ref. 1's results at higher values of C_{II}. The limiting case of C_{II} = 1/2 (ref. 8) is an interesting one, by the way, partly because it has no discontinuity in velocity across the bounding vortex sheet, so that the speed driving the body-scale flow is reduced by an order of magnitude, probably (but not necessarily) producing an asymptotically small drag c_D. Further, ref. 3's calculations are tending to point towards that limiting case, or near it, as regards the eddy width/length ratio.

The tasks of finalizing these inviscid solutions and then using them as input for the viscous-layer problems, to determine $c_D(\bar{\Omega})$ and $L(\bar{\Omega})$, have still to be addressed seriously. So has the body-scale problem fixing c_D, L, $\bar{\Omega}$. The overall picture so far, however, looks reasonably promising.

SECTION 4. SMALLER-SCALE SEPARATIONS

Another view of the physical mechanisms associated with massive-scale eddies comes from examining smaller-scale separations, and this can be more directly useful in practical terms, for many separated airfoil flows for example. Smaller-scale separations can be described completely, within an interactive framework of the triple-deck kind or similar, and so they provide a platform for moving outwards, as it were, towards the massive-scale separated cases of interest.

This means adopting an interacting boundary-layer or similar approach at finite Re or going back to triple-deck sizes of flow configurations, as the principal examples for Re >> 1. These latter sized flows are governed by the boundary-layer equations

$$\frac{\partial U}{\partial \bar{X}} + \frac{\partial V}{\partial \bar{Y}} = 0, \quad U \frac{\partial U}{\partial \bar{X}} + V \frac{\partial U}{\partial \bar{Y}} = -\frac{dP}{d\bar{X}} + \frac{\partial^2 U}{\partial \bar{Y}^2} \qquad (4.1)$$

in the lower deck closest to the surface, subject to the boundary conditions

$$U = V = 0 \text{ at } \bar{Y} = 0 \text{ (or other surface conditions)} \qquad (4.2)$$

$$U \sim \bar{Y} + A(\bar{X}) + f(\bar{X}) \text{ as } \bar{Y} \to \infty \qquad (4.3)$$

and to the pressure-displacement interaction law, for subsonic motion,

$$P(\overline{X}) = \frac{1}{\pi} \oint_{-\infty}^{\infty} \frac{dA}{d\xi} \frac{d\xi}{(\overline{X}-\xi)} . \qquad (4.4)$$

In (4.3) $f(\overline{X})$ denotes the scaled surface shape, be it a hump, a step, a corner or the trailing-edge shape: see figure 13.

Nonlinear solutions (see refs. 5, 6) for separating flows can be derived numerically for increasing disturbance size, α say. Here $\alpha = 0(|f|)$ is the typical height of the hump or step, or the angle of incidence in trailing-edge flow, or the ramp angle in the central example of supersonic ramp flow, for instance (figure 13). In supersonic flow (4.4) is replaced by Ackeret's law,

$$P(\overline{X}) = -\frac{dA}{d\overline{X}} . \qquad (4.5)$$

The numerical solutions are useful for finite values of α anyway, for example in investigating trailing-edge stall, and as α increases they should move towards the massive separation cases, creating much larger-scale eddies. The calculations become more difficult and prone to divergence as α increases, however, and windward differencing and multi-sweeping, in the forward and possibly the reversed sense also, seem essential in finite-difference methods. Much further investigation is required numerically into these increased-scale separating motions.

Some reasonable theoretical conjectures may be made about the large-α behavior of the separating motions. One such is the extended Kirchhoff form (see refs. 5, 6) but that appears to hit an inconsistency at eddy closure much like the one described in Section 2. The other main contender is a massive-eddy account, like that in Section 3. This is especially encouraging since the system of triple-deck equations (4.1)-(4.4) admits the Benjamin-Ono equation (3.6a) again, for the eddy shape S in _subsonic_ flow, under the assumption of a predominantly inviscid eddy or eddies much greater than the typical obstacle size α. So the previous inviscid, thin-eddy, solution of Section 3 and figures 10 and 11 applies.   It can also be extended to incorporate the surface-shape effect $f(\overline{X})$ if the eddy size is reduced to $0(\alpha)$, in which case we have a more Prandtl-Batchelor type of flow. The viscous parts of the large-α solution still need to be investigated carefully, to pin the solution down, and the join to the massive-separation cases studied previously still has to be followed through, but overall the large structure seems to be not inconsistent at this stage.

By constrast, the <u>supersonic</u> or <u>hypersonic-limit</u> regimes do not work in the same way, even at the inviscid level (ref. 5). For, in supersonic flow for instance, with (4.5) holding, we obtain for the massive-eddy balance the steady Burger equation,

$$\frac{1}{8} \bar{\Omega}^2 S^2(\bar{X}) - c_{II} = \frac{dS}{d\bar{X}} \tag{4.6}$$

in place of (3.6a). The equation (4.6) admits no closed-eddy solutions and neither does its hypersonic-limit counterpart where $dS/d\bar{X}$ is replaced by $S(\bar{X})$. On the other hand, analysis (ref. 9) shows that if the effect of the surface shape is included then the inviscid eddy can be closed, as in figure 14. So the implication is that for these regimes the eddy size is probably much less than in the subsonic regime. Moreover, the structure takes on the Prandlt-Batchelor form in essence. There is a difficulty, even so, because the start of the eddy ahead of a supersonic ramp requires a large pressure rise, on inviscid reasoning, and this contradicts the $0(1)$ pressure rise ($P \approx 1.8$) known to occur in the viscous free-interaction separation upstream (figure 14). Therefore, in this flow, either the proposed inviscid structure should be modified (it clearly needs to incorporate the secondary separations which could be expected in any case), or the free-interaction separation is nonunique and can lead (e.g.) to an arbitrarily large pressure rise. The latter prospect is an intriguing one and has still to be addressed. Again, the alternative of large-α separating flow produced by a backward-facing step or a ramp of finite length (figure 14) shows more immediate hope of fitting together with the above account, since a very large pressure variation can be built up, before the start of the step or ramp, by means of the alternative attaching-flow free-interaction.

SECTION 5. FURTHER COMMENTS

The following points are in addition to those in the previous Sections.

(i)   Despite the comments in Sections 2 and 3, free-streamline
        theory of the extended Kirchhoff kind still "works" in the
        numerical and practical sense (see figure 4) at the medium
        Reynolds numbers where external flow remains laminar, and
        moreover it appears to be self-consistent in internal
        motions through pipes and channels and in cascade flows.

(ii)   For axisymmetric separating flow past a blunt axisymmetric
        body the analogue of Sections 2 and 3 can also be con-
        structed. This is being considered by Mr. R. Avis in Ph.D
        research at University College. In particular the slen-
        der-eddy limit, analogous to the thin-eddy limit in Sec-
        tion 3, yields a surface pressure

$$p_I \propto (S^2)'' \tag{5.1}$$

        from axisymmetric linearized-flow properties, with unknown
        eddy radius $S(X)$, while the vorticity argument inside the
        slender eddy now gives

$$p_{II} = c_{II} - \bar{a}^2 S^4, \tag{5.2}$$

        where $\bar{a}$ is a constant. Equating the two pressures then
        produces an integrable nonlinear differential equation for
        the area function $\propto S^2$. Solutions with closed eddies are
        again found to exist.

(iii)   The stability properties of both the small- and the large-
        scale separated flows are of much theoretical and practi-
        cal concern. Further, there is a clear need to develop
        the theory to encompass nonlinear unsteady flows, and
        three-dimensionality, with a view to making a connection

with the Karman vortex trail observed in practice at higher Re, among other things. Some aspects of this are discussed in ref. 7.

(iv)   There is also much still to be done for the inclusion of turbulent-modelled boundary layers as opposed to laminar ones. This inclusion is significant because otherwise it is tempting to appeal to turbulence effects to justify the use of rather arbitrary inviscid outer solutions.

(v)   In all the flows discussed, although there is still quite a long way to go to complete the theory, it seems clear that viscous forces play a most important role. Inviscid solutions can be obtained fairly readily in principle, and there is often an infinity of them available; but viscous effects narrow the choice considerably. This applies throughout small- or large-scale separated flow theory, whether the flow is governed by the Navier-Stokes or the interactive boundary-layer equations.

REFERENCES

1.   Sadovskii, V. S., 1971, Prikl. Math. Mech., Vol. 35, p. 773 (Transl. Appl. Math. Mech., Vol. 35, 1971, p. 729).

2.   Smith, F. T., 1984, United Technologies Research Center, East Hartford, CT, Report UTRC84-31; also J. Fluid Mech., 55 (1985), 175-191, Cambridge University Press.

3.   Fornberg, B., 1985, J. Comp. Phys. (in press).

4.   Ingham, D. B., Private communications, 1984-85.

5.   Smith, F. T., 1984, AIAA Paper No. 84-1582, presented at AIAA Conf., Snowmass, Col., July 1984.

6.   Messiter, A. F., 1983, Trans. Am. Soc. Mech. Eng., Vol. 50, p. 1104.

7.   Smith, F. T., 1985, Proc. Symp. Stability of Spatially-Varying and Time-Dependent Flows, NASA Langley Research Center, August 1985, to be published by Springer-Verlag.

8.   Pierrehumbert, R. T., 1980, J. Fluid Mech., Vol. 99, p. 129.

9.   Burggraf, O. R. and Smith, F. T., 1984-5, work in progress.

10. Smith, F. T., 1985, United Technologies Research Center, East
     Hartford, CT, report in preparation.

ACKNOWLEDGEMENT

Thanks are due to Dr. M. J. Werle (United Technologies Research
Center, East Hartford, Connecticut) and to Dr. R. E. Whitehead (Office
of Naval Research) for their interest and partial support of this
research.

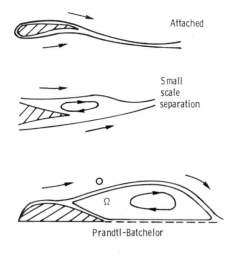

Fig. 1.   Various types of inviscid solutions with small- or large-
scale separation present.

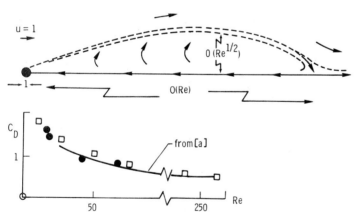

Fig. 2.   (a)   The extended Kirchhoff model.   (b) The drag coefficient
c_D, for flow past a cirular cylinder, versus Re, according
to experiments (o), numerical solutions of the Navier-Stokes
equations (X), and the extended Kirchhoff theory (——).

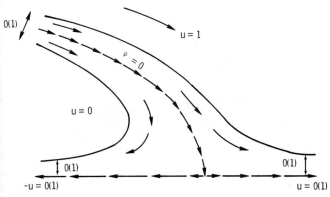

Fig. 3. Diagram of the eddy-closure process, with the incoming free
shear layer being turned by inviscid means.

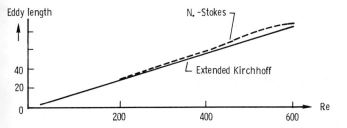

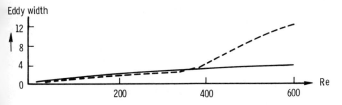

Fig. 4. Numerical Navier-Stokes results (---, ref. 3) for the eddy
length and eddy width behind the circular cylinder, including
results at higher Re. Extended Kirchhoff theory predicts the
solid curves, for comparison.

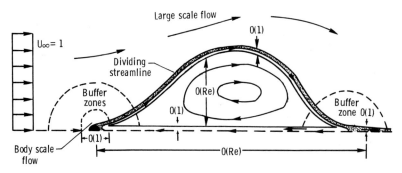

Fig. 5. The suggested massively-separated flow structure for large Re.

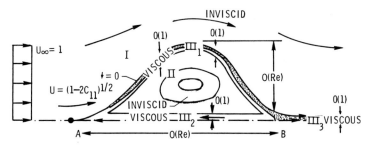

Fig. 6. The massive eddy-scale flow.

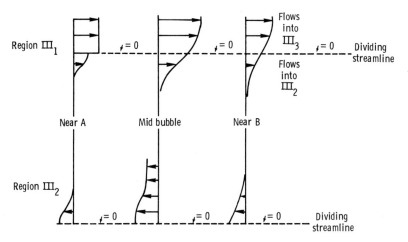

Fig. 7. Typical velocity profiles in the viscous layers III_1, III_2. Note the left-over jet profile in III_2 near point A.

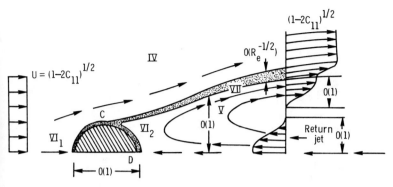

Fig. 8.   The structure of the body-scale flow, including the left-over jet (from figure 7) entering the turning-flow zone V.

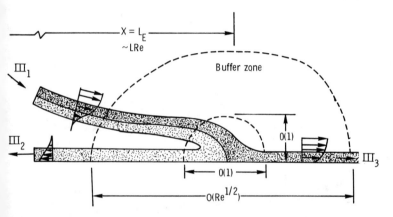

Fig. 9.   The flow structure during closure of the eddy, with buffer zones. Note the return jet produced, which forms the start of the viscous backward layer III_2.

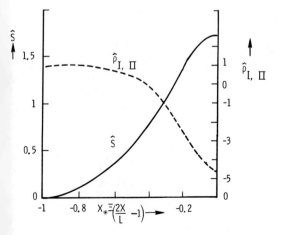

Fig. 10. Solutions of the thin-eddy problem (3.6a, b), in scaled terms (ref. 2), with symmetry about the eddy center $X = L/2$.

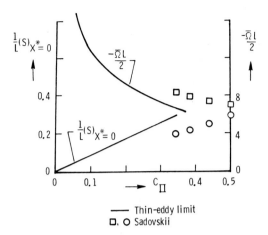

Fig. 11. Comparison between the thin-eddy predictions (———) (see also figure 10) and ref. 1's results ($\odot$,X), for the maximum eddy width, max-S/L, and the vorticity measure, $-\bar{\Omega}L/2$, versus the pressure parameter C_{II}.

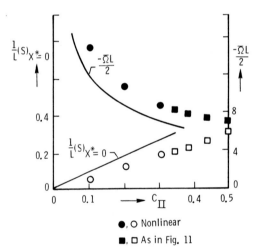

Fig. 12. Full computational solutions ($\bullet$, 0), from the present work, in addition to those in figure 11, for the massive-eddy problem of (3.1)-(3.3).

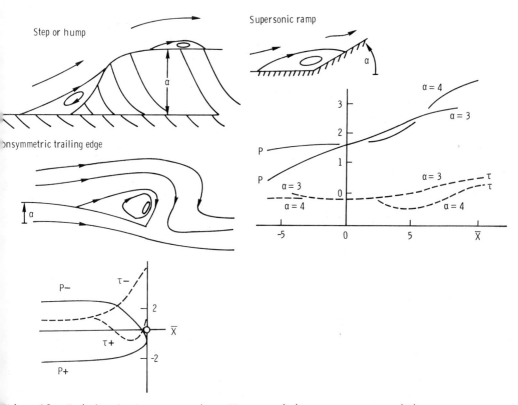

Fig. 13. Triple-deck separating flows: (a) over a step; (b) near a
nonsymmetric trailing edge (± refer to upper and lower
surface values, respectively); (c) past a supersonic ramp.
Here τ denotes the skin friction, and α is the typical
disturbance amplitude.

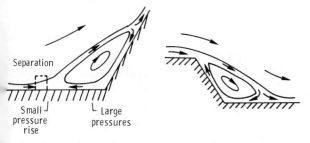

Fig. 14. Increased-scale separation for large α in supersonic motion:
(a) for the concave ramp; (b) for a backward-facing geometry.

Progress on the Calculation of Large-Scale Separation at High Reynolds Numbers

A. P. ROTHMAYER AND R. T. DAVIS

1. Introduction

The computation of high Reynolds number laminar separation has been one of the central issues in fluid mechanics over the past two decades. Laminar separation has eluded a description based on Prandtl's boundary layer theory primarily due to the presence of the Goldstein [10] singularity at separation (see also Stewartson [28]). However, this state of affairs changed abruptly with the development of the triple-deck theory by Stewartson & Williams [27] and Neiland [14] for supersonic flows. Drawing upon Lighthill's [12] earlier work with viscous sublayers, Stewartson & Williams [27] were able to show that a boundary layer in a supersonic flow could "spontaneously" separate by setting up a local interaction between a viscous sublayer lying within the depths of the boundary layer and the local inviscid flow just outside the boundary layer. The work of Neiland [14] and Stewartson & Williams [27] introduced a very new concept to viscous flow theory, that of a free-interaction. Prior to this time it had generally been thought that separation occurs due to a gradual reaction of the boundary layer to an externally imposed adverse pressure gradient. In fact, this turns out to be the exception rather than the rule. Viscous separation more often occurs in a very sudden and local fashion that is, to a large degree, independent of the geometry under consideration.

The application of the triple-deck free-interaction concept to subsonic flows is complicated by the fact that the leading order inviscid flow is elliptic, and so the triple-deck free-interaction must be embedded in the correct inviscid solution. There are a number of inviscid limit solutions which may be used for the leading order subsonic flow. In 1972, Sychev [30] proposed a theory of incompressible laminar separation based on the Kirchhoff [11] free-streamline model. In Sychev's [30] proposal, separation on the body scale occurs through a local triple-deck free-interaction embedded in a Kirchhoff [11] solution which is an $O(Re^{-1/16})$ perturbation from the Kirchhoff [11] smooth separation solution (see Figs. 1 and 2). In 1977, Smith [21] showed that a unique solution to Sychev's [30] proposed model exists, which indeed leads to a compressive free-interaction and separation. This solution of Smith [21] fixes the local inviscid behaviour about the separation point and is crucial in the development of a mechanism of cross-over from large-scale to small-scale separation (see Cheng & Smith [3]). This asymptotic structure is very interesting since the Kirchhoff [11] model has infinitly many solutions. The viscous effects in the triple-deck act as a mechanism for selecting the correct inviscid solution. The only global limitation imposed by the Kirchhoff [11] model is that the velocities entering the eddy from downstream infinity must be small (at least $O(Re^{-1/2})$ or smaller). It has been known for some time that the Kirchhoff [11] description of bluff body separation is valid in confined flows (for an example see Smith's [24] work on the cascade problem). This is because in a confined flow the viscous shear layer can diffuse into the inviscid region as the distance downstream of the airfoil increases. This allows for a fully viscous eddy closure and prevents large return velocities from entering the body scale flow. Unfortunately, the picture is not so simple in external flows. A number of researchers have attempted to achieve global consistency in the Kirchhoff [11] model (for example see Smith [24] and Cheng [4]). So far there has been little success using this approach primarily because of the inability to properly account for the re-entrant jet, which is present due to an inviscid reattachment mechanism in external flows. However, based on Fornberg's [9] full Navier-Stokes calculations at low Reynolds numbers, Smith [25] has proposed a new external flow structure which appears to be internally consistent. This new structure is based on the Sadovskii [20] constant vorticity eddy model and allows for a re-entrant jet in a consistent fashion. The primary complication that this new model introduces to the calculation of the body scale flow is that the re-entrant jet must be imposed as a downstream boundary condition on the infinite, non-constant vorticity, body scale eddy. More importantly, the flow near separation has a non-trivial dependence on the wake closure process. Cheng & Smith [3] have also provided a description of how the large-scale bluff body separation bubble undergoes a cross-over to small scale separation as the airfoil thickness is decreased. This asymptotic theory of Cheng & Smith [3] will be reviewed in more detail when the results are presented in section 3. It should be noted that the analogue of this cross-over theory for the Smith-Sadovskii [25,20] model of external separation has not been developed yet.

The procedure we favor for calculating high Reynolds number viscous flows is the interacting boundary layer model. In interacting boundary layer theory the higher order displacement interaction effect is assumed to be a leading order effect. This model can be justified for use at high Reynolds numbers if it can reproduce all of the asymptotic structure pertinent to the problem in question. It has been recognized for some time that interacting boundary layer theory is a composite of boundary layer and triple-deck theory, in the sense that it contains these structures and can reproduce them in the limit as Re→∞. Therefore, interacting boundary layer theory should be able to calculate small separation bubbles based on the triple-deck structure as well as Sychev-Smith [30,22] massive separation problems based on the Kirchhoff [11] inviscid model. Interacting boundary layer theory has had considerable success in calculating small separation bubbles. In fact, the number of applications of interacting boundary layer theory to small-scale separation are so numerous that they will not be listed here. However, there has been little success in calculating large-scale separation, particularly the extensive separation structures which have been predicted by the asymptotic theories. Recently we have developed an interacting boundary layer procedure which is capable of calculating large-scale separation (see Rothmayer & Davis [17]). It should be stressed that this model, which will be presented later in this paper, is only valid within the context of a body-scale Kirchhoff [11] description of separation. Such a solution is expected to be internally self-consistent. However, global consistency in an external flow requires that a re-entrant jet be imposed as a downstream boundary condition on the body scale flow. For now we choose to neglect this effect and instead we will concentrate on the local Kirchhoff [11] solution. The purpose of this paper is to review this new interacting boundary layer approach for large-scale separation, present some new results on the cross-over from large-scale to small-scale separation and to present a preliminary study which we feel will lead to a general interacting boundary layer formulation applicable to most two-dimensional incompressible flows. This formulation, which is presented in section 4, will allow lifting airfoils bluff body and cascade flows to be calculated using the same interacting boundary layer model. In addition, we have preliminary computational evidence which suggests that the re-entrant jet may not be present in finite eddies. Unfortunately, the evidence that we have at this time is inconclusive and does not point firmly towards Cheng's [4] account of the finite eddy nor to Smith's [25] re-entrant jet model. Even if the re-entrant jet is present in a finite length eddy, our results tend to suggest that this jet will not have a significant impact on the overall qualitative predictions of the Kirchhoff [11] finite eddy model of Cheng & Smith [3] and Cheng [4].

2. The Interacting Boundary Layer Models

The interacting boundary layer formulation and the numerical procedure used here substantially follows that of Davis & Werle [8] and Rothmayer & Davis [16,17]. The governing boundary layer equations are non-dimensionalized and scaled according to boundary layer scaling laws. The resulting equations are then rewritten using the Prandtl transposition theorem and Gortler variables referenced to an arbitrary baseline coordinate curve (see Fig. 3):

$$N = Re^{1/2}(y - f) \quad , \tag{1a}$$

$$\xi = \int_0^s Ue_0(t) \, dt \qquad , \qquad \text{(1b)}$$

and

$$\eta = \frac{Ue_0(s)}{\sqrt{2\xi}} N \qquad , \qquad \text{(1c)}$$

where Re is the Reynolds number, defined in the normal fashion by

$$Re = \frac{\rho U_\infty L}{\mu} \qquad . \qquad \text{(1d)}$$

Here, s and N are nondimensionalized arc-length and normal coordinates along the baseline coordinate curve, f is the distance from the baseline coordinate curve to the airfoil surface and Ue_0 is the inviscid surface speed over the baseline coordinate curve (see Fig. 3). A detailed discussion of this coordinate system is given by Davis & Werle [8] and Rothmayer [19]. For almost all the cases presented in this study, the baseline coordinate curve is taken to be the x-axis and so $Ue_0=1$ everywhere. Using the standard definitions of the streamwise and normal velocity components in Gortler variables (F and V respectively), the final form of the interacting boundary layer equations are found to be

$$V_\eta + F + \omega_1 2\xi F_\xi = 0 \qquad , \qquad \text{(2)}$$

$$F_{\eta\eta} - \omega_2 VF_\eta + \omega_3 FV_\eta + \beta_1 + \omega_4(1-\beta_0)F^2 = 0 \qquad \text{(3)}$$

and

$$\beta_{1_\eta} = 0 \qquad . \qquad \text{(4)}$$

In these equations, additional factors ω_1 through ω_4 have been introduced and their use will be discussed later. Unless otherwise stated, ω_1 through ω_4 are all taken to be equal to 1. The no-slip and wake center-line boundary and the numerical treatment of the boundary conditions are discussed by Davis & Werle [8] and Rothmayer & Davis [16]. This discussion will not be repeated here suffice for the very important displacement interaction effect, and its coupling into the boundary conditions. This procedure was first suggested by Veldman [32] and implemented in its final form by Davis [8]. The outer edge boundary condition on the u-velocity component is given by

$$F \to \frac{Ue}{Ue_0} \qquad \text{as } \eta \to \infty \qquad , \qquad \text{(5)}$$

where Ue is the inviscid velocity at the edge of the boundary layer. Using the normal formulation of interacting boundary layer theory, Ue is found from thin airfoil theory to be

$$Ue = 1 + \frac{1}{\pi} \int_{-\infty}^{\infty} \frac{T'(t)}{x - t} \, dt \qquad , \qquad (6)$$

where

$$T = Re^{-1/2}(Ue\delta + f) \qquad . \qquad (7)$$

Here $Ue_0 = 1$, δ is the boundary layer displacement thickness and f is the distance from the baseline coordinate curve to the airfoil surface. Note that when $Ue_0 = 1$, f is simply the scaled airfoil thickness. Substituting Eqn. (6) into Eqn. (5) gives F in terms of the boundary layer displacement thickness δ. At this point the V-matching condition

$$V - \eta V_\eta \rightarrow (2\xi)^{1/2} \frac{d}{d\xi}[Ue\delta] \qquad \text{as } \eta \rightarrow \infty \qquad (8)$$

is used to express δ in terms of the normal velocity component V. This yields a boundary condition solely in terms of unknowns in the boundary layer. The system of governing equations (2) through (4) and the associated boundary conditions are then solved in a completely coupled fashion in a direction normal to the body using a block tri-diagonal algorithm. If a spatial quasi-linearization is used, then the only iteration that is needed is a spatial sweeping of the strong-interaction region in order to converge the displacement thickness δ.

The Hilbert integral form of interacting boundary layer theory (i.e. using Eqn. (6)) has been applied to attached and slightly separated flows by a number of authors. It is now desired to apply this model to large-scale separation of the Sychev-Smith [30,22] form based on the Kirchhoff [11] model. In the Sychev-Smith [30,22] body scale theory of bluff body separation, the pressure is expected to be constant over a large region of the flow. In addition, the displacement thickness will eventually become proportional to $x^{1/2}$ as the distance, x, increases downstream of the airfoil. In order to directly apply the Hilbert integral of Eqn. (6), this asymptote would have to be naturally achieved in the calculation. This, in turn, would require that the downstream boundary condition be placed several airfoil chords downstream of the separation point. This situation is undesirable for a number of reasons, which will be discussed later. Fortunately there is an alternative to the Hilbert integral of Eqn. (6). In the massive separation interacting boundary layer model, we simply make the assumption that the pressure is constant past some arbitrarily prescribed point x_0. The eddy pressure for the single body is taken to be zero for simplicity. Typically, x_0 is chosen to be slightly downstream of the separation point and within the triple-deck plateau region (see

Stewartson [29] for a discussion of the plateau structure). The thin airfoil problem is now a mixed boundary value problem. Upstream of x_0 the displacement thickness is "prescribed", whereas downstream of x_0 the pressure is prescribed. The generalized method of Cheng & Rott [2] for mixed boundary value problems in thin airfoil theory is then used to solve for the inviscid flow. If $w(z)=u-iv$ is the complex velocity in the outer inviscid flow, we seek an expansion of the form $w(z) \sim 1 + \text{Re}^{-1/2} w_1(z) + \dots$. The problem for $w_1(z)$ is a mixed boundary value problem. $w_1(x,0)$ satisfies the injection condition (with $Ue_0 = 1$)

$$v_1(x,0) = Ue_0 \frac{d}{d\xi}\left[Ue\delta + f\right] \qquad (9)$$

on its imaginary part for $x < x_0$, and $u_1(x,0) = 0$ for $x > x_0$. All that is required is that $w_1(z)$ be an analytic function which satisfies these boundary conditions along the x-axis and the zero-flow condition at infinity. If a direct thickness problem is solved for $w_1(z)$, an integral equation results for an unknown singularity distribution along the x-axis. To avoid this, a direct thickness problem is solved for a constructed function $F(z)$, where

$$F(z) = \frac{w_1(z)}{H_0(z)} \qquad . \qquad (10)$$

$H_0(z)$ is chosen to switch from pure real to pure imaginary across x_0 in such a way as to make the imaginary part of $F(z)$ along the x-axis correspond to the prescribed boundary conditions for $w_1(z)$. $H_0(z)$ is chosen to be

$$H_0(z) = i(z-x_0)^{1/2} \qquad (11)$$

to give the correct form of the free-streamline at downstream infinity. The direct thickness problem is then solved for $F(z)$, with the boundary conditions for the imaginary part of $F(z)$ given by

$$\text{Im}\left[F(z)\right]\Big|_{y=0} = \frac{-1}{\sqrt{x_0-x}} \frac{d}{dx}\left[Ue\delta+f\right] \qquad , \text{ for } x < x_0 \qquad (12)$$

and

$$\text{Im}\left[F(z)\right]\Big|_{y=0} = 0 \qquad , \text{ for } x > x_0 \qquad . \qquad (13)$$

The resulting solution for $F(z)$ can be rearranged to give

$$Ue = 1 + \frac{Re^{-1/2}}{\pi} \sqrt{x_0-x} \int_0^{x_0} \frac{\frac{d}{dt}\left[Ue\delta+f\right]}{(x-t)\sqrt{x_0-t}} dt \qquad \text{for } x<x_0 \qquad (14)$$

and

$$\frac{d}{dx}\left[Ue\delta+f\right] = \frac{\sqrt{x-x_0}}{\pi} \int_0^{x_0} \frac{\frac{d}{dt}\left[Ue\delta+f\right]}{(x-t)\sqrt{x_0-t}} dt \qquad \text{for } x>x_0 \qquad . \qquad (15)$$

These equations assume that δ and $f(x)$ are zero from $-\infty$ to 0. As $x_0 \to \infty$, Eqn. (14) reduces to the standard form of Eqn. (6). These equations are handled numerically in a fashion similar to the original method of Davis & Werle [8]. The singularity near x_0 is removed analytically by subtracting out a Taylor series expansion of the integrand about x_0. some of the resulting integrals may then be integrated analytically. An integration by parts of the Cauchy principal value of the remaining integral is then used along with the midpoint rule to form element expressions. Details of this procedure may be found in Rothmayer & Davis [17] and Rothmayer [19]. A major advantage of using Eqn. (14), as opposed to Eqn. (6), is that the interacting boundary layer problem is solved only upstream of x_0. After the interacting boundary layer problem is converged upstream of x_0, Eqn. (15) is used to find the displacement thickness downstream of x_0.

The ability to avoid calculating the flow downstream of x_0 is a major advantage of this new technique, and in fact this was our initial motivation for considering this method. However, this advantage will almost certainly be lost in calculating a Smith-Sadovskii [25,20] flow with a re-entrant jet or a finite non-constant pressure eddy in a fully computational fashion. In fact if x_0 is taken to be relatively far downstream of the airfoil then the convergence properties of the method are found to suffer. In addition, if this method is to be generalized to three-dimensional flows an understanding of the basic mechanisms operating in this solver must be developed, since the complex variable approach obviously cannot be generalized to three-dimensions.

It should be stated at this point that even though much room remains for improvement, this method works surprisingly well. Initially in the iteration cycle the method causes separation to occur just upstream of x_0, this is on the first iteration and irrespective of the location of x_0! The local solution about the separation point (i.e. the free-interaction) then establishes itself usually within five to ten iterations. After that, the rest of the iterations (anywhere from 30 to 200, which reduce the residual approximately four orders of magnitude) appear to be taken up by a gradual "settling in" of the solution to the final answer. This "settling in" is also accompanied by a motion of the separation point along the airfoil surface. Based on the above observations, we feel that there are three important mechanisms operating in this method. The first is that the initial guess perceived by the boundary layer is the inviscid singular Kirchhoff [11] solution with the separation point at x_0. In all our calculations the initial guess was taken to be $\delta=0$. Setting $\delta=0$ in Eqn. (14) gives the inviscid Kirchhoff [11] solution (also see Cheng & Smith [3]). The inviscid Kirchhoff [11] solution almost always has a pressure gradient singularity just upstream of x_0. Therefore, built into this model is a mechanism for initiating the triple-deck free-interaction at an arbitrary point along the airfoil surface. This intrusive calculation is philosophically quite different from most computational fluid dynamics methods and, to our knowledge, has never been attempted before this time. Due to the basic physics of the free-interaction mechanism, once the free-interaction is initiated it tends to establish itself quite rapidly. In fact, it appears that the free-interaction can be established locally at any point on the airfoil surface without regard for the rest of the flow structure. Once the free-interaction has been established it then seeks to establish the global inviscid flow solution which can support the triple-deck separation structure. A second important feature of our solution method is that the initial inviscid Kirchhoff [11] solution is closer to the final answer than the initial guess in the Hilbert integral method. In fact, if δ is taken to be zero in Eqn. (6) then the initial guess perceived by the boundary layer is the attached flow solution. This initial guess may or may not cause separation to readily occur depending on the nature of the trailing-edge of the airfoil. Even if separation occurs within the first few iterations (as say with a wedged or blunt trailing-edge airfoil), it will be a small separation bubble centered at the trailing edge of the airfoil. This flow would then have to undergo considerable changes to reach the large-scale separation solution. In the Kirchhoff [11] integral of Eqn. (14), the final answer is close to the inviscid Kirchhoff [11] solution when $\delta=0$, and so viscous-inviscid interaction only has to locally modify the singular behaviour in the inviscid solution near the separation point. Note that this would also be true even if we allowed for a slight deviation from a constant pressure within the eddy to allow for a re-entrant jet. The third important point is that our method is internally consistent. An initial pressure gradient singularity exists in the method which allows us to intrusively pick the initial location of separation. However, as viscous-inviscid interaction takes over, this initial singularity is smoothed out and the final solution is found to be regular (see section 3 for numerical examples of this and Rothmayer & Davis [17] for further discussion on this point). This does not occur by accident, but is due to the fact that the method was formulated in a consistent fashion as a mixed boundary value problem. Further discussions on the interpretation of this method may be found in Rothmayer & Davis [17] and Rothmayer [19] and will not be repeated here.

Another question that immediatly arises concerns the zero-pressure downstream boundary condition and ultimately the validity of the Kirchhoff [11] model itself.

Smith [22] has indicated that the Kirchhoff [11] infinite eddy solution is a completely self-contained structure. As pointed out in the introduction, the only global requirements that this structure imposes on the flow field is that the velocities beneath the shear layer as $x \to \infty$ must be small. As a preliminary step to examining the consistency of the Kirchhoff [11] model, we chose to impose a constant non-zero pressure on the infinite eddy. The only boundary condition that was changed was that for $u_1(x,0)$ downstream of x_0. The equation which results for $x < x_0$ is

$$Ue = Ue_s + \frac{Re}{\pi}^{-1/2} \int_0^{x_0} \frac{\frac{d}{dt}[Ue\delta + f]}{\sqrt{x_0 - t}(x - t)} \, dt \qquad , \qquad (16)$$

where Ue_s is some constant imposed velocity past x_0. Changing the eddy pressure simply shifts the entire local pressure field by a constant amount, which suggests that it is consistent to choose a zero pressure on the eddy. This choice of the eddy pressure essentially defines the level of the free-stream velocity at infinity. However, the infinite eddy solution acts as a leading-edge region on a much larger eddy. It is known from thin airfoil theory (see Van-Dyke [31]) that the leading-edge region on a blunt leading-edge airfoil does not perceive the true free-stream velocity but rather a velocity that has been modified by the full airfoil geometry. In the massive separation problem, in the context of a Kirchhoff flow, the eddy pressure on the body scale will be non-zero. However, the actual pressure level must be set from a knowledge of the overall eddy properties. We choose not to address this problem, but instead will concentrate on the local solution. The overall consistency of the Kirchhoff [11] model now rests on the numerical results provided by the model. In section 3, it is convincingly demonstrated that changing x_0, even by a relatively large amount, has absolutely no effect on the solution. In addition, the leading order solution is shown to be completely insensitive to the behaviour in the slow reversed flow region beneath the shear layer. A further discussion on these points will be deferred until the results are presented.

3. Results and Discussion

The results for large-scale separation will be presented in terms of displacement thickness δ, pressure coefficient c_p and wall shear stress τ_w, where

$$\tau_w = \frac{2}{\sqrt{2\xi}} \frac{\partial F}{\partial \eta} (\xi, 0) \qquad (17)$$

and

$$c_p = \frac{1}{2} (1 - Ue^2) \qquad . \qquad (18)$$

In addition, the physical airfoil geometry is given by

$$y = \Omega f_A(x) \qquad . \qquad (19)$$

The wall shear stress for the sine airfoil,

$$f_A(x) = 1 + \sin\left[2\pi(x - 1/4)\right] \qquad , \qquad (20)$$

is shown in Fig. (4). The local free-interaction about the separation point and the asymptote to an almost stagnant reversed flow can be seen in this figure. The effect of changing the location of x_0 is shown in Figs. (5) and (6) for the pressure coefficient and the sum of the airfoil surface and the displacement thickness. These figures not only show the smooth asymptote to the constant pressure downstream boundary condition, but Fig. (6) also shows that the patching between the viscous solution of Eqn. (14) and the analytic solution of Eqn. (15) is regular. A typical velocity profile at x_0 is shown in Fig. (7) for the $\Omega=.025$ case. This velocity profile is compared with the Blasius profile which would exist at the same point on the airfoil if the airfoil were to be replaced by a flat plate. The patch between the analytic solutions downstream of x_0 and the computed solutions upstream of x_0 are shown in Fig. (8). An expanded view of this figure is shown in Fig. (9). As the airfoil thickness decreases, the separation point is forced towards the trailing-edge of the airfoil. This causes a local collapse of the shear layer which in turn causes the infinite, $O(Re^{3/4})$, eddy to undergo a rapid, but in this case steady-state, collapse towards the airfoil. This collapse will be referred to as wake deflation and the inverse process as wake inflation. This process is also called wake cut-off by Cheng & Smith [3]. The basic mechanism for the cross-over from a flow with massive separation to an attached flow, or at least a finite length separation bubble, was first set down by Cheng & Smith [3] for single non-lifting airfoils in a uniform stream. Extensions to this basic work were later made by Cheng [4,5] for the lifting case and asymmetric bifurcating solutions for a symmetric problem and by Rothmayer & Smith [18] for symmetric cascades of airfoils.

The Sychev-Smith [30,22] model is valid when an $O(1)$ length airfoil is a bluff body. That is, when the airfoil thickness is $O(1)$. In this case the inviscid solution is slightly perturbed from the Kirchhoff [11] smooth separation solution. As the thickness of the airfoil, say $O(\epsilon)$, tends to zero, the Kirchhoff [11] smooth separation solution can still be maintained. The separation point will then tend to some finite location on the airfoil surface. An extreme case is the blunt leading-edge airfoil. Here, the separation point actually tends towards the leading-edge as $\epsilon\to0$. This seems to be fundamentally incompatible with trailing-edge separation. On a blunt leading-edge/wedge trailing-edge airfoil for example, the flow would first separate at the trailing-edge when $\varepsilon \sim O(Re^{-1/4})$. This is very different from the tendency of the Kirchhoff [11] smooth separation point to move towards the leading edge as $\varepsilon \to 0$. The mechanism by which the trend towards the leading-edge is reversed, allowing the flow to cross-over from massive separation to an attached flow is provided by Cheng & Smith [3]. The pressure being generated on the airfoil surface is $O(\varepsilon)$ from classical thin airfoil theory. However, the pressure perturbation to the Kirchhoff [11] smooth separation solution is fixed at an $O(Re^{-1/16})$ level by the triple-deck. When ε decreases to $O(Re^{-1/16})$ the pressure being generated over the airfoil surface is comparable with the pressure perturbation being imposed by the triple-deck, and so the Kirchhoff [11] smooth separation solution can no longer hold. The triple-deck requirement on the local inviscid flow is given by Smith [21] and Cheng & Smith [3]:

$$p - p_e \sim - \alpha Re^{-1/16} \lambda(x_0)^{9/8}(x_0-x)^{1/2} + \ldots \qquad \text{as } x \to x_0^-, \qquad (21)$$

where x_0 is the separation point, p_e is the eddy pressure (which is zero for external flows), $\lambda(x_0)$ is the non-interacting boundary layer skin friction parameter at x_0 and $\alpha=0.44$ from Smith's [21] triple-deck free-interaction calculation. The inviscid solution is now the Kirchhoff [11] singular solution with the level of the singularity being set at a constant value by the triple-deck. Thus when $\varepsilon \sim O(Re^{-1/16})$, viscous effects have come into the picture and can modify the location of the separation point.

The basic problem to be addressed here is the Kirchhoff [11] flow over a thin airfoil with the thickness distribution $y = \varepsilon \tilde{h} f(x)$, where $\tilde{h}$ is the $O(Re^{-1/16})$ scaled airfoil thickness. Since the airfoil is thin, the non-interacting boundary layer flow will be a small perturbation on a Blasius boundary layer. Therefore, to leading order, $\lambda(x_0)$ takes on the Blasius value $\lambda(x_0) = \hat{\lambda} x_0^{-1/2}$, where $\hat{\lambda} = 0.332$. The Kirchhoff [11] problem is then solved for the pressure $p(x,y)$ using the mixed boundary value method of Cheng & Rott [2]. The resulting equation for the surface pressure $p(x,0)$ is then expanded for $x \to x_0$ and the coefficient of the $(x_0-x)^{1/2}$ term is fixed from the triple-deck criterion of Eqn. (21). This gives the equation derived by Cheng & Smith [3] for the location of the separation point x_0:

$$\int_0^{x_0} \frac{f'(t) - f'(x_0)}{(x_0-t)^{3/2}} \, dt - 2f'(x_0)x_0^{-1/2} = \frac{\bar{\alpha}\pi}{x_0^{9/16}\tilde{h}}, \qquad (22)$$

where $\bar{\alpha} = 0.127$. Equation (22) can be integrated by parts twice and solved numerically. The above equation will give a value of $\tilde{h}$ for every value of x_0 between the leading and trailing-edge. However, in order for the Kirchhoff [11] solution to be valid, the wake must be open and infinite. This means that the coefficient of the parabolic growth of the free-streamline as $x \to \infty$ must be positive (this coefficient will be called b for later reference). The point at which the wake ceases to be open and the free-streamline asymptotes to the x-axis (i.e. $b \to 0$) has been called the wake cut-off point by Cheng & Smith [3], but will be referred to here as the wake deflation, or inflation, point. It must be emphasized that this inflation/deflation point is inevitably reached as x_0 tends towards the trailing-edge, irrespective of the problem being solved. Solutions to Eqn. (22) for a biconvex (see Fig. 10a) and a sine (see Fig. 11a) airfoil are shown in Figs. 10b and 11b respectively. These two airfoils illustrate the two basic phenomenon encountered in the cross-over process. The solution for the wedge trailing-edge airfoil is unique and allows $b \to 0$ at some finite attainable point $(x_{0_c}, \tilde{h}_c)$. The approach to $(x_{0_c}, \tilde{h}_c)$ initiates the wake deflation process. As $b \to 0$ the infinite eddy collapses locally and in turn forces a rapid, but steady state, deflation of the $O(Re^{3/4})$ infinite eddy. Cheng & Smith [3] have shown that this deflation process continues until the eddy length shortens to $O(1)$ and the pressure within the eddy increases to $O(Re^{-1/16})$. The solution for the cusped trailing-edge sine airfoil shows the other basic type of behaviour which is possible in the cross-over from massive separation to an attached flow. For the sine airfoil, the solution is non-unique with wake deflation occuring on the uppermost branch. In a "realistic" problem the separation point would be moving towards the trailing-edge along the lower branch (having started at the smooth separation point as $\tilde{h} \to \infty$). Therefore, as the airfoil thickness $\tilde{h}$ is continually decreased, the separation point will eventually reach the unsteady wake cross-over point $(x_{0_u}, \tilde{h}_u)$ where $dx_0/d\tilde{h}(x_{0_u}) = 0$. In Fig. 11b $x_{0_u} \approx 0.65$ and $\tilde{h}_u \approx 0.00895$. If $\tilde{h}$ is decreased below $\tilde{h}_u$ then no massive separation solution exists. The flow is then expected to undergo an unsteady cross-over to some other flow. In the case of the cusped trailing-edge airfoil, an attached flow solution is always possible for $\epsilon \ll 1$. The flow will most likely make an unsteady cross-over to the attached flow solution, although the possibility of another solution with a small separation bubble should not be entirely discounted.

A qualitative account may now be given for these two types of cross-over processes. For thin airfoils, the strength of the singularity is, roughly speaking, proportional to the product of the airfoil thickness and the local body slope. This can be seen from Eqn. (22), although there is a distributed effect due to the integral, and can be much more clearly seen in the analytic solutions of Rothmayer & Smith [18] for the cascade problem. There are two possible mechanisms for the cross-over from massive separation to an attached flow. As the airfoil thickness decreases, the separation point tends towards regions of increasingly negative slope in order to maintain the level of the square root singularity. Typically, this tends to move the separation point towards the trailing-edge. In the first cross-over process, most typical of cusped trailing-edge airfoils, the separation point reaches the vicinity of an inflection point. At this point, the slope can no longer be negatively increased in order to compensate for the decreasing airfoil thickness. Below this critical thickness it would be expected that an unsteady breakdown of the Kirchhoff [11] solution would ensue, ultimately leading to either an attached flow or a smaller separation bubble. In the second cross-over process, most typical of wedged and blunt trailing-edge airfoils, an inflection point is never encountered, and the separation point tends towards the trailing-edge. However, before the trailing-edge is reached, the Kirchhoff [11] eddy collapses at infinity, thereby forcing the long $O(Re^{3/4})$ eddy to collapse towards the airfoil. This collapse is eventually halted when the eddy reaches a streamwise length scale of $O(1)$ and a thickness of $O(Re^{-1/16})$. In addition, the eddy pressure increases to $O(Re^{-1/16})$ and is no longer small in comparison with the pressure being generated by the airfoil ahead of the separation point.

Cheng & Smith [3] have tended to suggest that the type of cross-over encountered depends largely on the nature of the trailing-edge. The solutions of Cheng & Smith [3] were unique for wedge and blunt trailing-edges, implying a steady state cross-over, and were nonunique for cusped trailing-edges, implying an unsteady cross-over. However, as is seen later, the cross-over properties tend to be governed more by the local curvature of the airfoil near the separation point. In fact, wedge trailing-edge airfoils which have interior curvature similar to cusped trailing-edge airfoils are found to have nonunique solutions as well.

The airfoil geometry and triple-deck results are shown in Fig. (11) for the sine airfoil of Eqn. (20) and in Fig. (10) for the biconvex airfoil

$$f_A(x) = 8x(1-x) \tag{23}$$

These results are presented in terms of an $O(Re^{-1/16})$ scaled airfoil thickness $\tilde{h}$. The finite eddy portion of the results were calculated using the finite eddy Kirchhoff [11] model of Cheng & Smith [3]. The principle untested assumption of this model is the cusped reattachment condition which is required to yield a small drag on the airfoil. The basic steady-state wake deflation process is shown in Fig. (10b) for the biconvex airfoil. It should be noted that this solution is unique for all airfoil thicknesses. Figure (11b), for the sine airfoil, shows the nonunique steady-state massive separation solution which leads to the unsteady cross-over from large-scale separation to an attached flow. These triple-deck results are compared with the interacting boundary layer results in Fig. (12). Even though triple-deck theory can predict the qualitative behaviour quite accurately at this Reynolds number, it does not fare well in a quantitative comparison with interacting boundary layer theory. Indeed, it was found by Rothmayer & Davis [17] that even though there was a definite tendency towards the local triple-deck free-interaction solution of Smith [21] about the separation point as the Reynolds number increases, the limit solution was still inaccurate even at a Reynolds number as high as one hundred million.

It was desired to verify the downstream asymptote of the interacting boundary layer results in a more rigorous fashion. In the numerical calculations presented for the sine airfoil, no approximations were needed in order to stabilize the calculations in a reversed flow region. The region of reversed flow was treated in exactly the same fashion as the rest of the flow-field. It should be noted that this is justified by the fact that a unique downstream plateau structure in the reversed flow region is being forced to occur by the displacement thickness interaction (see Stewartson [29]). However, if x_0 is taken to be too far downstream of the separation point, an instability will set in within the reversed flow region. For example, in the sine calculation with $\Omega=.025$, an instability was encountered if x_0 was approximately .75 or larger. Therefore, in order to verify the downstream asymptote a means of suppressing this instability within an extensive region of reversed flow was needed. This stabilization was accomplished by introducing the factors ω_1 through ω_4 into the governing equations (2) through (4). Using a standard Fourier stability analysis, it was found that only ω_1 and ω_3 could be used to stabilize the calculations in an extensive region of reversed flow. In addition, two special cases were found to stabilize the calculations. The first involves setting $\omega_1=1$ and $\omega_3=0$ downstream of the point of initial instability in the reversed flow region for $\Omega=.025$ (taken here to be $x=.6$). This amounts to neglecting a portion of the uu_x term in certain regions of the reversed flow. The other approximation involves setting $\omega_1=0$ and $\omega_3=1$ downstream of $x=.6$ in the reversed flow region for $\Omega=.025$. The space marching scheme will then be stable only if $|F| < 2\Delta\eta$. However, this condition is seldom violated in an extensive region of reversed flow. This second approximation amounts to an assumption of local self-similarity. The results using both of these stabilizing approximations, for $x_0=2.0$, along with the interacting boundary layer model with no stabilizing approximation, for $x_0=.6$, are shown in Figs. (13) and (14). The pressure distribution is completely insensitive to the approximations made within the reversed flow region. In addition, the constant pressure downstream asymptote is achieved quite naturally over a fairly extensive region. Note that this figure also verifies the existence of the local Kirchhoff [11] solution. If the re-entrant jet were to be included in this calculation it must actually be imposed as a downstream boundary condition (i.e. it does not emerge naturally in the local body scale calculation). Figure (14) shows the overall insensitivity of the displacement thickness distribution to the flow within the eddy. A detailed picture of the streamline distribution for both stabilizing approximations are shown in Figs. 15 and 16. It should be noted from these figures that the flow within the separation bubble changes drastically from one stabilizing approximation to the next. However, this difference has a negligible effect on the solution external to the separation bubble. This can be seen more clearly in Figs. 17 and 18. Figure 17 shows that the velocity profile at the last grid point, and external to the separated region, are virtually identical irrespective of which stabilizing approximation is used. However, a blow-up of the velocity profiles within the separation bubble, shown in Fig. 18, shows that the reversed flow velocity profiles are significantly different. This is consistent with the Sychev-Smith [30,22] model based on the Kirchhoff [11] free-streamline theory. In the Sychev-Smith [30,22] theory, the reversed flow within the eddy is forced by the entrainment through the shear layer and is expected to be a higher order effect. In fact, the pressure correction is only $O(Re^{-1})$. This would not be the case in a Smith-Sadovskii [25,20] eddy calculation. Here, the reversed flow within the eddy would be expected to exert a leading order effect.

The steady-state wake cross-over process was examined for the biconvex airfoil of Eqn. (23) with a semi-infinite splitter plate. A series of Hilbert integral calculations were performed for increasing airfoil thickness and starting from an initially attached flow. These results for wall shear stress and pressure coefficient are shown in Figs. (19) and (20). These figures clearly show the beginnings of an almost stagnant, constant pressure, reversed flow region. The local triple-deck free-interaction can also be seen to be developing around the separation point as the airfoil thickness increases. However, a similar localized structure does not appear to be developing about the reattachment point. It should also be noted that the level of the constant pressure just downstream of separation appears to be asymptoting to a finite level (i.e. at least comparable with the pressures upstream of separation). This would seem to point towards a Smith-Sadovskii [25,20] account of the finite eddy problem. However, a Smith-Sadovskii [25,20] model would also be expected to possess a re-entrant jet. Interestingly enough, no such jet was found in the streamline and velocity profile plots near reattachment of Figs. (21) and (22), for the largest finite eddy of Fig. (20). In fact, these two figures appear to be more in line with the supersonic triple-deck reattachment calculations of Daniels [6]. In addition, the Smith-Sadovskii [25,20] account should have an expanding region of non-constant pressure downstream of the constant pressure region as the size of the separation bubble increases. No such trend was observed here. Unfortunately, the results appear to be inconclusive. The re-entrant jet was not found to appear within our calculations, but it may well be that the size of the separation bubble was still to small to discern such a jet. The streamline and velocity profiles tend to point more towards Cheng's [4] account of reattachment. But the asymptote of the pressure level to a value comparable with the pressures upstream of separation and the failure of a localized structure to appear about reattachment is disturbing and apparently contradicts Cheng's [4] account of the reattachment mechanism. Irrespective of which finite eddy model is correct, it should be noted that the Kirchhoff [11] finite eddy model of Cheng & Smith [3] and Cheng [4] does appear to give accurate qualitative predictions. The infinite eddy solutions for the biconvex airfoil are shown in Fig. (23). Note that as the airfoil thickness is decreased, a steady-state wake deflation is initiated. The location of the separation and reattachment points for this airfoil are shown in Fig. (24). This figure should be compared with the triple-deck results of Fig. (10b). As the airfoil thickness is increased from initially low values, separation first develops at the trailing-edge for an airfoil thickness of $\Omega \sim 0.005$. As the airfoil thickness is further increased the separation bubble gradually grows about the trailing edge. At a certain critical airfoil thickness, this separation bubble begins to undergo a rapid, but steady-state, wake inflation to the "infinite" eddy solution.

The next wedged trailing-edge problem to be considered is the quartic-arc airfoil with k=0.1 (shown in Fig. 25a), and given by

$$f_A(x) = 2\left\{\frac{12(x_1 x_2^2 + kx_1)x - 6(x_2^2 + 2x_2 x_1 + k)x^2 + 4(2x_2 + x_1)x^3 - 3x^4}{x_1^4 - 4x_1^3 x_2 + 6(k+x_2^2)x_1^2}\right\} , \qquad (24)$$

where

$$x_1 = \frac{1}{2} - \frac{1}{6}\left[\frac{3}{1+12k}\right]^{1/2} \qquad (25)$$

and

$$x_2 = \frac{2 - 3x_1}{3(1-2x_1)} . \qquad (26)$$

A more detailed discussion concerning the cross-over solutions for this airfoil is given by Rothmayer & Smith [18]. Solutions were initially sought for a predominantly attached flow, for the k=0.1 case, using the Hilbert integral method. The results for the locations of the separation and reattachment points are shown in Fig. (26). The steady state finite eddy solutions were found to terminate at a certain critical value of Ω. The last steady state solution which was found was Ω=.034, no steady state solution could be found for Ω=.035. It should be emphasized that no stabilizing approximation was needed in these calculations. In addition, neither the stabilizing approximations nor upwind differencing would stabilize the Ω=.035 case. The extreme sensitivity of the location of the separation and reattachment points to changes in Ω, near the point where the steady state solutions cease to exist, can be seen in Fig. (26). This behaviour is characteristic of a local region of nonunique solutions. The large-scale separation solutions for this airfoil are shown in Fig. (27). Again, the steady state solutions for infinite eddy massive separation terminate well before the point of wake deflation is reached. In fact, Ω=.03 was the last steady state solution which could be calculated, no steady state solutions could be found for Ω=.028. The global results calculated for this airfoil using triple-deck theory are shown in Fig. (25b). The corresponding results using interacting boundary layer theory are shown in Fig. (28). It should be noted that there is a small region of nonunique solutions for this airfoil. The nonuniqueness is believed to be the fundamental cause of stall hysteresis at high Reynolds numbers. The cross-over process from massive separation to an attached flow, and vice-versa, is an unsteady process for this airfoil as a result of the un-uniqueness. Whereas for the biconvex airfoil, of Fig. (25), the cross-over is a steady state process (albeit an extreamely rapid one). For this airfoil, the region of nonuniqueness should decrease with increasing Reynolds number. On the other hand, a cusped trailing-edge airfoil is expected to possess an enlarging region of nonuniqueness for increasing Reynolds number.

4. A Model for Blunt Leading-Edge Airfoils

The airfoils examined so far in this study possessed either cusped or wedged leading and trailing-edges. This allowed the thin airfoil approach to be used with the baseline coordinate curve chosen to be the x-axis. However, true bluff bodies and blunt leading-edge airfoils require a linearization of the thin airfoil equation about more complicated curves in space. The model presented in this section treats blunt leading-edge airfoils by choosing the baseline coordinate curve to be the osculating parabola to the nose of the airfoil. This formulation is suggested by the form of the leading-edge correction in thin airfoil theory (see Van-Dyke [32]). A more general class of problems which are solved along the same lines are discussed by Rothmayer [15], within the context of a Hilbert integral formulation for an airfoil problem. In fact, the following derivation is based on the inviscid work of Rothmayer [15]. The osculating parabola to the nose of the airfoil is given by

$$y = (2r_0 x)^{1/2} \tag{27}$$

where r_0 is the nose radius of curvature of the airfoil. This parabola was mapped from the z=x+iy plane to the $\bar{\xi}$-axis of the $\zeta=\bar{\xi}+i\bar{\eta}$ stagnation flow plane using the transformation

$$z = \zeta(\zeta + i(2r_0)^{1/2}) \tag{28}$$

or

$$\zeta = (z - r_0/2)^{1/2} - i(r_0/2)^{1/2} \tag{29}$$

The branch cut is chosen to be within the parabola, and so

$$(z - r_0/2)^{1/2} = |z - r_0/2|^{1/2} e^{i\theta/2} \tag{30}$$

where $\theta=0$ corresponds to the x-axis and θ is constrained to lie in the range $0<\theta<2\pi$. The flow in the transformed plane is then a stagnation flow with the complex velocity $w(\zeta)$ given by

$$w(\zeta) \sim w_0(\zeta) + Re^{-1/2} w_1(\zeta) + \dots \tag{31}$$

where

$$w_0(\zeta) = 2\zeta \qquad . \tag{32}$$

It is then found that the Gortler variable is given by

$$\xi = \int_0^s Ue_0(t) \, dt = \bar{\xi}^2 \tag{33}$$

and the arclength-parameter along the baseline coordinate curve is

$$s(\xi) = 2\left[\xi\left(\xi + \frac{r_0}{2}\right)\right]^{1/2} + r_0 \ln\left\{\left[\frac{2\xi}{r_0}\right]^{1/2} + \left[\frac{2\xi}{r_0} + 1\right]^{1/2}\right\} \qquad . \tag{34}$$

The base-flow over the parabola in the z-plane is then given by

$$Ue_0 = \left[\frac{2\xi}{2\xi + r_0}\right]^{1/2} \tag{35}$$

and

$$\beta_0 = \frac{r_0}{2\xi + r_0} \qquad . \tag{36}$$

Since the parabola is "folded out" in the ζ-plane there will be two free-streamlines, one tending to $\bar{\xi} \to -\infty$ and the other to $\bar{\xi} \to \infty$. The method of Cheng & Rott [2] must be modified accordingly and so we choose

$$H_0(\zeta) = i(\zeta^2 - \bar{\xi}_0^2)^{1/2} \tag{37}$$

where $\pm\bar{\xi}_0$ are the positions at which the changes in the boundary conditions are imposed. The imposition of symmetry and anti-symmetry conditions for the u and v velocity components in the z-plane leads to the analogue of Eqns. (14) and (15) in the ζ-plane. These equations can then be transformed back to the z-plane to give

$$Ue = Ue_0\left\{1 + \frac{\sqrt{\xi_0 - \xi}}{\pi} \int_0^{\xi_0} \frac{T'(t)}{(\xi - t)\sqrt{\xi_0 - t}} \, dt\right\} + Ue_0 I \qquad \text{for } 0 < \xi < \xi_0 \tag{38}$$

and

$$T'(\xi) = \frac{\sqrt{\xi_0 - \xi}}{\pi} \int_0^{\xi_0} \frac{T'(t)}{(\xi - t)\sqrt{\xi_0 - t}} \, dt \qquad \text{for } \xi > \xi_0 \tag{39}$$

In these equations the function $I(\xi)$ is given by

$$I(\xi) = \frac{1}{\pi}\left[Ue_s\left(1 + \frac{r_0}{2\xi}\right)^{1/2} - 1\right]\left\{\frac{\pi}{2} + \sin^{-1}\left(\frac{2\xi}{\xi_0} - 1\right)\right\} \tag{40}$$

where Ue_s is an imposed constant velocity (i.e. eddy pressure) for $\xi > \xi_0$. In the above equations $I(\xi)$ is generated from the fact that the base flow along the parabola has not reached the free-stream value when $\xi = \xi_0$. That is, the pressure along the parabola for $\xi > \xi_0$ is not zero. Therefore, a non-zero u-velocity boundary condition is generating a contribution to Ue for $\xi > \xi_0$. For consistency, these equations require that $r_0 << \xi_0$, thereby restricting the applicability of Eqn. (38) to thin blunt leading-edge airfoils. It is interesting to note that if the pressure on the parabola for $\xi > \xi_0$ had been the free-stream pressure then Eqn. (38) would not be restricted to thin airfoils and would be given by

$$Ue = Ue_0 \left\{ 1 + \frac{\sqrt{\xi_0-\xi}}{\pi} \int_0^{\xi_0} \frac{T'(t)}{(\xi-t)\sqrt{\xi_0-t}} \, dt \right\} \qquad . \qquad (41)$$

This equation is simply the single body thin airfoil equation written in Gortler variables, referenced to the baseline coordinate curve, and multiplied by Ue_0. A similar formulation is suggested by Rothmayer [15] for the symmetric Hilbert integral formulation for bluff bodies. Since the flow in the z-plane should not depend on which ζ-plane the problem is transformed to, these equations tend to suggest a very simple yet very general formulation for arbitrary two-dimensional incompressible interacting boundary layer problems. In particular since the boundary condition on the streamwise Gortler velocity $F(\xi)$ is

$$F(\xi,\eta) \rightarrow \frac{Ue}{Ue_0} \qquad \text{as } \eta \rightarrow \infty \qquad , \qquad (42)$$

the quantity that is dealt with in solving the governing equations is Ue/Ue_0 not Ue. Therefore, the numerical solution procedure for any two-dimensional interacting boundary layer problem should be virtually identical to the single body method. It should be pointed out that this is only speculation. However, work is currently underway to provide this general formulation in symmetric and asymmetric external and internal flows. It should be noted that the asymmetric problem will not have an airfoil equation of the form of Eqn. (41). However, no serious difficulties are anticipated in going to the asymmetric problem. It is the authors opinion that the solution of most high Reynolds number steady state incompressible laminar flows are now feasible through the use of the Hilbert and the Cheng & Rott [2] airfoil integrals coupled with the conformal mapping technique. The only serious difficulty which is anticipated in the two-dimensional incompressible problem is the incorporation of the re-entrant jet into the body scale flow for large-scale separation. Most likely this would be accomplished through a downstream asymptotic calculation (i.e. an inviscid calculation coupled with a non-interacting boundary layer calculation). Actually, the re-entrant jet does not present serious conceptual difficulties, it only complicates the numerical task.

5. Results

The airfoils examined so far in this study have possessed either cusped or wedged leading and trailing-edges. This allowed the thin airfoil approach to be used with the baseline coordinate curve chosen to be the x-axis. The model presented in the last section treated blunt leading-edge airfoils by choosing the baseline coordinate curve to be the osculating parabola to the nose of the airfoil. This model will be briefly examined using the ellipse (see Fig. 29)

$$f_A(x) = 4[x(1-x)]^{1/2} \quad . \tag{43}$$

For this airfoil, the nose radius of curvature is

$$r_0 = 8\Omega^2 \tag{44}$$

In the following problems a prescribed pressure c_{p_s} was chosen in the eddy. This can easily be related to the velocity Ue_s in Eqn. (40). Figure (30) shows the results for the ellipse $\Omega = .15$, given in terms of the arclength $s(\xi)$ along the baseline coordinate curve. These results look very similar to the results for the other airfoils examined in this chapter. However, it is seen that the eddy pressure can have a sizeable impact on the location of the separation point and the results upstream of separation. At this point, the interacting boundary layer technique needs to be honed to a finer degree of accuracy by including: more accurate, possibly adaptive, baseline coordinate curves, higher order boundary layer composite models, an account of any re-entrant jet which may be present, pre and post-separation turbulence modeling and perhaps a finite eddy Cheng-Rett [2] approach to more accurately model the turbulent eddy. The modeling of the downstream eddy and wake closure process should also provide a rational basis for determining the eddy pressure.

6. Conclusion

In this study, we have presented a new interacting boundary layer model which is capable of calculating large-scale laminar separation. Both unique and nonunique solutions were computed for airfoils at high Reynolds numbers. The nonunique steady-state solution was seen to provide a purely aerodynamic basis for the important phenomenon of stall hysteresis. A preliminary generalization to this basic model was also presented for calculating large-scale separation over thin blunt leading-edge airfoils. This new model is believed to be capable of an extension to a very general formulation of two-dimensional incompressible interacting boundary layer theory. The general formulation should be capable of calculating a wide variety of problems including: bluff bodies, lifting airfoils and cascade flows. At the same time, this general formulation would preserve the computational efficiency of the thin airfoil approach to calculating strongly interacting boundary layers.

Acknowledgment

The authors wish to thank Dr. M.J. Werle and Dr. J.E. Carter, of the United Technologies Research Center in East Hartford Connecticut, for their support, encouragement and suggestions throughout the course of this research. a special thanks is due to Dr. J.M. Verdon, also of the UTRC Research Center, whose insightful comments motivated some of our calculations and led to a deeper understanding of the method presented in this study. This research was supported by the General Electric Corp. Fellowship 2-37000-9306-17 and by the Office of Naval Research Grant No. N00014-76C-0364, monitored by Dr. R. E. Whitehead. In addition, the blunt leading-edge work was supported by the United Technologies Resarch Center under the guidance of Dr. J. E. Carter and Dr. M. J. Werle. I would like to dedicate this work to the memory of Professor R. Thomas Davis, who recently passed away in Cincinnati, Ohio. The untimely death of Professor Davis has deprived the aerospace engineering community of a truly great researcher, teacher and humanist.

Alric P. Rothmayer

REFERENCES

1. Achenbach, E., "Distribution of Local Pressure and Skin Friction Around a Circular Cylinder in Cross-Flow up to Re=5×10^6," J. Fluid Mech., vol. 34, 1968.

2. Cheng, H.K., and Rott, N., "Generalizations of the Inversion Formula of Thin Airfoil Theory," J. Rat. Mech. An., No. 3, 1954.

3. Cheng, H.K., and Smith, F.T., "The Influence of Airfoil Thickness and Reynolds Number on Separation," J. of Applied Maths. and Physics, vol. 33, 1982.

4. Cheng, H.K., "Laminar Separation from Airfoils Beyond Trailing-Edge Stall," AIAA 84-1612 presented at the AIAA 17th Annual Fluid Dynamics, Plasma Dynamics and Lasers Conference, 1984.

5. Cheng, H.K., and Lee, C.J., "Laminar Separation Studied as an Airfoil Problem," Numerical and Physical Aspects of Aerodynamic Flows, ed. T. Cebeci, Springer-Verlag, 1985.

6. Daniels, P.G., "Laminar Boundary-Layer Reattachment in Supersonic Flow. Part 2. Numerical Solution," J. Fluid Mech., 97, part 1, 1980.

7. Davis, R.T., "Numerical Methods for Coordinate Generation Based on Schwarz-Christoffel Transformations," AIAA79-1463 presented at the 4th Computational Fluid Dynamics Conference, 1979.

8. Davis, R.T., and Werle, M.J., "Progress on Interacting Boundary Layer Computations at High Reynolds Number," Numerical and Physical Aspects of Aerodynamic Flows, Springer-Verlag, 1982.

9. Fornberg, B., "Steady Viscous Flow Past a Circular Cylinder up to Reynolds Number 600," submitted to J. Comp. Phys., 1985.

10. Goldstein, S., "On Laminar Boundary Layer Flow Near a Position of Separation," Quart. J. Mech. Appl. Math., vol. 1, 1948.

11. Kirchhoff, G., "Zur Theorie freier Flussigkeitsstrahlen," J. Reine Angew. Math., vol. 70, 1869.

12. Lighthill, M.J., "On Boundary Layers and Upstream Influence II. Supersonic Flows Without Separation," Proc. Roy. Soc., Series A 217, 1953.

13. Messiter, A.F., and Enlow, R.L., "A Model for Laminar Boundary-Layer Flow Near a Separation Point," SIAM J. Appl. Math., vol. 25 no. 4, 1973.

14. Neiland, V. Ia., "Supersonic Flow of a Viscous Fluid Around a Separation Point," presented at the 3rd pan-Soviet meeting on Theoretical and Applied Mechanics, 1968.

15. Rothmayer, A.P., "The Development of a Comprehensive Two-Dimensional Linearized Airfoil Theory," AIAA 9th Annual Mini-Symposium on Air Science and Technology, Wright Patterson AFB, Dayton OH, 1983.

16. Rothmayer, A.P., and Davis, R.T., "An Interacting Boundary Layer Model for Cascades," AIAA83-1915 presented at the AIAA 6th Computational Fluid Dynamics Conference, 1983.

17. Rothmayer, A.P., and Davis, R.T., "Massive Separation and Dynamic Stall on a Cusped Trailing-Edge Airfoil," to be published in Numerical and Physical Aspects of Aerodynamic Flows, Springer-Verlag, 1985.

18. Rothmayer, A.P., and Smith, F.T., "Large Scale Separation and Hysteresis in Cascades," to be published in Proc. Roy. Soc., Series A, 1985.

19. Rothmayer, A.P., Ph.D. Dissertation, Univ. of Cincinnati, 1985.

20. Sadovskii, V.S., "Vortex Regions in a Potential Stream with a Jump of Bernoulli's Constant at the Boundary," Prikl. Math. Mech, vol. 35, transl. Appl. Math. Mech., vol. 35, 1971.

21. Smith, F.T., "The Laminar Separation of an Incompressible Fluid Streaming Past a Smooth Surface," Proc. Roy. Soc., Series A 356, 1977.

22. Smith, F.T., "Laminar Flow of an Incompressible Fluid Past a Bluff Body: The Separation, Reattachment, Eddy Properties and Drag," J. Fluid Mech., vol. 92, 1979.

23. Smith, F.T., "On the High Reynolds Number Theory of Laminar Flows," IMA J. Appl. Math., vol. 28, 1982.

24. Smith, F.T., "Large-Scale Separation and Wake Closure/Reattachment - The Cascade Problem," to appear in J. Maths. Phys. Sci., 1985.

25. Smith, F.T., "A Structure for Laminar Flow Past a Bluff Body at High Reynolds Number," J. Fluid Mech., 155, 1985.

26. Smith, F.T., and Merkin, J.H., "Triple-Deck Solutions for Subsonic Flow Past Humps, Steps, Concave or Convex Corners and Wedged Trailing Edges," _Computers and Fluids_, vol. 10 no. 1, 1982.

27. Stewartson, K., and Williams, P.G., "Self-Induced Separation," _Proc. Roy. Soc._, Series A 312, 1969.

28. Stewartson, K., "Is the Singularity at Separation Removable?," _J. Fluid Mech._, vol. 44, 1970.

29. Stewartson, K., "Multistructured Boundary Layers on Flat Plates and Related Bodies," _Adv. Appl. Mech._, vol. 14, Academic Press, 1974.

30. Sychev, V.V., "On Laminar Separation," _Meck. Zhid. i Gaza_, vol. 3, 1972.

31. Van-Dyke, M.D., _Perturbation Methods in Fluid Mechanics_, The Parabolic Press, 1975.

32. Veldman, A.E.P., "New Quasi-Simultaneous Method to Calculate Interacting Boundary Layers," _AIAA Journal_, vol. 19, 1981.

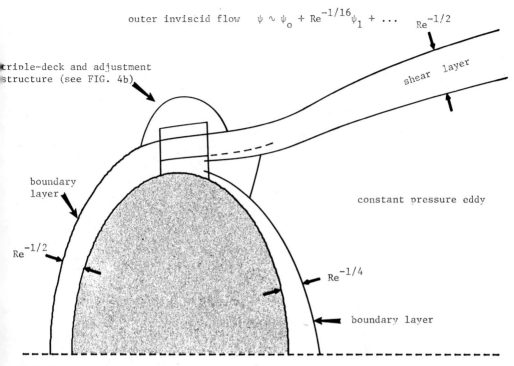

outer inviscid flow $\psi \sim \psi_o + Re^{-1/16}\psi_1 + \ldots$ $Re^{-1/2}$

triple-deck and adjustment structure (see FIG. 4b)

shear layer

boundary layer

$Re^{-1/2}$

constant pressure eddy

$Re^{-1/4}$

boundary layer

FIG. 1. THE BLUFF BODY ASYMPTOTIC STRUCTURE BASED ON THE INVISCID KIRCHHOFF WAKE MODEL.

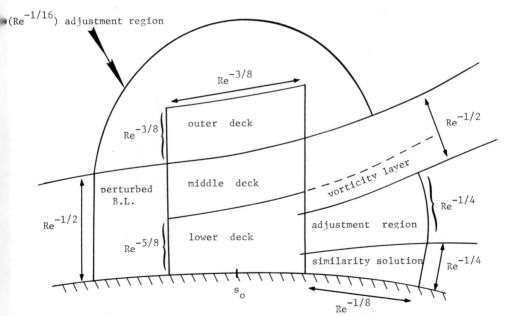

$(Re^{-1/16})$ adjustment region

$Re^{-3/8}$

$Re^{-3/8}$ outer deck

$Re^{-1/2}$

perturbed B.L.

middle deck

vorticity layer

$Re^{-1/2}$

$Re^{-5/8}$ lower deck

$Re^{-1/4}$

adjustment region

similarity solution $Re^{-1/4}$

s_o

$Re^{-1/8}$

FIG. 2. THE TRIPLE-DECK STRUCTURE AT THE SEPARATION POINT.

132

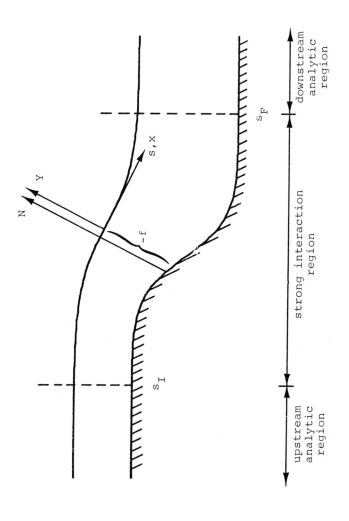

FIG. 3. THE BASELINE COORDINATE SYSTEM GEOMETRY FOR INTERACTING BOUNDARY LAYER THEORY.

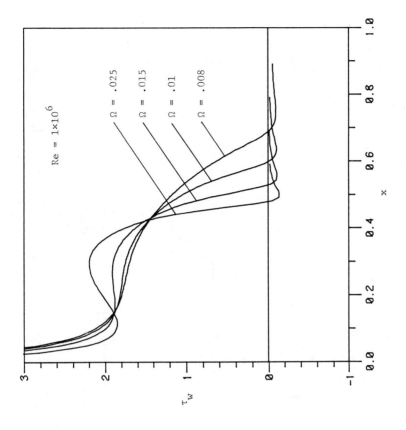

FIG. 4. WALL SHEAR STRESS FOR MASSIVE SEPARATION ON THE SINE AIRFOIL,
FOR VARIOUS AIRFOIL THICKNESSES.

134

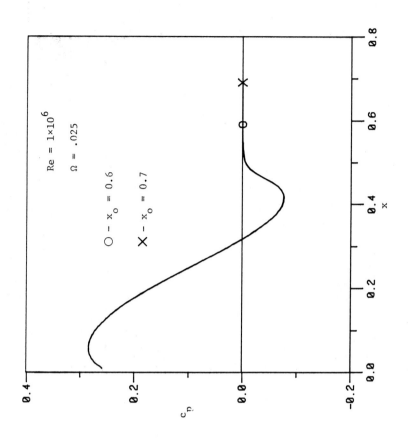

FIG. 5. A SUPERPOSITION OF THE PRESSURE COEFFICIENT OVER THE SINE AIRFOIL
FOR TWO DIFFERENT VALUES OF x_o.

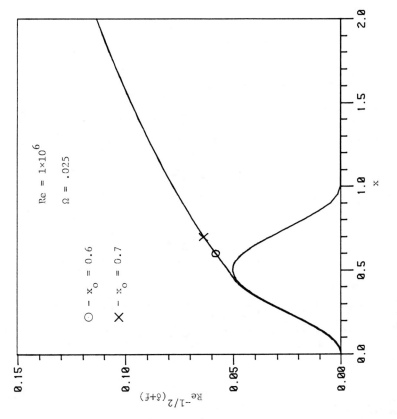

FIG. 6. A SUPERPOSITION OF THE DISPLACEMENT THICKNESS OVER THE SINE AIRFOIL FOR TWO DIFFERENT VALUES OF x_o.

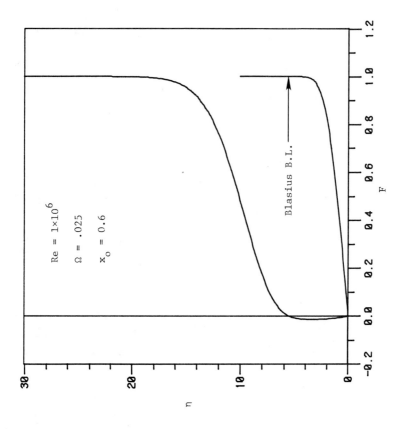

FIG. 7. A TYPICAL VELOCITY PROFILE AT THE LAST GRID LINE OF THE
CALCULATION DOMAIN.

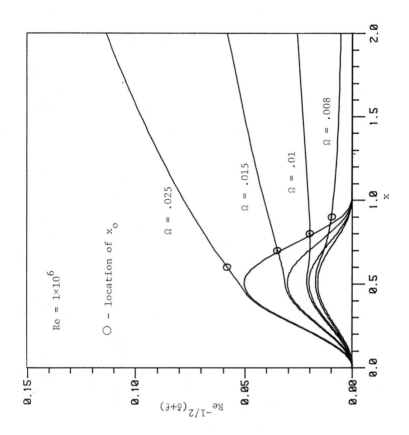

FIG. 8. THE DISPLACEMENT THICKNESS SUPERIMPOSED UPON THE AIRFOIL GEOMETRY
FOR THE SINE AIRFOIL; A PATCH BETWEEN THE NUMERICAL RESULTS AND
THE ANALYTIC MODEL.

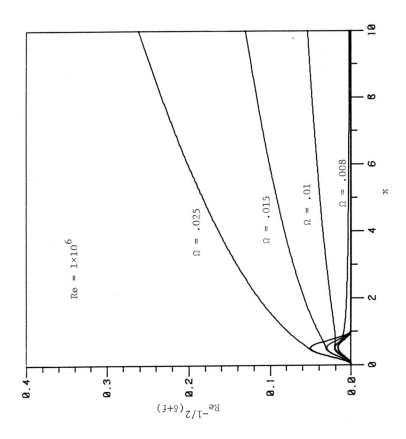

FIG. 9.   THE DISPLACEMENT THICKNESS SUPERIMPOSED UPON THE AIRFOIL GEOMETRY,
SHOWING THE APPROACH TO WAKE DEFLATION (THE $\Omega=.008$ CASE IS WELL
BEYOND THE POINT OF THE WAKE DEFLATION.).

FIG. 10a. THE BICONVEX AIRFOIL $f(x) = 8x(1-x)$.

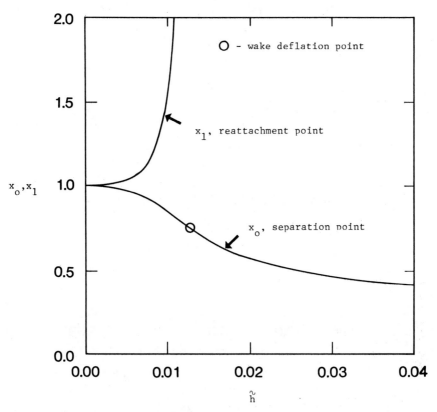

FIG. 10b. THE TRIPLE-DECK RESULTS FOR THE SINGLE BICONVEX AIRFOIL.

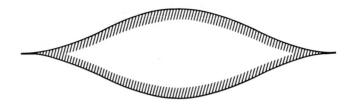

FIG. 11a.   THE SINE AIRFOIL $f(x) = 1 + \sin[2\pi(x-1/4)]$.

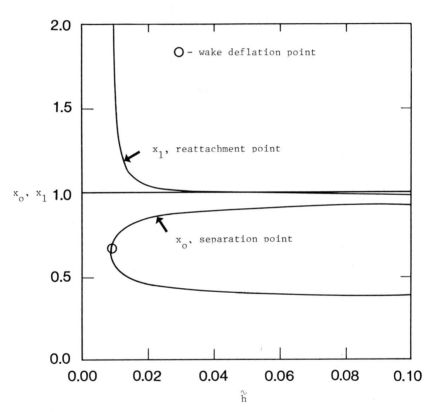

FIG. 11b.   THE TRIPLE-DECK RESULTS FOR THE SINGLE SINE AIRFOIL.

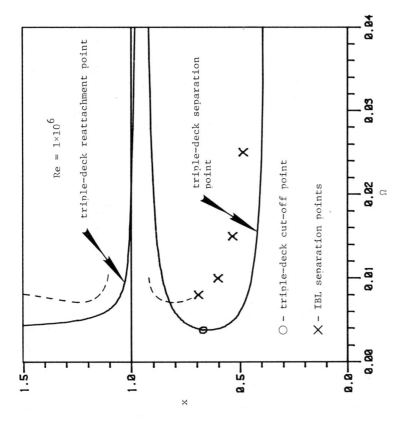

FIG. 12. A GLOBAL COMPARISON WITH TRIPLE-DECK THEORY; THE TRAILING-EDGE IS AT x=1.0 and - - - - INDICATES THE POSSIBLE INTERACTING BOUNDARY LAYER RESULTS PAST WAKE DEFLATION.

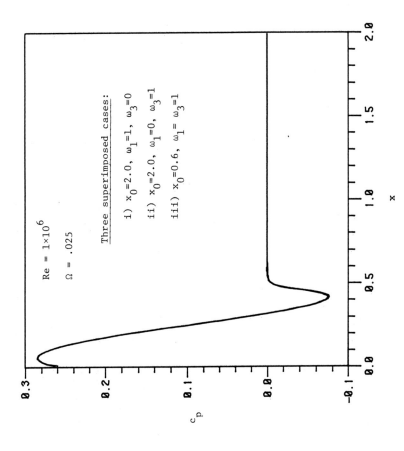

FIG. 13. PRESSURE COEFFICIENT FOR MASSIVE SEPARATION ON THE SINE AIRFOIL WITH A SEMI-INFINITE SPLITTER PLATE FOR TWO DIFFERENT VALUES OF x_o.

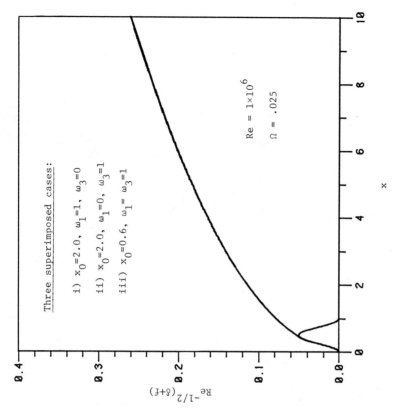

FIG. 14. A SUPERPOSITION OF THE DISPLACEMENT THICKNESS AND AIRFOIL
GEOMETRY FOR TWO DIFFERENT VALUES OF x_o.

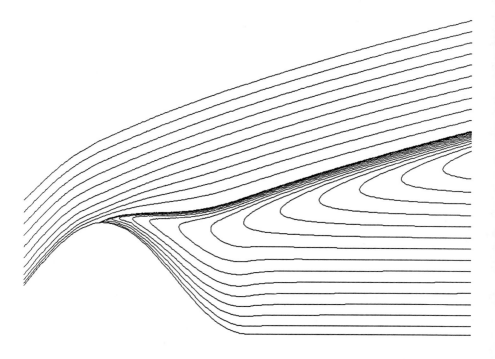

FIG. 15. THE STREAMLINE PATTERN IN THE EDDY FOR A SINE AIRFOIL WITH
$Re = 1 \times 10^6$, $\Omega = 0.025$, $\omega_1 = 1$ and $\omega_3 = 0$.

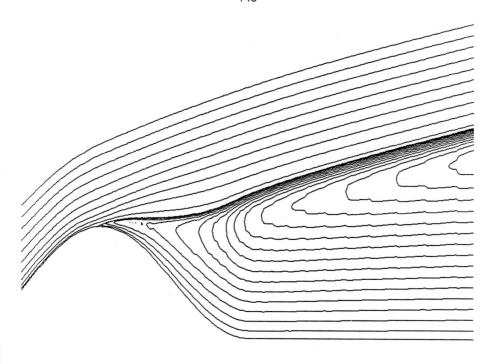

FIG. 16.   THE STREAMLINE PATTERN IN THE EDDY FOR A SINE AIRFOIL WITH
$Re = 1 \times 10^6$, $\Omega = 0.025$, $\omega_1 = 0$ AND $\omega_3 = 1$.

146

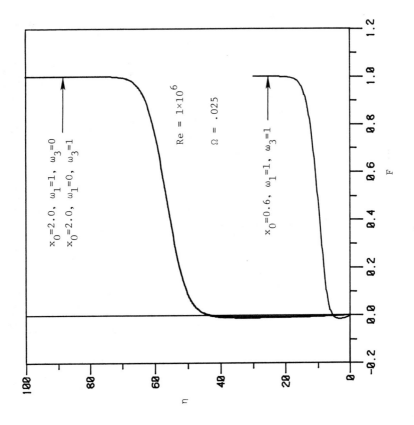

FIG. 17.   TYPICAL VELOCITY PROFILES AT THE LAST GRID LINE OF THE
CALCULATION DOMAIN FOR $x_o = 2.0$ AND $x_o = 0.6$.

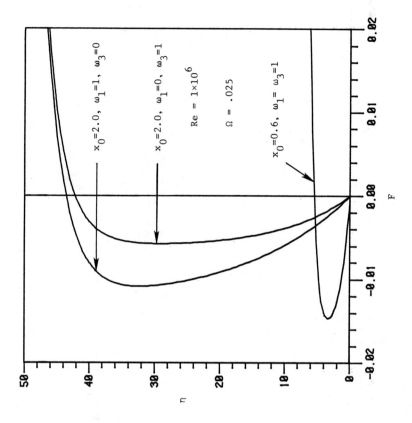

FIG. 18.   TYPICAL REVERSED FLOW VELOCITY PROFILES AT THE LAST GRID LINE
OF THE CALCULATION DOMAIN FOR $x_o = 2.0$ AND $x_o = 0.6$.

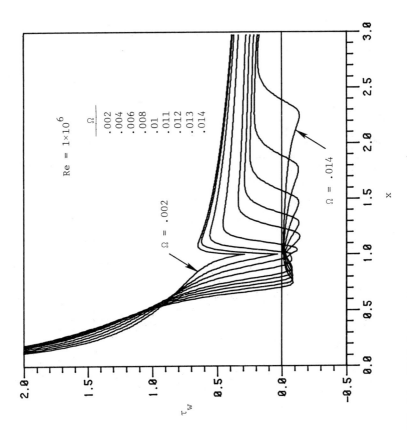

FIG. 19. WALL SHEAR STRESS FOR THE FINITE EDDY PROBLEM ON A BICONVEX AIRFOIL; THE APPROACH TO WAKE INFLATION.

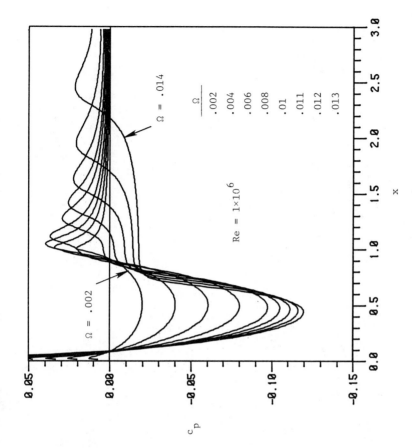

FIG. 20. PRESSURE COEFFICIENT FOR THE FINITE EDDY PROBLEM ON A BICONVEX AIRFOIL; THE APPROACH TO WAKE INFLATION.

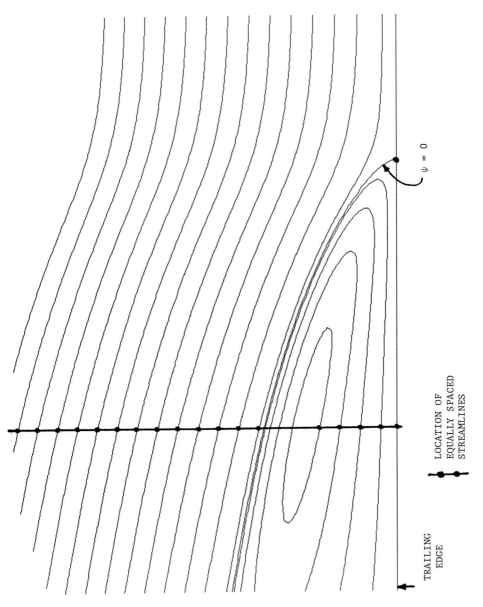

$\psi = 0$

LOCATION OF
EQUALLY SPACED
STREAMLINES

TRAILING
EDGE

FIG. 21. WAKE STREAMLINES FOR THE FINITE EDDY PROBLEM ON A BICONVEX AIRFOIL.

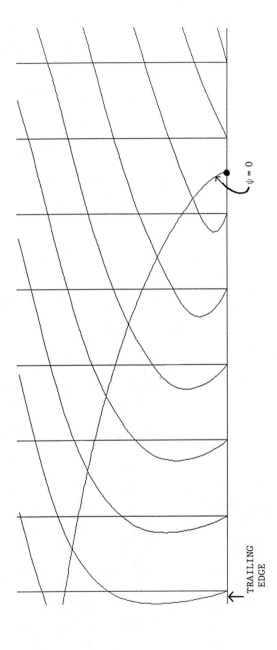

FIG. 22. WAKE VELOCITY PROFILES FOR THE FINITE EDDY PROBLEM
ON A BICONVEX AIRFOIL.

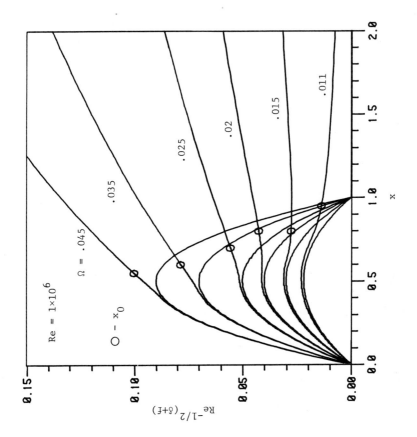

FIG. 23. A SUPERPOSITION OF THE DISPLACEMENT THICKNESS AND AIRFOIL GEOMETRY FOR MASSIVE SEPARATION OVER A BICONVEX AIRFOIL.

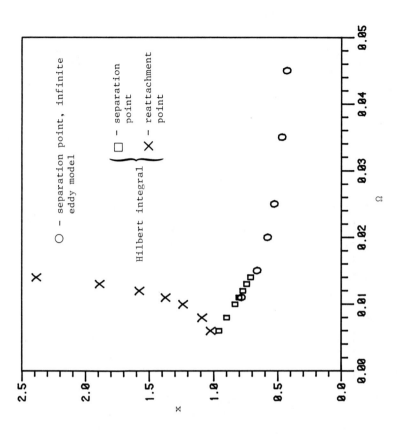

FIG. 24. THE MASSIVE SEPARATION POINTS AND THE FINITE EDDY SEPARATION AND
REATTACHMENT POINTS AS A FUNCTION OF AIRFOIL THICKNESS;
A STEADY-STATE WAKE CROSS-OVER PROCESS.

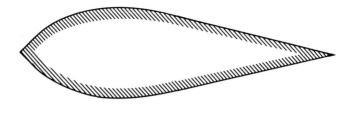

FIG. 25a.   THE AIRFOIL GEOMETRY OF EQN. (24) WITH k = 0.1.

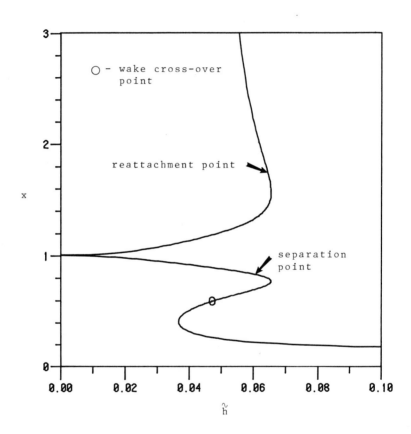

FIG. 25b.   THE LOCATION OF THE SEPARATION AND REATTACHMENT POINTS FOR THE
AIRFOIL OF EQN. (24) AS PREDICTED BY TRIPLE-DECK THEORY
(SEE FIG. 25a FOR THE AIRFOIL GEOMETRY).

155

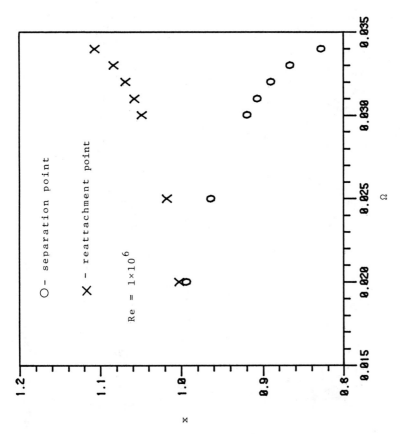

FIG. 26. THE LOCATION OF THE SEPARATION AND REATTACHMENT POINTS AS A FUNCTION OF AIRFOIL THICKNESS FOR THE FINITE EDDY PROBLEM ON THE AIRFOIL OF EQN. (24) WITH k = 0.1; THE APPROACH TO UNSTEADY WAKE CROSS-OVER.

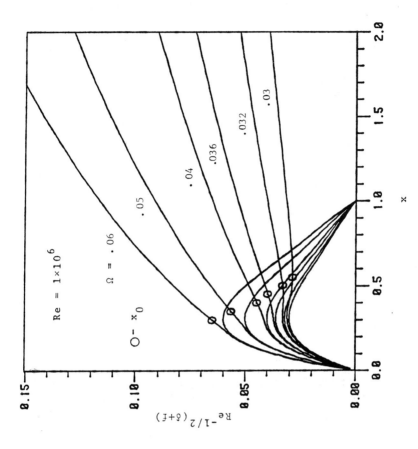

FIG. 27. THE DISPLACEMENT THICKNESS SUPERIMPOSED UPON THE AIRFOIL GEOMETRY FOR THE AIRFOIL OF EQN. (24) WITH k = 0.1.

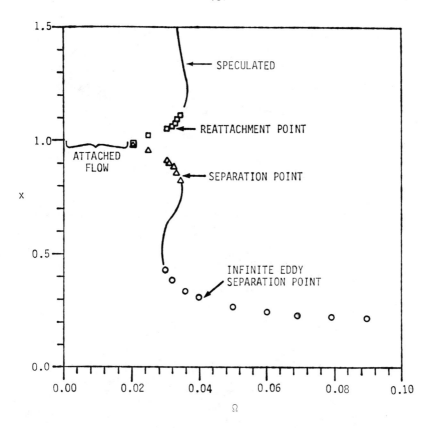

Fig. 28. The location of the massive separation points and the finite eddy separation and reattachment points for the airfoil of equation (24) with k = 0.1.

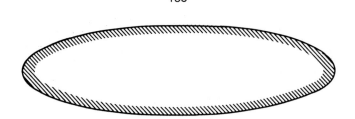

FIG. 29.   THE ELLIPSE $f(x) = 4[x(1-x)]^{1/2}$.

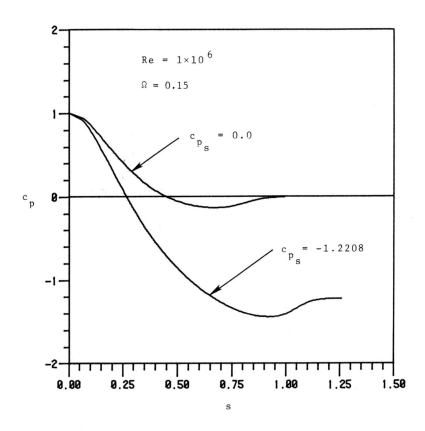

FIG. 30.   PRESSURE COEFFICIENT FOR MASSIVE SEPARATION OVER THE ELLIPSE
$f_A(x) = 4[x(1-x)]^{1/2}$ (SHOWN IN FIG. 29) FOR TWO DIFFERENT
PRESCRIBED EDDY PRESSURES.

Viscous-Inviscid Interaction Solvers and Computation of Highly Separated Flows

J. C. LE BALLEUR

INTRODUCTION

Progress in computation of high Reynolds numbers flows is currently based, either on "Direct" Navier-Stokes (NS) solvers, or on "Indirect" Viscous-Inviscid Solvers (VIS), [1 to 4].

The Indirect VIS methods, which differentiate simply from the Direct NS methods by splitting the numerical treatment into viscous, plus inviscid-like, plus coupling solvers, were more or less designed, in the past, to be used within the approximations of the boundary layer theory. The recent advances have removed this limitation and have shown that the VIS methods are acquiring capacities that are constantly approaching those of the Direct NS methods, including the resolution of separations, shock wave-boundary layer and trailing-edge interactions, see for example Melnik [5 to 7], Carter [8 to 11], Werle [12], Lock, Firmin [13 to 15], Yoshihara [16], Veldman [17,18], Whitfield [19], Steger, Van Dalsem [20], Le Balleur et al. [21 to 29].

Recently in addition, the possibility to recover a full Navier-Stokes-like capability with the VIS-methods has been shown also for time-consistent solutions, see Le Balleur, Girodroux-Lavigne [30,31]. In this case, an unsteady Defect integral method, "strongly" interacted with an inviscid solver, was found capable to describe self-oscillatory flows at buffeting conditions, with transonic shock-induced unsteady separation.

Although the use of Interacting Defect Integral (IDI) equations, which are somewhat less restrictive than the interacting Prandtl equations, is only to be considered as the first step of approximation when developing Indirect VIS-Solvers based on the full "Defect Formulation" [3,23,4], the question arises now to understand how far a Navier-Stokes-like capability will be recovered in this way for the computation of massive separations.

Obviously, a full answer to this question is still difficult within the state of the art. However, the following preliminary theory and computations suggest that the VIS-Solvers with IDI equations do provide such a capability, not only when the separation is massive at the viscous scale, with for example slender reverse flows bubbles and high shape parameters, but also when the separation is massive at the inviscid scale, with for example the thick reverse flow bubbles generated by airfoils beyond stall or with spoiler-induced separations.

1 - FULL DEFECT FORMULATION AND APPROXIMATE DEFECT INTEGRAL EQUATIONS.

1.1 - Defect Formulation.

This viscous-inviscid splitting approach, that we suggested previously [3,23,4], assumes that both computational domains fully overlap, and that both solutions are fully matched in the far field. The pseudo-inviscid problem is selected to satisfy the Euler equations (or some approximation of them) even inside the viscous layers, in such a way that the viscosity controls its boundary conditions. The viscous problem estimates the "Defect" between the real and pseudo-inviscid flows.

In two-dimensions for example, we consider the usual curvilinear orthogonal coordinates (x,y) linked to a moving body-contour $y{=}0$, whose curvature is $K(x)$. Denoting $\overline{u},\overline{v}$ the velocity components, $\overline{\rho}$ the density, $\overline{p}$ the pressure, $\overline{\gamma}_x$, $\overline{\gamma}_y$ the entrainment and Coriolis accelerations, $u,v,\rho,p,\gamma_x,\gamma_y$ the corresponding quantities of the pseudo-inviscid problem, $\overline{V}$ the viscous stresses terms, $h{=}1{+}Ky$, the Navier-Stokes equations

$$\frac{\partial \overline{U}h}{\partial t} + \frac{\partial \overline{F}}{\partial x} + \frac{\partial \overline{G}h}{\partial y} - K\overline{H} - \overline{I} = \overline{V} \quad (1)$$

$$\overline{U} = \begin{bmatrix} \overline{\rho} \\ \overline{\rho}\,\overline{u} \\ \overline{\rho}\,\overline{v} \end{bmatrix}, \overline{F} = \begin{bmatrix} \overline{\rho}\,\overline{u} \\ \overline{\rho}\,\overline{u}^2 + \overline{p} \\ \overline{\rho}\,\overline{uv} \end{bmatrix}, \overline{G} = \begin{bmatrix} \overline{\rho}\,\overline{v} \\ \overline{\rho}\,\overline{uv} \\ \overline{\rho}\,\overline{v}^2 + \overline{p} \end{bmatrix}, \overline{H} = \begin{bmatrix} 0 \\ -\overline{\rho}\,\overline{uv} \\ \overline{\rho}\,\overline{u}^2 + \overline{p} \end{bmatrix}, \overline{I} = \begin{bmatrix} 0 \\ \overline{\rho}\overline{\gamma}_x \\ \overline{\rho}\overline{\gamma}_y \end{bmatrix}$$

and the boundary conditions are split into an equivalent system of Defect Formulation, involving an Euler system, plus a viscous Defect system, plus strong coupling relations :

$$\frac{\partial Uh}{\partial t} + \frac{\partial F}{\partial x} + \frac{\partial Gh}{\partial y} - KH - I = 0 \quad (2)$$

$$\frac{\partial}{\partial t}(U-\overline{U})h + \frac{\partial}{\partial x}(F-\overline{F}) + \frac{\partial}{\partial y}(G-\overline{G})h - K(H-\overline{H}) - (I-\overline{I}) = -\overline{V} \quad (3)$$

$$0 = \lim_{y\to\infty} (U-\overline{U}) \quad (4)$$

The splitting operates on the "Defects" between the viscous fluxes $\overline{U},\overline{F},\overline{G},\overline{H},\overline{I}$ and the inviscid fluxes U,F,G,H,I. The strong matching (4) couples $v(x,y)$ and $\overline{v}(x,y)$, and is free of any influence of the outer-edge $y{=}\delta(x,t)$ of the viscous layers. After a y-integration of the Defect Continuity equation, and denoting $<f>$ the discontinuity of the inviscid quantity f along the wake-cut, the "exact" boundary conditions of the pseudo-inviscid problem provide the generalized coupling conditions (displacement and curvature effects) :

$$wall \;:\; [\rho v]_{(x,0,t)} = \frac{\partial}{\partial t} \int_0^\infty (\rho - \overline{\rho}) h \, dy + \frac{\partial}{\partial x} \int_0^\infty (\rho u - \overline{\rho u}) \, dy \qquad (5)$$

$$wakes \;:\; <\rho v>_{(x,0,t)} = \frac{\partial}{\partial t} \int_0^\infty (\rho - \overline{\rho}) h \, dy + \frac{\partial}{\partial x} \int_0^\infty (\rho u - \overline{\rho u}) \, dy$$

$$<p>_{(x,0,t)} = \int_{-\infty}^{+\infty} - \frac{\partial (p - \overline{p})}{\partial y} dy \qquad (6)$$

1.2 Approximate Defect Integral Equations.

Within the approximations of Thin-Layer-NS equations, the viscous tensor $\overline{V}$ reduces to the single shear term $\overline{\tau}$. At first order with respect to δ , $h=1$ and the defect system (3) reduces to :

$$\frac{\partial}{\partial t}[\rho - \overline{\rho}] + \frac{\partial}{\partial x}[\rho u - \overline{\rho u}] + \frac{\partial}{\partial y}[\rho v - \overline{\rho v}] = 0$$

$$\frac{\partial}{\partial t}[\rho u - \overline{\rho u}] + \frac{\partial}{\partial x}[\rho u^2 - \overline{\rho u^2}] + \frac{\partial}{\partial y}[\rho uv - \overline{\rho uv}] = -\frac{\partial \overline{\tau}}{\partial y} + \gamma_x(\rho - \overline{\rho}) \qquad (7)$$

$$\frac{\partial}{\partial y}[p - \overline{p}] = 0$$

No approximation (except $h=1$) is introduced in the strong coupling (5),(6), which has to be satisfied (at each time-step) and to be discretized consistently in space (and time), in order to maintain a full NS-like capability in strong interaction (unsteady) phenomena.

The corresponding first order defect integral equations, involving continuity (8), momentum (9),(10), plus the local x-momentum (defect) equation at $y=\delta(x,t)$ or entrainment equation (11), are :

$$\left\{ \frac{\partial}{\partial t}[\rho \delta_0] + \frac{\partial}{\partial x}[\rho q \delta_1] = [\rho v - \overline{\rho v}] \right\}_{(x,0,t)} \qquad (8)$$

$$\left\{ \frac{\partial}{\partial t}[\rho q \delta_1] + \frac{\partial}{\partial x}[\rho q^2(\theta_{11}+\delta_1)] = [\rho uv - \overline{\rho uv}] + \rho q^2 [\frac{Cf}{2} + \frac{\delta_0}{q^2}\gamma_x] \right\}_{(x,0,t)} \qquad (9)$$

$$\overline{p}_{(x,y,t)} \simeq p_{(x,y,t)} \qquad (10)$$

$$\left\{ \frac{\partial}{\partial t}[\rho(\delta - \delta_0)] + \frac{\partial}{\partial x}[\rho q(\delta - \delta_1)] = \rho q [E + \frac{\overline{\rho v}}{\rho q}] \right\}_{(x,0,t)} \qquad (11)$$

$$\delta_{0\,(x,t)}[\rho]_{(x,-H,t)} = \int_{-H}^\infty [\rho - \overline{\rho}]_{(x,y,t)} h \, dy \qquad , \qquad H(x) = 0$$

$$\delta_{1\,(x,t)}[\rho q]_{(x,-H,t)} = \int_{-H}^\infty [\rho u - \overline{\rho u}]_{(x,y,t)} \, dy$$

$$[\delta_1 + \theta_{11}]_{(x,t)}[\rho q^2]_{(x,-H,t)} = \int_{-H}^{\infty} [\rho u^2 - \overline{\rho}\,\overline{u}^2]_{(x,y,t)}\,dy$$

$$E(x,t) = \left[\frac{\dfrac{\partial \overline{\tau}}{\partial y}}{\rho q \dfrac{\partial}{\partial y}(u - \overline{u})}\right]_{x,\delta,t} \qquad , \qquad q^2 = u^2 + v^2 \qquad (12)$$

Equation (8) is an exact integral, identical to (5). Equations (9),(11) assume $\overline{p}(x,y,t) = p(x,y,t)$, which is less restrictive than the boundary layer theory. They remain simple because they govern the "Defect" integral thicknesses $\delta_0, \delta_1, \theta_{11}$, whose definitions include the y-variations of the inviscid quantities ρ, u .

At second order, the y-momentum equation (10) is replaced by the approximate modelling :

$$\frac{\partial}{\partial y}[p - \overline{p}] = K^{*}_{(x)}[\rho u^2 - \overline{\rho}\,\overline{u}^2]$$

$$[p - \overline{p}]_{(x,0\pm,t)} = [-K^{*}\rho q^2(\delta_1 + \theta_{11})]_{(x,0\pm,t)} \qquad (13)$$

For strong interactions, the induced curvature $K^{*}_{(x,0\pm)}$, which is an averaged streamline curvature across the shear layer, cannot reduce to the wall or wake-cut curvature $K(x)$, as it would be assumed in boundary layer theory. A first approximation is to reduce the induced curvatures $K^{*}_{(x,0\pm)}$ to the upper and lower curvatures of the displacement body. This can be deduced from the following section.

1.3 - Massive Separation. Coordinate adjustement.

For massive separations, the y-direction is no more approximately normal to the shear layer, and the v-component of the pseudo-inviscid field is no more small. In order to maintain "Thin-layer" approximations, these limitations can be removed by using a new coordinate system (x,y) , which is better adjusted to the coordinate system defined by the pseudo-inviscid streamlines, figure 1.

The same form as previously is obtained for the NS-equations, system (1), when choosing the usual curvilinear orthogonal coordinates (x,y) linked to a contour $y=0$, different from the body and the wake-cut, where the curvature $K(x)$ is no more the body curvature, but an induced curvature $K=K^{*}$. The body and the wake deviate from the contour $y=0$, and are such that $y= -H(x)$, $\theta=(\vec{0s},\vec{0x})$, if we are denoting now (s,n) the coordinate system which is tangent and normal to the wall (or wake-cut), fig. 1.

After a y-integration of the Defect NS-equations (3), the new set of integral equations is more complex than (8),(9),(10),(11). This set involves the new term $\dfrac{\partial H}{\partial x}$. The space derivatives are versus x , instead of s . The integral thicknesses are defined with a y-integration, instead of a n-integration.

Denoting v_n the inviscid velocity "normal" to the wall, and $\overline{v}_n$ the viscous-one, the new integral equations are simplified when expressing them with v_n , $\overline{v}_n$, and with space-derivatives versus s , by using the following relations :

$$\frac{\partial}{\partial x} = \frac{h}{\cos\theta}\frac{\partial}{\partial s} \quad , \quad \frac{\partial H}{\partial x} = \frac{\sin\theta}{\cos\theta}h \quad , \quad h = 1 - K^*H$$

$$v_n = u\sin\theta + v\cos\theta \tag{14}$$

The generalized Defect integral equations for continuity, x- and y-momentum, entrainment are then, in the body coordinate system (s,n,t) :

$$\left\{\frac{\cos\theta}{h}\frac{\partial}{\partial t}[\rho\delta_0] + \frac{\partial}{\partial s}[\rho q\delta_1] = \rho v_n - \overline{\rho v_n}\right\}_{(s,0,t)} \tag{15}$$

$$\left\{\frac{\cos\theta}{h}\frac{\partial}{\partial t}[\rho q\delta_1] + \frac{\partial}{\partial s}[\rho q^2(\delta_1 + \theta_{11})] = [\rho u v_n - \overline{\rho u v_n}] + \rho q^2[\frac{Cf}{2} + \frac{\delta_0}{q^2}\gamma_x]\frac{\cos\theta}{h}\right\}_{(s,0,t)} \tag{16}$$

$$\overline{P}_{(s,n,t)} = P_{(s,n,t)} \tag{17}$$

$$\left\{\frac{\cos\theta}{h}\frac{\partial}{\partial t}[\rho q(\delta - \delta_0)] + \frac{\partial}{\partial s}[\rho q(\delta - \delta_1)] = \overline{\rho v_n} + \rho q E\frac{\cos\theta}{h}\right\}_{(s,0,t)} \tag{18}$$

Equation (15) is still an exact integral, whatever is the choice of the induced curvature $K^*(x)$ that defines the contour $\theta(x), H(x)$.

Equations (16),(18) assume the first order approximation for the pressure defect (17), and for neglecting the term $K^*(\rho uv - \overline{\rho uv})$. The latter point requires that v is small and leads to choose $K^*(x), \theta(x)$ such that :

$$[v_n \simeq q\sin\theta]_{(s,0,t)} \quad , \quad [u \simeq q]_{(s,0,t)} \tag{19}$$

Within the usual thin-layer approximations, this choice is equivalent to defining the contour $y=0$ whose curvature is $K^*(x)$ as the steady displacement body. In the steady case, as $h \simeq 1$ at first order, the integral equations (15),(16),(18) with $u \simeq q$ provide a set of differential equations in the s-direction which is very similar to the usual one (8),(9),(11), except for the friction and entrainment terms, which are multiplied by $\cos\theta$, as we have suggested previously, see [23]. This conclusion makes the coordinate ajustment very easy for an integral method. We have however to keep in mind that the generalized integral equations govern defect thicknesses which depend on the choice of the coordinate system. With the present choice, the closure modelling will have to be based on the Defect in the modulus of the velocity. This modelling is the more appropriate, and may be different from the Defect in the s-component of the velocity.

At second order, the present theory justifies the equation (13) for the pressure defect, that we have suggested first empirically, see [21,23].

2 - INTEGRAL TURBULENT CLOSURE.

2.1 - Velocity profile modelling.

The defect integral equations (15),(16),(17),(18) are closed with the velocity profile modelling of Le Balleur [23,25,26,27], suggested for attached or separated turbulent layers $(1 < Hi < \infty)$ in two- or three-dimensions. It is assumed that the closure relations deduced from this boundary-layer or mixing-layer modelling are still valid for the "Defect" integral thicknesses in the (x,y) coordinate system. The steady profile modelling is assumed for the instantaneous velocity profile, in the moving reference frame linked to the body, in the case of unsteady flows.

In two-dimensions, a two-parameter description of the mean velocity profiles $[\delta, C_2]_{(s,t)}$ is assumed :

$$\overline{q} = q\left[1 - C_2\tilde{F}(\eta) + C_1 Log\,\eta\right]$$

$$\tilde{F}(\eta) = F\left[\frac{\eta - \eta^*}{1 - \eta^*}\right] \;,\quad F(\eta) = \left[1 - \eta^{3/2}\right]^2 \;,\quad y = \delta\eta \qquad (20)$$

$$C_1 = \frac{1}{0.41}\frac{Cf}{2}\quad\sqrt{2/|Cf|}$$

$$1 - C_2 = C_1\left[Log(R_\delta\quad\sqrt{|Cf|/2} - 5.25 \times 0.41\right] \qquad (21)$$

$$\eta^* = 0 \qquad\qquad\qquad ,if \qquad 0 < \frac{\delta_{1i}}{\delta} < 0.44$$

$$\eta^* = 4.598\,(\frac{\delta_{1i}}{\delta} - 0.44)^2 \;,if \qquad 0.44 < \frac{\delta_{1i}}{\delta} < 0.69 \qquad (22)$$

$$\eta^* = 2.229\,(\frac{\delta_{1i}}{\delta} - 0.565) \;,if \qquad 0.69 < \frac{\delta_{1i}}{\delta} < 1$$

$$\frac{\delta_{1i}}{\delta} = \int\limits_0^\infty (1 - \frac{\overline{q}}{q})\,d\eta$$

The description, with $\tilde{F}=1$ if $\eta < \eta^*$, generalizes the Coles profile by using a new wake function. The same modelling, with $C_1 = 0$, describes also the half-wake velocity profiles. For adiabatic walls, the density profiles are deduced from the velocity profiles with the isoenergetic approximation.

2.2 - Turbulence modelling.

For equilibrium turbulence, a mixing length model using the scale $\tilde{\delta} = (1 - \eta^*)\delta$ of the main shear layer $(\eta^* < \eta < 1)$ provides an algebraic entrainment and dissipation closure, E_{eq} and Φ_{eq}, deduced from the velocity profile modelling $(20),(21),(22)$, see Le Balleur [23]. $\lambda_1, \lambda_2, \lambda_3$ are correction terms near unity for usual boundary layers; λ_1 is near 2 for usual wakes :

$$E_{eq} = \left[0.053\,(1 - \tilde{u}_w) - 0.182\frac{Cf}{\sqrt{|Cf|}}\right]\lambda_1\lambda_2\lambda_3$$

$$\Phi_{eq} = \left[|Cf|.\tilde{u}_w + 0.018\,(1 - \tilde{u}_w)^3\right]\lambda_1\lambda_2\lambda_3 \qquad (23)$$

$$\tilde{u}_w = \frac{2.22}{1 + 1.22\eta^*}\frac{\delta_{1i}}{\delta} \;,\quad \Phi_{(x,t)} = \frac{2}{q^3}\int\limits_0^\infty \tilde{\tau}\,\frac{\partial\overline{u}}{\partial y}dy$$

For taking into account the out-of-equilibrium turbulence, in addition, the final closure involves the two-equation model previously suggested for integral methods, see Le Balleur [23,25]. Denoting $\tilde{k}(x,t), \tilde{\varepsilon}(x,t), \tilde{\tau}(x,t)$ the averaged levels across the viscous layer of the turbulent kinetic energy, dissipation, and Reynolds shear stress, denoting $\tilde{k}_{eq}(x,t), \tilde{\varepsilon}_{eq}(x,t), \tilde{\tau}_{eq}(x,t)$ the corresponding values deduced from the mean velocity profile and the mixing length model, the departure from equilibrium of the entrainment and dissipation is calculated by solving two additionnal integral equations for $\tilde{k}$ and $\tilde{\tau}$:

$$\tilde{\delta}\frac{\partial \tilde{k}}{\partial t}+q(\tilde{\delta}-\delta_{1i})\frac{\partial \tilde{k}}{\partial x}=\frac{\tilde{\tau}}{\tilde{\tau}_{eq}}\frac{q^3}{2}\Phi_{eq}-\tilde{\epsilon}\tilde{\delta}$$

$$\tilde{\delta}\frac{\partial \tilde{\tau}}{\partial t}+q(\tilde{\delta}-\delta_{1i})\frac{\partial \tilde{\tau}}{\partial x}=1.5\frac{\tilde{\epsilon}\tilde{\delta}}{\tilde{k}}\left[\left(\frac{\tilde{k}}{\tilde{k}_{eq}}\right)^2\frac{\tilde{\epsilon}_{eq}}{\tilde{\epsilon}}\tilde{\tau}_{eq}-\tilde{\tau}\right] \qquad (24)$$

$$\frac{\tilde{\epsilon}}{\tilde{\epsilon}_{eq}}=\left(\frac{\tilde{k}}{\tilde{k}_{eq}}\right)^{3/2} \quad , \quad \begin{bmatrix} E(x,t) \\ \Phi(x,t) \end{bmatrix}=\frac{\tilde{\tau}_{(x,t)}}{\tilde{\tau}_{eq}(x,t)}\begin{bmatrix} E_{eq}(x,t) \\ \Phi_{eq}(x,t) \end{bmatrix} \qquad (25)$$

$$\tilde{\delta}\tilde{\epsilon}_{eq}=0.5\Phi_{eq}q^3$$

$$\tilde{k}_{eq}=\left[0.045\,\lambda_1\lambda_2\lambda_3(1-\tilde{u}_w)\Phi_{eq}\right]^{1/2}q^2$$

$$\tilde{\tau}_{eq}=\lambda_1\lambda_2\lambda_3\left[0.09\,(1-\tilde{u}_w)\right]^2q^2 \qquad (26)$$

$$2.222\,(\tilde{\delta}-\delta_{1i})=\tilde{\delta}\,(1.222+\tilde{u})$$

2.3 - Massive separation.

If we except the model for out-of equilibrium turbulence (24),(25),(26), which introduces history effects in the Reynolds stresses and entrainment, the previous modelling has been designed to satisfy the constraints which are already present in equilibrium layers.

The main qualitative constraint is to insure that the mathematical domain of dependence of the integral equations never violates the physics, nor the domain of dependence of the Prandtl's equations, and recovers an upstream influence in the whole reverse flow range. The main quantitative constraint is to describe accurately the two limiting test-cases for equilibrium layers, the flat plate boundary layer in the attached flow range, and the isobaric mixing layers in the highly separated flow range $(\delta_1\rightarrow\delta, H\rightarrow\infty)$.

The main qualitative conclusions are summarized on Figure 2, which sketches the diagrams $\tilde{H}(H)$, and θ_{11}/δ versus δ_1/δ. As these diagrams are dependent on Mach and Reynolds numbers, we consider here for simplicity the incompressible case at very high Reynolds number.

As we have shown previously, see Le Balleur [21], the characteristic cone of dependence of the unsteady integral equations, when solving the problem in direct mode, is such that no upstream influence is present if $\dfrac{\partial \tilde{H}}{\partial H}<0$, and that the upstream influence is possible if $\dfrac{\partial \tilde{H}}{\partial H}>0$. The limiting case $\dfrac{\partial \tilde{H}}{\partial H}=0$ degenerates, for steady flows, into a Goldstein-like singularity. It indicates the separation point in integral methods (beginning of upstream influence). The physics then requires, in order to obtain a proper domain of dependence, that the diagrams $\tilde{H}(H)$ exhibit a minimum, and only one, for any Mach and Reynolds numbers.

The present closure provides the separation minimum always around $Hi=2.7$. The limiting possibility $\tilde{H}(H)=$constant beyond separation is in principle excluded, since generating only a marginal upstream influence. The occurence of several minima is also excluded, since any area with $\dfrac{\partial \tilde{H}}{\partial H} < 0$ in the separated flow range $(H > 2.7)$ would eliminate there the upstream influence.

When considering the diagram $[\theta_{11}/\delta](\delta_1/\delta)$, the closure locus in the incompressible case at infinite Reynolds number is approximately a parabola, for attached and slightly separated flows, figure 2. The constant values of $\tilde{H}$ are straigth lines issued from the point $\delta_1 = \delta, \theta_{11} = 0$, which corresponds to the mixing-layer limit. The extremal value of $\tilde{H}$, drawn with the mixed line on figure 2, shows that the separation is near $\delta_{1i} = 0.40\delta$. For massive separation, if the Coles-like modelling with $\eta^* = 0$ was maintained, the maximum reverse velocity in the turbulent velocity profile would tend to infinity, and the closure locus would follows the parabola until $\theta_{11} = 0$ (dashed line on fig.2). Such a non-physical behaviour for the velocity profile would then never allow to reach the mixing-layer limit $(\delta_1 \to \delta, \theta_{11} \to 0)$, which is believed to represent the infinitely-separated boundary layers $(H \to \infty)$.

The closure locus in the diagram θ_{11}/δ (δ_1/δ) has then to join smoothly the mixing-layer phase-point $(\delta_1 = \delta, \theta_{11} = 0)$, with the parabola of the attached and slightly separated layers. The condition of upstream influence recovery in the separated flow range restricts the possible paths to the shaded area on figure 2, in the plane $(\theta_{11}/\delta)(\delta_1/\delta)$. The same condition of upstream influence recovery, which requires to have no other extremum of $\tilde{H}$ than the separation one, restricts still the possible locus to a narrow range of very regular curves in the diagram θ_{11}/δ versus δ_1/δ, such as the solid line on figure 2.

It was then found that the simplest way to obtain such regular closure loci was to assume a linear variation $\eta^*(\delta_1/\delta)$, in the range of high shape parameters, when modifying the wake function from $F(\eta)$ to $\tilde{F}(\eta)$ in the velocity profile modelling (20), see figure 2. This modelling is fruitful because the slope $d\eta^*/da$ $(a = \delta_1/\delta$ in incompressible flow) can be determined from the isobaric mixing-layer limit $(\delta_1 \to \delta, \theta_{11} \to 0)$, in order to fit the spreading rate σ, figure 3. Finally, the two piecewise-linear closures $\eta^*(\delta_1/\delta)$, respectively for attached or slightly separated flows, and for massively separated ones, have been joined smoothly, as shown by relations (22), the smooth junction being adjusted to fit the usual levels of maximal reverse velocities observed in different experiments [32].

2.4 - Isobaric mixing layer limit.

The adjustment on the isobaric mixing-layer limit is based on the following equations, figure 3 :

$$\frac{d\delta_1}{dx} = \frac{v}{q} = 0 \quad , \quad \frac{d(\delta - \delta_1)}{dx} = E \quad , \quad \frac{d\theta_{11}}{dx} = \frac{Cf}{2} = 0$$

$$\delta_1 \to \delta \quad , \quad \delta \to \infty \quad , \quad \theta_{11} \to 0 \quad , \quad \frac{1}{q}\frac{dq}{dx} \to 0 \tag{27}$$

Assuming a closure $\eta^*(a)$, $[\theta_{11}/\delta](a)$, $\delta_1/\delta = a$, we need :

$$\frac{d\delta}{dx}=E \quad , \quad \delta\frac{da}{dx}=-E \quad , \quad \frac{d}{da}\left[\frac{\theta_{11}}{\delta}\right]\to 0 \quad , \quad \delta\frac{d\eta^*}{dx}=-E\frac{d\eta^*}{da} \qquad (28)$$

From the velocity profile modelling (20), we assume :

$$\tilde{u}=\frac{\overline{q}}{q}=1-\left[1-\left(\frac{\eta-\eta^*}{1-\eta^*}\right)^{3/2}\right]^2$$

$$\tilde{u}=0.10 \qquad \eta_1=\eta^*+0.138\ (1-\eta^*) \qquad (29)$$

$$\tilde{u}=0.90 \qquad \eta_2=\eta^*+0.776\ (1-\eta^*)$$

If the conventional thickness of the mixing-layer is defined with η_1,η_2, the spreading rate σ is deduced from (28) :

$$y_{0.9}-y_{0.1}=(\eta_2-\eta_1)\delta$$

$$\frac{1}{\sigma}=\frac{d}{dx}\left[y_{0.9}-y_{0.1}\right]=(0.776-0.138)\ E\frac{d\eta^*}{da} \qquad (30)$$

With the present closure modelling, the relation $(22),(23)$ provide :

$$E=0.053 \quad , \quad \frac{d\eta^*}{da}=\frac{1}{0.435}\simeq 2.3 \quad , \quad \sigma=12.9 \qquad (31)$$

If the conventional thickness of the mixing-layer is defined with the maximal slope of the profile $\tilde{u}(\eta)$, the new spreading rate $\tilde{\sigma}$ is also deduced from $(28),(22),(23)$:

$$\left(\frac{\partial\tilde{u}}{\partial y}\right)_{max}^{-1}=0.7055\ (1-\eta^*)\ \delta$$

$$\frac{1}{\tilde{\sigma}}=\frac{d}{dx}\left[\left(\frac{d\tilde{u}}{dy}\right)_{max}^{-1}\right]=0.7055\ E\ \frac{d\eta^*}{da} \qquad (32)$$

$$\tilde{\sigma}\simeq 11.6$$

The two estimates σ, $\tilde{\sigma}$ of the spreading rate demonstrate that the present modelling is consistent, if $(d\eta^*/da)_{a=1}\simeq 2.3$, with the usual incompressible mixing layers where $\sigma\simeq 12$, within experimental accuracy. This value 2.3 was the first one suggested previously, see Le Balleur [23]. With some more analysis, a possible improvement would be perhaps $(d\eta^*/da)_{a=1}\simeq 2.4$, which would insure $\frac{d}{da}(\theta_{11}/\delta)=0$.

On the other hand, at last, the present modelling can be adjusted with the experimental data on the flat plate boundary layer. An improved tuning of the Reynolds number effect on the entrainment is possible.

3 - VISCOUS - INVISCID COUPLING SOLVERS

3.1 - Direct-Inverse viscous solvers

The defect equations of the viscous problem, in integral form $(15),(16),(17),(18)$:

$$\left\{\frac{1}{q}C_j^i{}_{(a,m,\delta)}\frac{\partial f^j}{\partial t}+A_j^i{}_{(a,m,\delta)}\frac{\partial f^j}{\partial x}=b^i\right\}_{(s,0,t)} \qquad (33)$$

$$[f^j]=\begin{bmatrix}\hat{m}\\\hat{h}_i\\\hat{p}_i\\\delta\\a\\\tilde{k}\\\tilde{\tau}\end{bmatrix}_{(s,0,t)} \quad,\quad [b^i]=\begin{bmatrix}\dfrac{\hat{v}}{q}\\[4pt]\dfrac{Cf}{2}+\dfrac{\delta_0}{q^2}\gamma_x\\[4pt]\dfrac{\tilde{\tau}}{\tilde{\tau}_{eq}}E_{eq}\\[4pt]b^4\\b^5\end{bmatrix} \quad,\quad \begin{array}{l}i=1,5\\[4pt]j=1,7\end{array}$$

have to be closed by the inviscid problem.

In the viscous problem, the unknowns are the reduced Mach number $\hat{m}_{(s,0,t)}=0.5(\gamma-1)M^2$, transpiration velocity $\hat{v}_{(s,0,t)}$, total enthalpy $\hat{h}_{i\,(s,0,t)}$, total pressure $\hat{p}_{i\,(s,0,t)}$, and the viscous parameters $[\delta,a,k,\tilde{\tau}]_{(s,t)}$.

The four unknowns $[\hat{m},\hat{v},\hat{h}_i,\hat{p}_i]_{(s,0,t)}$ have to be coupled with the corresponding variables which are computed in the pseudo-inviscid problem $[m,v,h_i,p_i]_{(s,0,t)}$:

$$\left\{\hat{m}\rightarrow m\ ,\ \hat{v}\rightarrow v\ ,\ \hat{h}_i\rightarrow h_i\ ,\ \hat{p}_i\rightarrow p_i\right\}_{(s,0,t)} \qquad (34)$$

These four coupling conditions (34), which have to be satisfied at each space- and time-step, are generally solved with an iterative numerical method (coupling solver), in order to dissociate the numerical techniques used for the viscous and pseudo-inviscid problems.

The rank-5 viscous system (33), with 8 unknowns, is then solved with only three coupling conditions, choosen in (34) :

$$\left\{\hat{h}_i=h_i\right\}_{(s,0,t)}\quad,\quad\left\{\hat{p}_i=p_i\right\}_{(s,0,t)}\quad,$$
$$\left\{\epsilon\frac{\hat{v}}{q}+(1-\epsilon)\frac{1}{\hat{m}}\frac{\partial\hat{m}}{\partial x}=\epsilon\frac{v}{q}+(1-\epsilon)\frac{1}{m}\frac{\partial m}{\partial x}\right\}_{(s,0,t)} \qquad (35)$$

If $\epsilon=0$, the viscous problem is solved in the "Direct" mode (pressure-prescribed), compatible with a space-marching solution for attached flows ($a<0.40$). The switch $\epsilon=1$ provides a solution in "Inverse" mode, which is necessary to remove the Goldstein singularity at separation in steady flow , and to maintain a space-marching solution in unsteady flow for the reverse flow range ($a>0.40$), see Le Balleur [21], and Cousteix, Le Balleur, Houdeville [33].

Let us notice that the viscous upstream influence, which is eliminated at the viscous step by this space-marching solution, is fully recovered at the coupling step, where the fourth condition of (34) is satisfied (strong coupling). This coupling step can never then be reduced to a non-iterative space-marching solution, even when simultaneously solved with the pseudo-inviscid problem.

3.2 - Viscous influence function.

In order to build the coupling step, if we consider the discretized problems with the unknowns $[\hat{m},\hat{v},\hat{h}_i,\hat{p}_i]_{i,1}^n$, $[m,v,h_i,p_i]_{i,1}^n$ at $s=s_i, n=0, t=n\Delta t$, we need to analyse both the viscous and the pseudo-inviscid operators, which are connecting these discrete unknowns.

The viscous operator along $n=0$ has to be deduced from the full Defect and coupling equations (3),(4). In the case of the integral equations (33), which give a rank-5 system, it is possible to use 4 equations to eliminate the 4 purely viscous unknowns $\delta, a, \bar{k}, \bar{\tau}$.

The residual equation provides the viscous influence function that operates on $[\hat{m},\hat{v},\hat{h}_i,\hat{p}_i]_{i,1}^n$. As we have discussed previously, this elimination is an analytical-one in the steady case, and involves the numerical integration schemes of (33) in the unsteady case, see Le Balleur [21,22,1] and Le Balleur, Girodroux-Lavigne [30,31]. The viscous influence function :

$$\left[\frac{\hat{v}}{q} + \alpha_j^* \frac{\partial \hat{f}^j}{\partial t} + \beta_j^* \frac{\partial \hat{f}^j}{\partial x}\right]_{j,1}^n = R_{i,1}^n\left(\frac{\hat{v}}{q}, \hat{f}^j\right) = 0$$

$$\hat{f}^j = \left[\hat{q}, \hat{h}_i, \hat{p}_i\right] \quad , \quad j = 1,2,3 \tag{36}$$

is highly non-linear, the terms α_j^*, β_j^* growing to infinity at the separation or reattachment stations. It displays the different roles played by the unknowns $\hat{v}$ and $\hat{f}^j$.

3.3 - Inviscid operator. Steady flows.

The analysis of the inviscid operator along the boundary $n=0$ is more difficult. We have to estimate the influence coefficients that connect the discrete unknowns of the boundary conditions at $n=0$. This involves the dependency of the inviscid field on perturbations of the boundary conditions, which is generally a complex non-linear problem.

However, when the inviscid operator is only involved as a numerical conditioning or a stability control in the relaxation techniques for coupling, we have suggested to use a linearized and local approximation, for small perturbations around a given solution $v_{0(s,0)}, q_{0(s,0)}$. This approximation is of the Prandtl-Glauert type, and is solved with a Fourier analysis, see Le Balleur [22,3,27] :

$$\left[\begin{matrix} v - v_0 \\ q - q_0 \end{matrix}\right]_{j,1} = \sum_\alpha \left[\begin{matrix} v' \\ q' \end{matrix}\right]_{\alpha)} \cdot e^{j\alpha x_i} \quad , \quad (j^2 = -1) \tag{37}$$

Within these approximations, where $\cos\theta \simeq 1$, it is found at any wave number α :

$$\left[v - v_0\right]_{j,1} \simeq j \cdot \beta_i \cdot \left[q - q_0\right]_{j,1} \quad , \quad \beta = \sqrt{\left(\frac{q}{u}\right)^2 - M^2} \tag{38}$$

This value of the Prandtl-Glauert constant β has been used succesfully to develop the direct and semi-inverse coupling methods in quasi-threedimensional separated flows over infinite swept wings, see Le Balleur [25,27]. The "local" constant β_i at each coupling node is used.

3.4 - Strong coupling solvers.

The following results have been computed by using three coupling solvers.

The steady flows are solved with the original form of the "Direct" and "Semi-Inverse" methods of Le Balleur [22], which includes the linear stability control required by their explicit character. This automatic stability control [22], deduced from the analysis of sections 3.2, 3.3, is believed to be important for computing shock wave-boundary layer interactions or stalled airfoils. Such a control is missing in the semi-inverse method suggested more recently by Carter [8]. Except for this stability control, Carter's method was however found linearly equivalent to the subsonic term of the present method, see [4].

The unsteady flows are solved with the more recent "Semi-Implicit" method of Le Balleur, Girodroux-Lavigne [30,31], whose discretization is fully time-consistent with a strong coupling. Although involving a step where the inviscid field is semi-simultaneously relaxed with a viscous influence equation, the "Semi-Implicit" method is different from Veldman's method [18], and maintain a step where an uncoupled "Direct" or "Inverse" viscous problem is solved. Roughly, Veldman's method solves a viscous problem plus an inviscid influence function. On the contrary, the "Semi-Implicit" method solves an inviscid problem plus a viscous influence function, deduced from a direct or inverse viscous solution.

3.4.1 - 'Direct' method .

Within areas of attached steady viscous layers, where the viscous solution is Direct ($\epsilon = 0$, $H_i < 1.8$), the transpiration velocity of the pseudo-inviscid problem $v_{i,1}$ is explicitly relaxed by using the estimate $\hat{v}_{i,1}$ of the viscous problem :

$$v_{i,1}^{\nu+1} - v_{i,1}^{\nu} = \omega \cdot \omega_i^1 \left[\hat{v}_{i,1} - v_{i,1}^{\nu} \right]$$

$$\omega_i^1 = \omega_{opt} \left[\mu(x_i, \alpha_{max}) \right] \quad , \quad \alpha_{max} = \frac{\pi}{\Delta x_i}$$

$$\mu_{(x,\alpha)} = \frac{\alpha \beta_1^*}{\beta} = R + jI \quad , \quad j^2 = -1 \qquad (39)$$

$$\omega_{opt}(\mu) = \frac{1 - R}{(1 - R)^2 + I^2}$$

$$0 < \omega < 2$$

The update is over-relaxation-like for the coefficient ω. The local coefficient ω_i^1 decrease however with Δx_i (explicit method), and with the vicinity of the sonic-like stations ($\beta \to 0$), or of the separation and reattachment stations ($\beta_1^* \to \infty$). The upstream influence in supersonic zones is recovered through $\hat{v}_i$, deduced from (36) with centered or downwind schemes for coupling $\left(\dfrac{\partial \hat{q}}{\partial x} \right)_{i,1}$ with $q_{i,1}^{\nu}$.

3.4.2 - 'Semi-Inverse' method.

Within areas of separated or separating steady viscous layers, where the viscous solution is Inverse ($\epsilon = 1$, $H_i > 1.8$), the transpiration velocity $v_{i,1}$ is explicitly relaxed with an update formula. This update uses the coupling error on the first and second derivatives of the two velocities, estimated in the pseudo-inviscid problem $q_{i,1}^\nu$, and in the viscous problem $\hat{q}_{i,1}$:

$$v^{\nu+1}_{i,1} - v^\nu_{i,1} = \omega \cdot \left[\omega_i^2 \left(\frac{q^\nu}{\hat{q}} \frac{\partial \hat{q}}{\partial x} \frac{\partial q^\nu}{\partial x} \right) + \omega_i^3 \left(\frac{q^\nu}{\hat{q}} \frac{\partial^2 \hat{q}}{\partial x^2} \frac{\partial^2 q^\nu}{\partial x^2} \right) \right]_{i,1}$$

$$\omega_i^2 = \omega_{opt} \left[\mu^{-1}(x_i, \alpha_{max}) \right] \quad , \quad \omega_i^3 = 0 \quad , \quad \text{if } M_i < |\frac{q}{u}|_i$$

$$\omega_i^2 = 0 \quad , \quad \omega_i^3 = \omega_{opt} \left[\mu^{-1}(x_i, \alpha_{max}) \right] \quad , \quad \text{if } M_i > |\frac{q}{u}|_i \tag{40}$$

$$0 < \omega < 2$$

The sonic-like switch of the formulae (40),(39) is valid for quasi-3D infinite swept wing flows, where $q \neq u$. With the present choice of ω_i^2, ω_i^3, the convergence is accelerated at supersonic nodes (for wave number α_{max}) by discretizing $\left(\frac{\partial^2 q^\nu}{\partial x^2} \right)_{i,1}$ downwind of $\left(\frac{\partial^2 \hat{q}}{\partial x^2} \right)_{i,1}$. This also improves the upstream influence recovery, which is however insured so long as a full upwind discretization of these two terms is avoided. As the stability is assessed on α_{max}, the convergence rate has an explicit character, and decreases with the mesh size Δx_i.

3.4.3 - 'Semi-Implicit' method.

When applying the viscous influence operator $R_{i,1}^n(\frac{\hat{v}}{q}, \hat{f}^j)$ of (36) to the variables of the pseudo-inviscid problem, non-zero residuals $R_{i,1}^n(\frac{v}{q}, f^j)$ are obtained at time n, and coupling iteration ν, so long as the strong coupling is not converged. The Semi-Implicit method is a relaxation technique for these coupling residuals at iteration ν, where the non-linear terms α_j^*, β_j^* and the residuals $R_{i,1}^n$ are provided by a previous viscous solution, in Direct or Inverse mode, and a previous inviscid solution :

$$\left[\Delta(\frac{v}{q}) + \alpha_j^{*\nu} \cdot \Delta\left(\frac{\partial f^j}{\partial t} \right) + \beta_j^{*\nu} \cdot \Delta\left(\frac{\partial f^j}{\partial x} \right) \right]_{j,1}^n = -\omega (R^\nu)_{i,1}^n$$

$$[\Delta f]_{i,1}^n = \left[f^{\nu+1} - f^\nu \right]_{i,1}^n$$

$$R_{i,1}^n = (1-\epsilon) \left[\frac{\hat{v}}{q} - \frac{v^\nu}{q} \right]_{j,1}^n + \epsilon \left[\alpha_j^* \left(\frac{\partial \hat{f}^j}{\partial t} - \frac{\partial f^j}{\partial t} \right)^\nu + \beta_j^* \left(\frac{\partial \hat{f}^j}{\partial x} - \frac{\partial f^j}{\partial x} \right)^\nu \right]_{j,1}^n \tag{41}$$

$$f^j = \left[q, h_i, p_i \right] \quad , \quad j = 1,2,3 \quad , \quad \omega = 1$$

The conditioning terms Δ on the left hand side are discretized with the variables of the inviscid solver. They use upwind schemes or centered schemes but, in the latter case, a Gauss-Seidel discretization based on iterations ν and $\nu+1$ (at the same time level n) is involved. Then the coupling relaxation sweeps of equation (41) can be performed space-marching in the free-stream direction, simultaneously with the y-sweep of an ADI inviscid solver, at iteration ν. The viscous upstream influence is fully recovered from the discretization of the residuals $R_{i,1}^n$, and from the iterative-marching relaxation ν for coupling, which is converged at time n.

4 - MASSIVE SEPARATION AT THE VISCOUS SCALE.

When the local viscous layer thickness $\delta_{(x,t)}$ is small with respect to the inviscid scale based on the overall geometry, a local wall-discontinuity or a shock-induced separation can provide massive recirculations in the viscous layer, which however are still rather slender (or even short) bubbles at the inviscid scale. In this case, the significant criterium for a massive separation is not the separation length, but the maximal (incompressible) shape parameter $H_i = (\frac{\delta_1}{\theta_{11}})_i$ of the viscous layer. The attached flow range of the boundary layer theory is roughly $1 < H_i < 2.7$, the usual flat plate value being near $H_i \simeq 1.4$.

In the following examples of applications, the shape parameter range is roughly $1.2 < H_i < 25$. The separations zones are slender, and the separation lenghts are significant at the inviscid scale. The shape parameters are yet in the massive separation range, although higher values have been solved for stall, backward-facing steps and spoilers $(50 < H_i < 500)$, see section 5.

4.1 - Steady shock wave-boundary layer interactions.

The following results are taken from Le Balleur, Blaise [29]. This method solves the viscous-inviscid interaction as explained in the upper sections, and involves an Euler solver for the pseudo-inviscid problem. The explicit Euler solver of Viviand, Veuillot [34] is presently used. The method solves the steady viscous equations and uses the Direct and Semi-Inverse coupling methods.

The figures 4-5-6 display typical results for shock wave-boundary layer interactions. The figure 4 shows a plane transonic nozzle. The prescribed back-pressure induces a normal shock wave-boundary layer interaction at M=1.45 that extensively separates the boundary layer, see the Isomach contours on figure 4. The Reynolds number $R_\delta \simeq 33000$ is based on the incoming boundary layer thickness. The computation is performed here with the equilibrium turbulent model. This is found to underestimate the separation trend, and to overestimate the pressure at separation point, which explains that the supersonic tongue at the root of the shock is slighty underestimated on figure 4.

The figure 5 shows the capability to compute the supersonic shock wave-boundary layer interactions. The high Mach number M=2.85 demontrates that the present theory always removes the "supercritical" breakdown of the Crocco theory, and never fails to propagate upstream the viscous influence, see Le Balleur [21]. A downstream boundary condition at the wall is directly prescribed to the coupling solver, the condition $\partial p / \partial x = 0$ being used for the supersonic ramp. The present theory allows the shock waves to penetrate within the viscous layers, figure 5. The computed wall pressure is compared to the experiment of Bogdonoff [35] on fig. 5. Due to the high normal pressure gradient accross the layer at M=2.85, the second order modelling (13) is here found to be necessary, because the first order defect approximation (10)(17) overestimates $\partial \bar{p} / \partial y$.

The figure 6 displays the computation results for a rather massive separation, induced by the transonic shock wave-boundary layer interactions in a plane dissymetrical channel, at Mach number M=1.36. The back-pressure is adjusted to fit the beginning of interaction on the lower wall. The use of the two-equation model for the turbulence is also found necessary to obtain an acceptable prediction of the extent of the separation. The rather good agreement with the experiment of Delery [36] on fig 6. for the mean velocity profiles shows that the separation is still slightly underestimated. Similar conclusions are obtained for the turbulent kinetic energy profiles and for the shear stress profiles, that are computed from the present two-equation model (24) and from the velocity profiles (20), see Le Balleur, Blaise [29]. The figure 6 compares with experiment the hysteresis effect computed for the maximal shear stress, and shows the large departure from the closure locus that would be assumed with the equilibrium model (mixing length).

In the previous calculations, adapted grids or self-adaptative grids, with clustering techniques, have been designed to have at least four cells per boundary layer thickness in the areas of beginning of interactions.

4.2 - Transonic unsteady separation.

The following unsteady computations are extracted from Le Balleur, Girodroux-Lavigne [30,31]. The unsteady viscous equations and the strong coupling are solved fully time-consistently. The full viscous upstream influence is recovered within a single time-step. The transonic small perturbations solver of Couston, Angelini [37] is presently used for the pseudo-inviscid problem. The coupling is solved with the Semi-Implicit method.

The figures 7-8 display typical results in forced oscillations, with unsteady shock-induced separation. As in the steady case, a grid clustering, fig.8, is used to resolve the viscous-inviscid interaction scale, at the root of the shock wave, which is moving during a cycle, and to obtain a continuous compression at the wall. A steady state solution over the NACA64A010 airfoil with extensive separation is shown on fig.7. The separation point is at mid-chord and the reattachment in the wake. The two-equation turbulent model is still necessary to recover an agreement of the separation length with the experiment of Davis [38]. The figure 8 displays the instantaneous Mach number contours during one cycle of forced oscillation in pitch, at reduced frequency $K = \omega c / u_\infty = 0.40$. The evolution of the boundary layer thickness at the trailing-edge shows the unsteadiness of the separation. The first harmonic of the unsteady pressure distributions, figure 8, displays an agreement with experiment similar to the agreement at steady state. A rather satisfactory prediction of the modulus is obtained, both at shock wave, for the intensity and amplitude of the motion, and beyond the shock, in the reverse flow area. In this area, the experimental phase-inversion with respect to an inviscid analysis is here very well recovered.

The figures 9-10 display the results obtained for the prediction of buffeting separated flows over a steady symmetrical airfoil. The viscous flow is computed fully time-consistently over a symmetrical circular-arc airfoil in free air, whose thickness is very rapidly growing from zero to 18 percent, starting from a parallel uniform flow with attached boundary layer. The computations are performed at Reynolds number $R = 11 \times 10^6$, at different Mach numbers in the transonic range. At low Mach number, without separation at the shock, steady symmetrical solutions are obtained, with a trailing-edge separation. Steady symmetrical solutions are also obtained at high Mach number M=0.788, with extensive shock-induced separation near 60 percent chord, without reattachment before the wake, figure 9. The agreement with the experiments of Marvin et al [39] is still improved by the two-equation model, figure 9.

On the contrary, at the intermediate Mach number M=0.760, the transitory computed solution, after starting a symmetrical evolution, deviates slowly toward a dissymetrical one. The computation never reaches a steady limit, but a self-oscillatory solution, with buffeting shock-induced separation, and with a 180° phase shift between upper and lower surfaces, as shown by the instantaneous Mach number contours, figure 10. The solution agrees with experiment [39]. At the same conditions, a steady solution is obtained with an inviscid calculation. The dominant frequency of the present self-oscillatory viscous solution (131 Hz) is of the same order that the experimental one (188 Hz), or than the solutions of Direct Navier-Stokes solvers (155 Hz) [39].

5 - MASSIVE SEPARATION AT THE INVISCID SCALE.

Examples of massive separation are provided by the computation of separated flows over airfoils, at hight-lift or beyond stall, or with deflected spoilers.

Although self-oscillatory solutions induced by viscosity are still possible, the following results have been obtained by assuming and solving only steady equations.

5.1 - Hight-lift and stalled airfoils.

The figure 11 shows a first result of trailing-edge separation, obtained with the full viscous transonic method developed for separated flows over airfoils, see Le Balleur [2,23], which has been extended also to the spoiler problem, see Le Balleur [24,25]. The boundary layer and wake are computed with the present method. The strong coupling is obtained with the Direct and Semi-Inverse methods. The full dissymmetry of the wake, see [23], is included in the viscous integral equations, as well as in the coupling and positioning of the wake-cut, thus giving a full visualization of the flow field based on the defect velocity profiles modelling (20), figure 11. The configuration is the "stalled airfoil" test-case of the STANFORD Conference on Complex turbulent flows, rather well predicted with the equilibrium turbulent model, fig. 11, but corresponding only to an incipient stall.

A simplified form of this viscous-inviscid interaction method, restricted to the symmetrical wake coupling, see [23], has been used to develop a method for multi-element airfoils in incompressible flow, see Le Balleur, Neron [40].

The following results for stall prediction have been obtained with a recent new version of this simplified method, developed by Le Balleur, Henry. The new method incorporates an adaptable mesh generation issued from the transonic method [23], that allows to cluster automatically the mesh near the separation point, and to resolve the viscous-inviscid interaction scale near the leading-edge, at stall conditions. The new method incorporates the two-equation turbulent model, and still involves the panel method of Neron [40] for the pseudo-inviscid problem.

The figure 12 shows that the deep stall range with massive separation is accessible to this method. An extensive plateau-pressure is obtained without special modelling. The calculation with the two-equation turbulent model is however believed here to overestimate the separation.

The computations at increasing angles of attack for the NACA0012 airfoil are shown on figure 13, still using the two-equation model that overestimates the separation. The stiffness of the stall on the NACA0012 airfoil, as compared to the smoother stall on the NACA4412 airfoil, is well reproduced by the present computations, fig. 11-12-13. At the higher incidence on figure 13, the massive separation involves roughly two chord lengths, with a separation point around 5 percent chord and a reattachment in the wake, one chord downstream the trailing-edge, fig. 13. The resolution at the viscous scale (local boundary layer thickness) near the leading-edge is crucial for convergence, and probably for accuracy, see fig. 15.

Altough the tuning of the model is still preliminary, the method indicates however that the accurate prediction of the stall is a stiff problem. The figure 14 shows that an important influence of the turbulent model has to be expected for the prediction of lift and stall angle.

For the massive separation case at $\alpha = 17$ ° on the NACA0012 airfoil, the streamlines of the pseudo-inviscid field are displayed on figure 15 (symmetrical wake coupling), showing the large difference between the airfoil and the displacement body. This case was found very severe for the convergence rate, more severe than higher incidences, and was uses to check the convergence of the present Semi-Inverse method.

The first conclusion was to find that the present Semi-Inverse method converges strictly, at each coupling node, even for such an extensive separation. This conclusion is obtained here with a panel solver for the inviscid subroutine, which probably minimizes the interferences due to the numerical dissipation of the inviscid solvers. The convergence history for the more severe test-case is shown on figure 16 for the lift, drag, position of the separation point, pressure near trailing-edge. The discontinuities on the convergence curves corresponds to the successive adaptations of the mesh. As the Semi-Inverse method is explicit in character, the convergence rate is only indicative, and can be reduced by decreasing the mesh size. The present results are obtained with 256 nodes on the airfoil and wake-cut, with a strong clustering near the leading-edge separation. The slowest convergence is obtained with 1000 iterations.

5.2 - Airfoils with deflected spoilers.

The following results are obtained with the method developed for transonic airfoils [23]. The method was extended to compute the airfoils with spoiler by simulating the spoiler deflexion with an equivalent deflexion of the velocity along the wall, superimposed with the deflexion induced by the strong viscous coupling, and by suddenly increasing the displacement thickness at the trailing-edge of the spoiler, considered as a backward-facing step, see Le Balleur [24,25].

The figure 17 shows a result obtained with the full transonic method, including the dissymetrical treatment of the wake, at moderate spoiler deflexion $\delta_{sp} = 10$ ° and transonic conditions. The large separation bubble is roughly 45 percent chord, and induces a negative lift, with a supersonic pocket on the lower surface, on the RA16SC1 supercritical airfoil. A rather very satisfactory agreement with the experiments of Consigny et al [41], Costes [42] is obtained for the pressure and drag prediction, including the plateau-pressure.

Figure 18 displays typical comparisons with experiment at different spoiler deflexions, obtained with a simplified version of the method where the symmetrical wake coupling is used. The different behaviours observed at small deflexions, without plateau-pressure, and at high deflexions are well recovered by the calculation. The comparison with the experiment at the higher deflexion $\delta_{sp} = 18$ ° is subject to wind-tunnel interferences. At moderate deflexion $\delta_{sp} = 6$ ° however, the discrepancy with experiment indicates an overestimate of the separation, without reattachment before the wake.

The figure 19 shows on the pseudo-inviscid streamlines, that the simplified method, with symmetrical wake coupling, has been found capable to compute massive separations at high spoiler deflexions, in subsonic flow. The strict convergence of the present Semi-Inverse method is still observed. The adaptation of the grid is still here a crucial point, especially to resolve the rear stagnation point R in the wake with an acceptable mesh size. This constraint is sensitive because this reattachment point R is moving quickly far downstream of the trailing edge when the spoiler deflexion is increased, fig 19. The solution at higher deflexions have been obtained with an initial guess of the flow obtained at lower deflexions. At the maximal spoiler deflexion $\delta_{sp} = 45$ °, a noticeable separation bubble is also computed at the hinge of the spoiler, and the massive separation beyond the spoiler has a rear stagnation point one chord downstream the trailing edge. The dotted lines on fig. 19 visualize the displacement body, still very different from the airfoil.

The figure 20 displays the viscous streamlines at high spoiler deflexions, deduced from the field of figure 19 by adding the turbulent velocity profile modelling (20). The trend toward the mixing-layer limit of the massively separated boundary layer, just downstream the spoiler, is clearly shown. The second bubble of the separation at the spoiler hinge is also shown at $\delta_{sp} = 45$ °.

CONCLUSIONS

The Indirect Viscous-Inviscid Solvers (VIS) based on IDI equations do provide a Navier-Stokes-like capability in computation of massive separations.

The approximate Defect integral equations have been generalized for massive separations by adjusting the coordinate system where the thin-layer approximations are involved. The coordinate system linked to the displacement body is found satisfactory. The final integral equations are however projected on the directions tangent and normal to the wall, where they are simpler.

The integral turbulent closure based on a velocity profile modelling has been generalized for massive separation. The guidelines are to obtain the proper mathematical domains of dependence for the integral equations, insuring the viscous upstream influence recovery in the whole reverse flow range, and to reach the description of the isobaric mixing-layer limit for infinitely-separated boundary layers.

The explicit "Semi-Inverse" coupling method of Le Balleur was found to converge for massive separations of two chord lengths over airfoils at deep stall conditions, or with highly deflected spoilers. The same method is capable to solve the supersonic and transonic shock wave-boundary layer interactions. The "Semi-Implicit" coupling method is capable to solve time-consistently the strong viscous-inviscid interaction at conditions of buffeting separation.

Very encouraging results are obtained for shock wave-boundary layer interactions, oscillating airfoils, transonic buffeting, computation of airfoils stall, and spoiler-induced separations.

For most of these stiff viscous-inviscid interactions, the sensivity to the hysteresis effect of the two-equation model for turbulence is found important, even with the approach of an integral method closure.

Acknowlegments.   *The author wishes to thank here his colleagues for kindly helping him, and especially S. Henry who has contributed to the results presented on stalled airfoils.*

-- REFERENCES --

[1] LE BALLEUR J.C, PEYRET R., VIVIAND H. - Numerical Studies in high Reynolds number aerodynamics - Computers and Fluids, Vol. 8,n° 1, p. 1-30, (March 1980).

[2] METHA U., LOMAX H. - Reynolds-averaged Navier-Stokes computations in transonic flows. The state of the art. - Proceedings Symp. Transonic perspective, ed. D. Nixon, Progress in Astronautics, vol. 81 (1982).

[3] LE BALLEUR J.C. - Calcul des écoulements à forte interaction visqueuse au moyen de méthodes de couplage. - AGARD-CP-291, General introduction, Lecture 1, Colorado-Springs, (1981), or ONERA-TP 1980-121.

[4] LE BALLEUR J.C. - Viscid-invicid coupling calculations for two- and three-dimensional flows. - Lecture series 1982-04, Von Karman Institute, Computional Fluids Dynamics, Belgium, (march 1982).

[5] MELNIK R.E., CHOW R., MEAD H.R. - Theory of viscous transonic flow airfoils at high Reynolds number. - AIAA-Paper 77-680, Albukerque (1977).

[6] MELNIK R.E. - Turbulent interactions on airfoils at transonic speeds. Recent development. - AGARD-CP-291, Lecture 10, Colorado-Springs (1981).

[7] MELNIK R.E., BROOK J.W. - The computation of Viscid/Inviscid Interaction on airfoils with separated flow. - Proceed. 3rd Symp. on Numerical and Physical Aspects of Aerodynamic Flows. Long Beach (Jan. 1985), T. Cebeci editor, Springer-Verlag to be published.

[8] CARTER J.E. - A new boundary layer interaction technique for separated flows . - AIAA Paper n° 79-1450, 4th Comp. Fl. Dyn. Conference, (July 1979).

[9] CARTER J.E. - Viscous inviscid interaction analysis of transonic turbulent separated flow. - AIAA Paper 81-1241, Palo-Alto (1981).

[10] CARTER J.E., EDWARDS D.E, DAVIS R.L, HAFEZ M.M. - Analysis of strongly interacting inviscid-viscid flows including separation. - Proceed 9th ICNMFD, Saclay (1984), Lecture Notes in Physics, 218, Springer Verlag (1985).

[11] CARTER J.E., EDWARDS D.E, HAFEZ M.M. - Analysis of transonic shock-induced separation flow including normal pressure gradient. - Report AFOSR-TR-83-1283 (Oct 1983).

[12] WERLE M.J. - Compressor and turbine blades boundary layer separation. -AGARD-CP-351, Paper 9, Copenhague (1983).

[13] LOCK R.C. - A review of method for predicting viscous effects on aerofoils at transonic speeds. - AGARD-CP-291, Lecture 2, Colorado-Springs (1981).

[14] LOCK R.C., FIRMIN M.C.P - Survey of techniques for estimating viscous effects in external aerodynamics. - RAE Tech. Memo. Aero. 1900 (1981).

[15] COLLYER M.R., LOCK R.C. - Prediction of viscous effects in steady transonic flow past an airfoil. - Aero. Quart., Vol. 30, Part 3 (1979).

[16] WAI J.C., YOSHIHARA H. - Planar transonic airfoil computations with viscous interactions. - AGARD-CP-291, Paper 9, Colorado-Springs (1981).

[17] VELDMAN A.E.P. - The calculation of incompressible boundary layers with strong viscous-inviscid interaction. - AGARD-CP-291, Paper 12, Colorado-Springs (1981)

[18] VELDMAN A.E.P. - New, quasi-simultaneous method to calculate interacting boundary layers. - AIAA J. Vol. 19, n°1 (1981), p.79-85.

[19] WHITFIELD D.L., THOMAS J.J., JAMESON A., SCHMITT W. - Computation of transonic viscous-inviscid interacting flow. - Proceed. 2nd Symp. Numerical and Physical Aspects of Aerodynamic Flow, Long-Beach, Springer-Verlag, ed. t. Cebeci, (1984).

[20] VAN DALSEM W.R., STEGER J.L. - Simulation of transonic separated airfoil by finite difference viscous-inviscid interaction. - Proceed. 9th ICNMFD, Saclay, France (1984), Lecture Notes in Physics, 218, Springer-Verlag (1985).

[21] LE BALLEUR J.C. - Couplage visqueux-non visqueux : Analyse du problme incluant décollements et ondes de choc. - La Recherche Aerospatiale 1977-6, p. 349-358 (Nov 1977).
   or English transl. ESA-TT-476 (Viscous-inviscid flow matching : Analysis of the problem including separation and shock waves).

[22] LE BALLEUR J.C. - Couplage visqueux-non visqueux : Méthode numérique et applications aux écoulements bidimensionnels transsoniques et supersoniques. - La Recherche Aerospatiale 1978-2, p. 67-76 (March 1978).
   or English transl. ESA-TT-496 (Viscous-inviscid flow matching : Numerical method and applications to two-dimensional transonic and supersonic flows).

[23] LE BALLEUR J.C - Strong matching method for computing transonic viscous flows including wakes and separations. Lifting airfoils. - La Recherche Aerospatiale 1981-3,p 21-45, English and French editions, (1981).

[24] LE BALLEUR J.C. - Calculation method for transonic separated flows over airfoils including spoiler effects. - Proceed. 8th ICNMFD, Aachen, Lecture Notes in Physics, 170, Springer-Verlag (1982), or ONERA-TP 1982-66.

[25] LE BALLEUR J.C. - Numerical viscid-inviscid interaction in steady and unsteady flows. - Proceed. 2nd Symp. Numerical and Physical Aspect of Aerodynamic Flows, Long-Beach, (1983), Springer-Verlag, t. Cebeci editor, chapt. 13, p. 259-284 (1984), or ONERA-TP 1983-8.

[26] LE BALLEUR J.C. - Progrès dans le calcul de l'interaction fluide parfait-fluide visqueux. - AGARD-CP-351, Paper 1, Copenhague, (1983).

[27] LE BALLEUR J.C. - Numerical flow calculation and viscous-inviscid interaction techniques. - Recent Advances in Numerical Method in Fluids, Vol 3. : Computational methods in viscous flows, p. 419-450, W. Habashi editor, Pineridge Press, (1984).

[28] LE BALLEUR J.C., LAZAREFF M. - A multi-Zonal-Marching integral method for 3D boundary layer with viscous-inviscid interaction. - Proceed. 9th ICNMFD, Saclay, France (1984) Lecture Notes in Physics, 218, Springer Verlag (1985) or ONERA-TP 1984-67.

[29] LE BALLEUR J.C., BLAISE D. - Computation of separated internal flows and shock wave-boundary layer interactions by viscous-inviscid interaction. - La Recherche Aerospatiale 1985-4, p. 211-227, English and French editions, (July 1985).

[30] LE BALLEUR J.C., GIRODROUX-LAVIGNE P. - A semi-implicit and unsteady numerical method of viscous-inviscid interaction for transonic separated flows. - La Recherche Aerospatiale 1984-1, p. 15-37, English and French editions, (1984).

[31] LE BALLEUR J.C., GIRODROUX-LAVIGNE P. -A viscous-inviscid interaction method for computing unsteady transonic separation. - Proceed. 3rd Symp. Numerical and Physical aspects of Aero. Flows, Long-Beach (Jan. 1985), T. Cebeci editor, Springer Verlag to be published. (or ONERA-TP 1985-5).

[32] LE BALLEUR J.C., MIRANDE J. - Etude expérimentale et théorique du recollement bidimensionnel turbulent incompressible - AGARD-CP-168, Göttingen, (May 1975).

[33] COUSTEIX J., LE BALLEUR J.C., HOUDEVILLE R. - Calcul des couches limites turbulentes instationnaires en mode direct ou inverse, écoulements de retour inclus. Analyse des singularités. - La Recherche Aerospatiale 1980-3, English and French editions (1980).

[34] VIVIAND H., VEUILLOT J.P. - Méthodes Pseudo-instationnaires pour le calcul d'écoulements transsoniques. - Publication ONERA n°1978-4. English translation ESA-TT-549.

[35] SETTLES T.J., FITZPATRICK T.J., BOGDONOFF S.M. - Detailed study of attached and separated compression corner flowfield in high number supersonic flow. - AIAA J., vol. 17, n°6,(1979),p579.

[36] DELERY J. - Experimental investigation of turbulence properties in transonic shock/boundary layer interaction. - AIAA J.,vol 21,n °2,(1983),p.180-185.

[37] COUSTON M., ANGELINI J.J. - Solution of nonsteady two-dimensional transonic small disturbances potentiel flow equation. - ASME Conf., San Francisco (1978) or J. of Fluids Engin., vol. 101, n °3, (1979).

[38] DAVIS S.S., MALCOLM G.N. - Transonic shock wave/boundary layer interactions on an oscillating airfoil. - AIAA J., vol. 18, n °11, (1980).

[39] MARVIN J.G., LEVY L.L., SEEGMILLER H.L. - Turbulence modelling for unsteady transonic flows. - AIAA J., vol. 18, n °5, (1980).

[40] LE BALLEUR J.C., NERON M. - Calculs d'écoulements visqueux décollés sur profils d'ailes par une approche de couplage. - AGARD-CP-291, Paper 11, (1981).

[41] CONSIGNY H., GRAVELLE A., MOLINARO R. - Aerodynamic characteristics of a two-dimensional moving spoiler in subsonic and transonic flow. - AIAA Paper 83-109, (1983).

[42] COSTES M. - Comparaison between experimental and computational results for airfoils equipped with a spoiler and a flap. - AIAA Paper 85-5008 , Colorado (Oct. 1985).

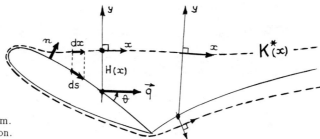

Fig.1 - Coordinate system.
Massive Separation.

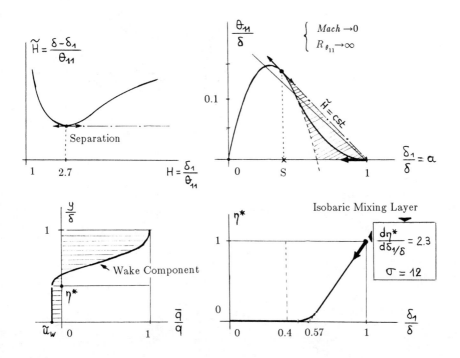

Fig.2 - Integral turbulent closure. Massive separation.

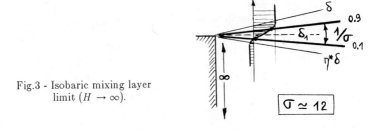

Fig.3 - Isobaric mixing layer
limit $(H \to \infty)$.

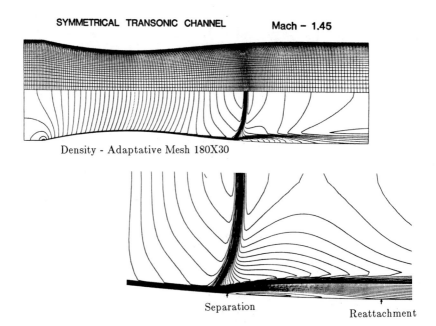

Fig.4 - Transonic shock wave. Boundary layer interaction.
(M =1.45, R_δ = 32900, Equilibrium turbulence model).

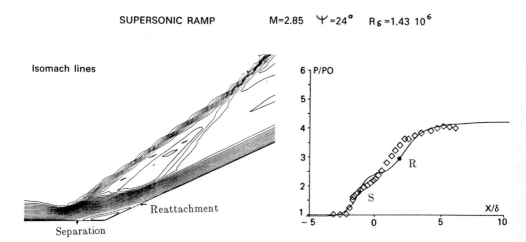

Fig.5 - Supersonic shock wave - boundary layer interaction.
(M =2.85, $\psi = 24°$, $R_\delta = 1.43\ 10^6$, 2 equation model).

DISSYMMETRICAL TRANSONIC CHANNEL   Mach − 1.36

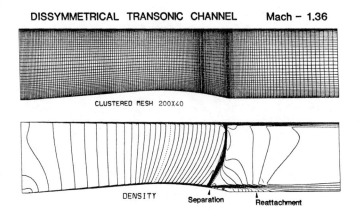

CLUSTERED MESH 200X40

DENSITY   Separation   Reattachment

Velocity Profiles

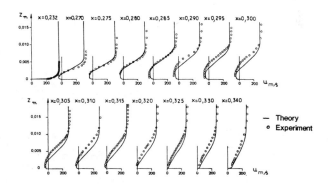

— Theory
o Experiment

M=1.36   SHAPE PARAMETER

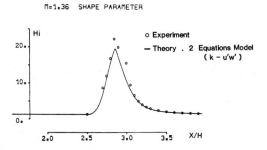

o Experiment
— Theory . 2 Equations Model
(k − u'w')

MAXIMAL SHEAR STRESS EVOLUTION

Versus   SHAPE PARAMETER H_i

$$C_\tau^{\frac{1}{2}} = \left[\frac{2\,\tau_{max}}{\rho_e\,\bar{u}_e^2} \right]^{\frac{1}{2}}$$

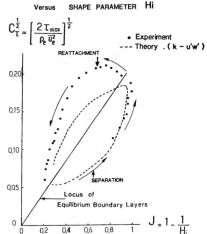

• Experiment
--- Theory . (k − u'w')

REATTACHMENT

Locus of
Equilibrium Boundary Layers

SEPARATION

$$J = 1 - \frac{1}{H_i}$$

Fig.6 - Transonic shock wave - boundary layer interaction.
(M=1.36, $R_\delta = 38400$, 2 equation model).
Mean flow and turbulent shear stress hysteresis.

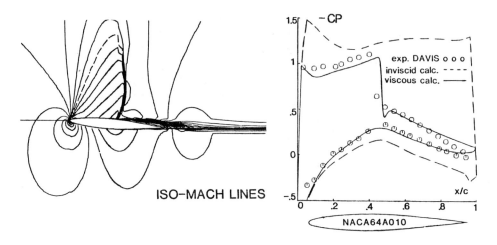

Fig.7 - Steady state solution - NACA64A010 airfoil
(M =0.789, $\alpha_0 = 4^o$, $R_e = 12\ 10^6$, 2 equation model).

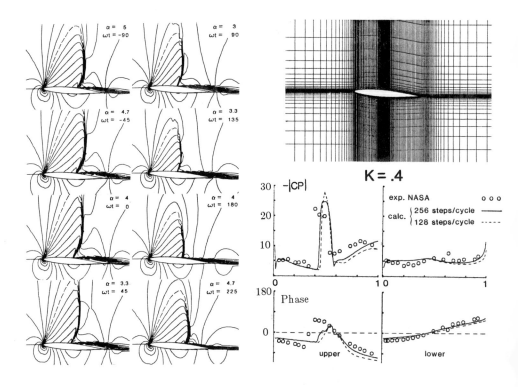

Fig.8 - NACA64A010 airfoil in pitch (M =0.789, $\alpha = 4^o - 1^o \sin\omega t$, $R_e = 12\ 10^6$,
$K = \omega c / u_\infty = 0.40$, $x_\alpha/c = 0.25$, 2 equation model). 180X60 grid.
Instantaneous Mach contours. First harmonic pressure distributions.

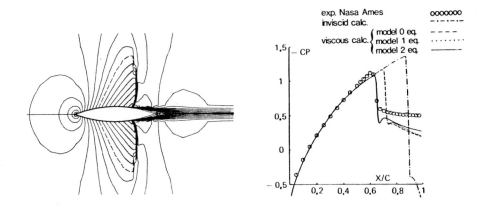

Fig.9 - Steady 18% circular-arc airfoil
$(M = 0.788, \alpha_0 = 0^o, R_e = 11 \ 10^6)$.
Symmetrical separated flow solution.

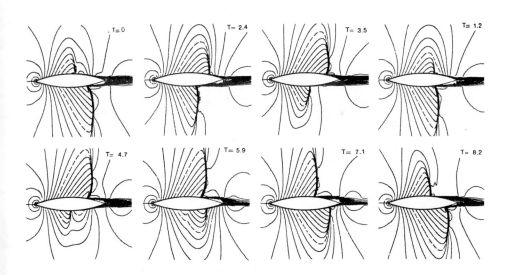

Fig.10 - Steady 18% circular-arc airfoil
$(M = 0.760, \alpha_0 = 0^o, R_e = 11 \ 10^6)$.
Oscillatory solution. Buffeting separation.

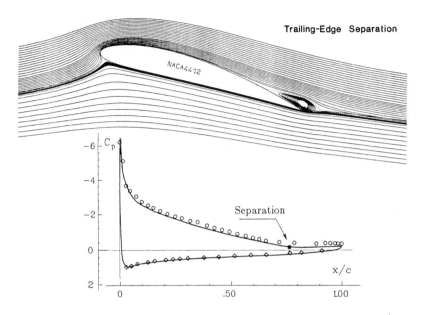

Fig.11 - Incipient stall. "Stalled airfoil" test-case of the 1981-Stanford conference.
(NACA4412,M=0.077, $\alpha = 13.6°$, $R = 1.5 \ 10^6$, equilibrium turbulence model).

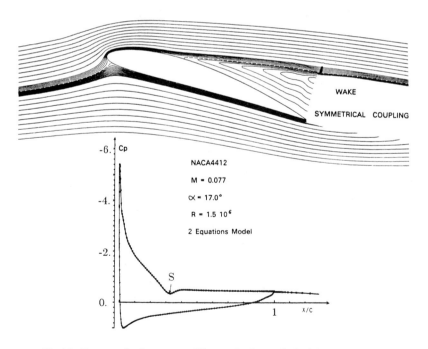

Fig.12 - Deep stall - Incompressible steady flow calculation.
(NACA4412, M=0.077, $\alpha = 17°$, $R = 1.5 \ 10^6$, 2 equation model).

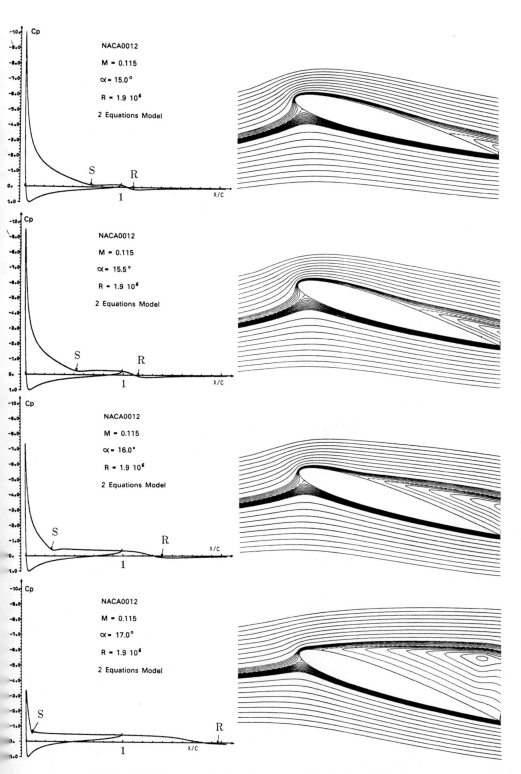

Fig.13 - Stall history prediction - Incompressible steady calculation.

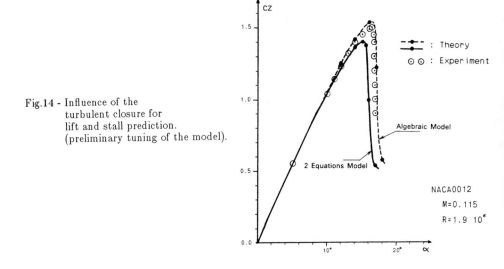

Fig.14 - Influence of the
turbulent closure for
lift and stall prediction.
(preliminary tuning of the model).

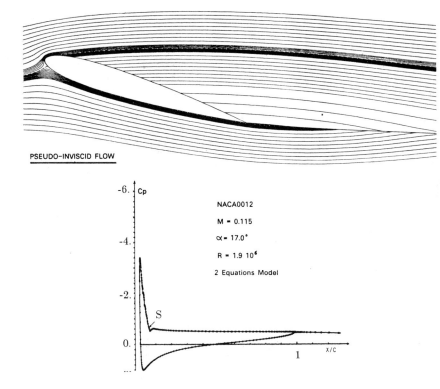

PSEUDO-INVISCID FLOW

Fig.15 - Deep-stall test-case for convergence history (Fig.16).

189

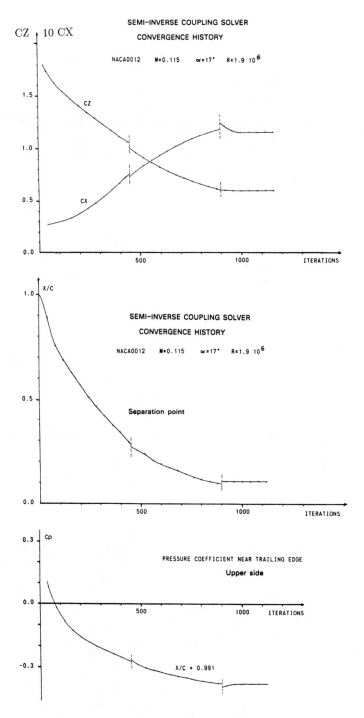

Fig.16 - Convergence history of a difficult test-case.
Explicit Semi-Inverse (Le Balleur). 256 nodes.

SPOILER FLAP

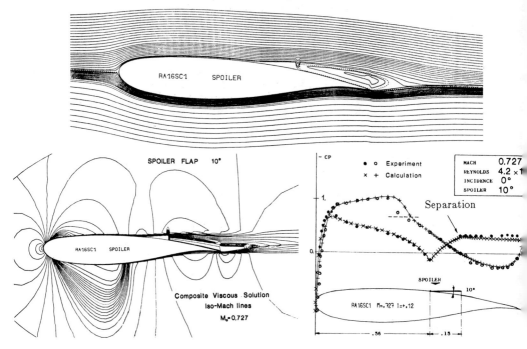

Fig.17 - Massive separation on a transonic airfoil with spoiler flap
(RA16SC1,M=0.727, $\alpha = 0^{o}$, $R = 4.2\ 10^{6}$,$\delta_{SP} = 10^{o}$, 2 equation model).

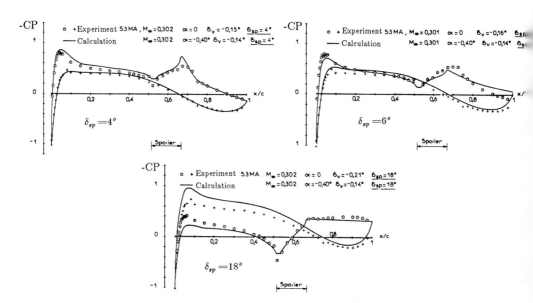

Fig.18 - Typical pressure computations at different spoiler deflexions $\delta_{SP} = 4^{o}$,6° ,18°
(RA16SC1,M=0.30, $\alpha = 0^{o}$, $R = 2\ 10^{6}$,2 equation model, symmetrical wake coupling).

Pseudo-Inviscid flow - Spoiler flap

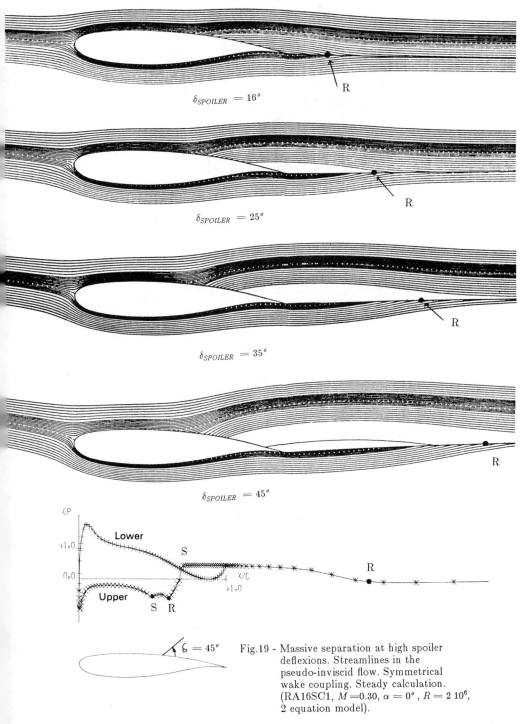

$\delta_{SPOILER} = 16^o$

$\delta_{SPOILER} = 25^o$

$\delta_{SPOILER} = 35^o$

$\delta_{SPOILER} = 45^o$

$\delta = 45^o$

Fig.19 - Massive separation at high spoiler deflexions. Streamlines in the pseudo-inviscid flow. Symmetrical wake coupling. Steady calculation. (RA16SC1, $M = 0.30$, $\alpha = 0^o$, $R = 2\ 10^6$, 2 equation model).

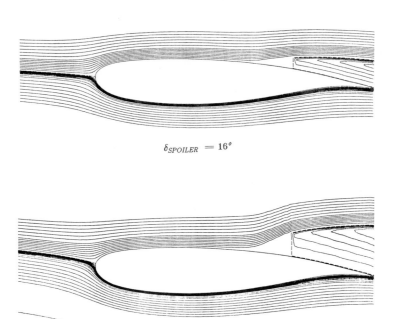

$\delta_{SPOILER} = 16^o$

$\delta_{SPOILER} = 30^o$

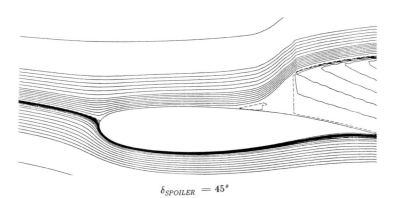

$\delta_{SPOILER} = 45^o$

Fig.20 - Computation of massive separation at high spoiler deflexions.
Viscous streamlines - Steady calculation (RA16SC1 airfoil,
$M = 0.30$, $\alpha = 0^o$, $R = 2\ 10^6$, 2 equation model).

Section IV

Vortex Shedding From Sharp Leading Edges

Simulation Studies of Vortex Dynamics of a Leading Edge Flap

H. K. CHENG, R. H. EDWARDS AND Z. X. JIA

ABSTRACT

The leading-edge flap is studied as an aerodynamic concept for generating and trapping vortices to enhance lift. For an incompressible
plane flow satisfying the Kutta condition at the trailing edges of both
the airfoil and the leading-edge flap, the position and strength of a
single vortex can be found, which is free and stationary, and gives
rise to a lift hysteresis, signifying multiple steady states with high
lift. The nonlinear stability of these bifurcating states are studied
using a separation-vortex model and the more computation-intensive,
point-vortex method. With the separation-vortex model which treats
vortices shed from a salient edge as a single vortex, solutions on the
hysteresis loop prove to be remarkably stable and are recoverable from
the evolution of time-dependent solutions. Preliminary study with the
point-vortex method show reasonable agreement with the separation-
vortex model up to periods comparable with the flow-transit time. In
the longer periods, vortex lift higher than anticipated by the latter
model is possible, depending on methods of generation. Features of
interacting vortex clusters familiar in laboratory and other simulation studies are evident.

1. INTRODUCTION

Nonlinear lift arising from leading-edge eddies is familiar in
works on slender wings with highly swept leading edges or side
edges.[1-6] To retain the full benefit of this lift enhancement, or the
vortex lift, the shed vortices must be kept from drifting away and be
counted as a part of the lift-carrying bound vortices. Similar may be
said about the vortex shedding and the temporal lift enhancement of
oscillating airfoils during "dynamic stall"[7,8] as well as some insects
and birds in certain phases of their hovering and climbing flight.[9-11]

As an aerodynamic design concept, vortex lift is most directly expressed, perhaps, in the Casper wing,[12-15] in which a free vortex,
with the help of a forward- and/or backward-facing flap, is rendered

stationary on top of the wing. Whether or not the resulting flow can be made steady, the major issues in the final analysis must deal in this case with the flow stability and its uniqueness, which cannot be resolved by the solution to a stationary problem. The present work will address these issues in the framework of a separation-vortex model adopted after Refs. (1-3,11) and the more computation-intensive, point-vortex method,[16-22] while exploring ways to capture shed vortices to enhance lift.

A flap designed to trap shed vortices for lift enhancement may be called a vortex flap;[23] the particular example studied below concerns mainly a flap with its hinge located at the leading edge. An unsuspected feature of this leading-edge vortex flap, to be brought out in Section 2, is the lift hysteresis rendered by a single (stationary, free, concentrated) vortex, signifying the existence of multiple steady states. Interestingly, the major solution branches giving the hysteresis loop is found to be remarkably stable on the basis of the separation-vortex model and are recoverable from the time-dependent solution at the end of an evolution process (cf. Section 3). A program of numerical simulations employing existing inviscid and viscous CFD algorithms is initiated to ascertain the significance of the aforementioned results and to study the stability and the long-time behavior of this vortex dominated flow. The results from the point-vortex method (Section 4) represents a preliminary study under this program.

One major aim in the application of point-vortex methods has been the discretized representation of a continuous vortex sheet, although it may also be used to describe a vorticity patch and a shear layer structure of variable intensity. The merits and shortcoming of the approach have been extensively reviewed.[21,22] Interesting details of vortex traces determined by the method exhibit nonlinear instability behavior resembling pairing, merging and the tearing of vortical clusters, blobs or segments observed in the laboratories.[24-26] Little can one conclude, however, on the convergence of the method to the continuous-sheet model; nor can one establish firmly that these intriguing trace behaviors are intrinsic of a certain discrete numerical algorithm. Obviously, the method is inadequate as a genuine approximation whenever neighboring point vortices are stretched too far apart, or whenever vortices belonging to different turns of a spiral are brought too close to one another in a tightly wound sheet. These inadequacies should become most serious at large time and may therefore affect the long-timer flow behavior of interest. [Neither could the existing Euler or Navier-Stokes codes reduce such an uncertainty at the very long time.] Though considerably more refined than the separation-vortex model (which considers only a single (growing) vortex from a salient edge), the point-vortex method used below should be viewed only as an alternative model in numerical experiments.

Unlike the studies of Refs. 16-18, in which the strength of each point vortex and their total number are prescribed, the vortices are generated in the present problem at the sharp trailing edge and at the sharp end of the leading-edge flap, as the time-dependent solution evolves. Essential to the vortex production is the Kutta condition which requires the flow velocity and pressure be finite at the sharp edges. Following the separation-vortex model,[1-4] the Kutta condition and the force-free requirement will be used to determine the strength and location of the nascent vortices in our point-vortex calculation. This procedure of generating vortices appear to differ (considerably) with those used in Refs. 19, 20.

Certain CFD codes based on the Euler equations[27,28] are capable to capture not only shocks but also (free) vortex sheets (as demonstrated in the recent works of Murman and Powell and others reported in this conference proceeding). It is uncertain if this, as well as the Navier-Stokes codes, could offer an approach to the vortex-sheet capturing far more adequately than the point-vortex method, inasmuch as the degree to which the numerical/artificial viscosity affects their solutions is far from being understood and their accuracy in representing an unsteady, two-dimensional contact discontinuity has not been critically tested. (One may recall in this connection the long-standing difficulty of the Euler codes in predicting correctly the induced drag of a high aspect ratio wing, owing to the poor resolution in describing a trailing vortex sheet.[29])

In view of these uncertainties, the Euler and Navier-Stokes codes cannot be considered far superior than the point-vortex algorithm for the present study. Their uses may nevertheless complement the point-vortex and other approaches. In this connection, we observe that all inviscid models fail to allow for the breaking away of a boundary layer from the smooth part of the (wing) surface, which may render the inviscid analysis of the vortex dynamics useless and misleading. In this case, a Navier-Stokes calculation (with appropriate turbulent model) or even an ad hoc boundary-layer analysis, should be helpful.

2. STATIONARY FREE VORTEX AND VORTEX LIFT

Stationary, Free Vortex Over a Flat-Plate Airfoil

Saffman and Sheffield[14] examine the steady 2-D flow about an inclined flat plate, which is dominated by a discrete, stationary, free vortex. Two types of solutions with greatly enhanced lift are found admissible. In one, the separation streamline reattaches at the trailing edge; the flow speed at the leading edge is, however, left unbounded. In the other type, a separation bubble can occur next to the leading edge. Stability analyses show that only the first type is stable;[14] the analyses omit, however, the shedding of vortices during the perturbation, and thus appear to be not self-consistent.

A more complete stability study of the aerodynamic flow in the presence of a discrete free vortex are made in Huang and Chow[31] and more recently in Chow, Huang and Yan[31] where vortex shedding at the trailing edge of a Joukowskii airfoil has been accounted for in a linearized fashion. The equilibrium solution for the free vortex again proves to be unstable in this case. [Prof. C.Y. Chow indicates to us that he succeeded in stabilizing the vortex recently with a combined pitching-heaving airfoil motion.]

Vortex Flaps

Rossow[32] has studied the lift achievable through a vortex trapped between the trailing edge and a raised flap on the upper wing surface. To obtain higher lift, the separation streamline in each example is designed to reattach at the trailing edge. This calls for an extra degree of freedom in the design parameters, and the requirement is met by introducing a sink at the equilibrium position along with the vortex. As a result, an inviscid drag penalty must be paid. However, numerical study of the vortex trajectory with slightly different initial vortex locations indicates that the equilbrium solutions may be unstable.[32] This conclusion may again be questioned, since vortex shedding was not allowed and the Kutta condition not maintained in Rossow's stability analysis.

The most exhaustive study of the vortex flap and related steady-state problems is, perhaps, covered in the more recent works of Saffman and Tanveer[33,34] in which the forward-facing flap with an overhang as in Figs. 1a,b has been the principal configuration considered. The separation streamline is required to leave the leading the trailing edges smoothly, and also to reattach at the leading-edge of the overhang smoothly; a Prandtl-Batchelor[35] type recirculating region with a uniform vorticity is stipulated (cf. Fig. 1b). Other alternatives of the vorticity distributions including a discrete stationary free vortex (cf. Fig. 1a) have also been studied.

Here, the flap angle and the vorticity level ω in Fig. 1b (or the

circulation and equilibrium location of the free vortex in Fig. 1a)
cannot be arbitrarily given but has to be determined for each combination of the incidence, the flap and overhang dimensions from a design
analysis. We observe that recirculating flows with patches of uniform
vorticities at different levels may also exist inside one another
(like onion rings, cf. Fig. 1c). Therefore, the flow represented by
Fig. 1b represents only one of the alternatives consistent with the
Prandtl-Batchelor idea (and with the steady-state Navier-Stokes equations for an infinite Reynolds number). In any event, the realizabili-
ty of the stipulated flow depends on the question of stability, which
has yet to be answered in each case.

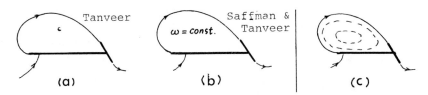

Fig. 1. Illustration of flow models for a vortex flap with a forward-
facing flap: (a) a trapped potential vortex, (b) a version of the
Prandlt-Batchelor model, (c) a more general version of the P-B model.

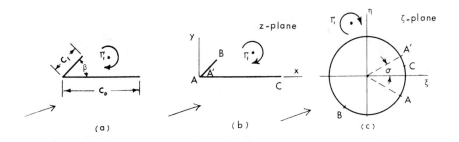

Fig. 2. Leading-edge flap with a single trapped potential vortex: (a)
definitions of β, c_0 and c_1, (b) the vortex flap in z-plane, (c) transformation to the ζ-plane.

As indicated, our study will address mainly the problem of a leading-edge flap. To provide a focus for subsequent investigations, we
consider here the family of solutions dominated by a single stationary
free vortex, as sketched in Fig. 2a. Like an L-shaped barrier (sea
wall) bracing an incident current (from the left), the leeside of the
leading-edge flap should be ideal for harboring eddies shed from the
barrier, intuitively speaking. Results from Rossow's[32] study also

suggest that a higher lift can be achieved with the flap placed closer
to the leading edge. A major difference of the present study from the
previous works on vortex flaps[32-34] lies in the approach to the problem
solving. Namely, we address it as a direct (instead of a design) problem: the aerodynamic quantities of interest are solved as a function
of the angle of attack α, for a fixed airfoil geometry. The latter is
defined in this case by the flap angle β and the flap-to-chord ratio
c_1/c_0. Solution to the direct problem brings out clearly the feature
of hysteresis which signifies the existence of multiple steady states
which does not appear to have been recognized in previous studies. The
following will discuss in detail an example of the leading-edge vortex
flap.

Leading-Edge Vortex Flap: Example

The complex potential W of the incompressible irrotational plane
flow can be readily written down in the transformed ζ-plane where the
L-shaped barrier in the original z-plane is mapped to a unit circle
(cf. Figs. 2b,c). The transformation needed, which also leaves the
flow at infinity unchanged, is

$$z = \zeta^{-1}(\zeta-e^{-i\sigma})^{\beta/\pi}(\zeta-e^{i\sigma})^{2-\beta/\pi} \qquad (2.1)$$

where the constant σ controls the flap-to-chord ratio c_1/c_0 and thus is
a function of c_1/c_0 and σ; z and ζ are normalized by the circle radius a.

Apart from requiring the vortex to be stationary and free, the
Kutta condition will be enforced at the sharp edges B and C in Figs.
2b. To minimize the analytical complexity, the Kutta condition is not
enforced at the leading edge A. This may not lead to a serious problem
in application and may in fact be mitigated with a small nose radius,
as practiced in airfoil design. We shall return to this point later.
In the ζ-plane, the aforementioned constraints are

$$\lim_{\zeta \to \zeta_1} \left[\frac{dW}{d\zeta} - \frac{i\Gamma_1}{2\pi} \frac{1}{\zeta-\zeta_1} - \frac{i\Gamma_1}{4\pi} \left(\frac{z''}{z'}\right)_1 \right] = 0 \qquad (2.2a)$$

$$\left|\frac{dW}{d\zeta}\right| = 0 \quad \text{at } \zeta = \zeta_B, \zeta_C \qquad (2.2b,c)$$

where the subscripts 1, B and C refers to the vortex, the edge B and
edge C, respectively. These conditions suffice to determine the
strength Γ_1 and the location ζ_1 of the stationary vortex, hence the
complex potential W. In passing, we point out that the circulation
Γ_1 so determined could take on complex values, and therefore the
existence of a stationary free vortex cannot be assured a priori.

Lift Hysteresis

Figure 3 presents the lift coefficient C_L as a function of the incidence α for a flap angle $\beta = 45°$ and a flap-to-chord ratio $c_1/c_0 = 1/3$. The C_L curves are neither monotone nor single-valued, but three adjoining monotone segments may separately be identified. There is a lower C_L branch with positive slope extending up to $\alpha \approx 8.25°$ and a higher C_L branch also with $dC_L/d\alpha > 0$ extending from $\alpha = 6.18°$; connecting the upper and lower branches is a third branch with negative slope (shown in dashes). Within $6.18° < \alpha < 8.25°$, three values of C_L corresponding to three alternative steady states are admissible. Accordingly, continuous C_L increases with α along the lower branch must be terminated and followed by a jump to a high C_L value of 6.5 belonging to the upper branch (if α is to increase beyond $8.25°$); whereas, a continuous C_L decrease with decreasing α along the upper branch must be followed by an abrupt drop to $C_L \approx 1.7$ of the lower branch at $\alpha = 6.18°$ (if α is to reduce to a value lower than $6.18°$). An anticlockwise hysteresis loop may thus be identified in the C_L-α domain (see arrows in Fig. 3). The open circles plotted next to the C_L curve in Fig. 3 are data recovered from time-dependent solutions to be explained and discussed in Section 3.

The vortex strength Γ_1 responsible for the lift shown in Fig. 3 is considerably larger than the total circulation Γ_0 around the airfoil and the trapped vortex, which is $Uc_0C_L/2$ and must also be equal to the total circulation of the starting vortex left behind. This indicates that the bounded vorticity on the wing in this case contributes only to a negative circulation, hence, being counter-productive. The normalized value of Γ_1 and Γ_0 (based on Uc_0) for $\beta = 45°$, $c_1/c_0 = 1/3$ are shown as functions of α in Fig. 4.

Reattachment and Break-Away

More revealing features are the stagnation-point locations on the wing surfaces. The stagnation location occurring on the upper surface must be a reattachment point of the streamline defining the boundary of the recirculation region; its distance from the leading edge (in fractional chord) is denoted by x_s (cf. sketch on the left side of Fig. 4). The first stagnation location on the lower surface downstream of the leading edge is the entry point for the surface streamline, its distance from the apex is designated as x_s'. Another stagnation point at a distance x_s'' from the apex may occur on the lower side, and will be recognized as a break-away location for the separation streamline leaving the flat plate (cf. sketch on the right side

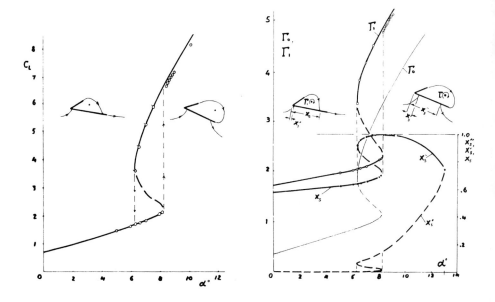

Fig. 3. Lift coefficient as function of incidence illustrated for $\beta = 45°$, $c_1/c_0 = 1/3$.

Fig. 4. Normalized equilibrium vortex strength Γ_1, total circulation Γ_0 and stagnation locations for $\beta = 45°$, $c_1/c_0 = 1/3$.

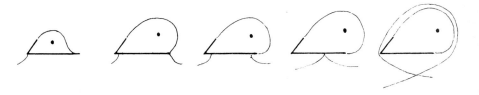

Fig. 5. Illustration of streamlines wetting the wing surface at different incidences.

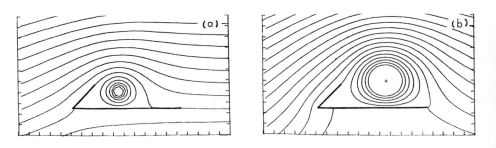

Fig. 6. Streamline pattern with a stationary, free vortex for $\beta = 45°$, $c_1/c_0 = 1/3$: (a) $\alpha = 8.2°$ (lower branch), (b) $\alpha = 8.27°$.

of Fig. 4). The results of x_s, x_s' and x_s'' as a function of α are also presented in Fig. 4, where the x_s curve featuring a hysteresis loop is seen to terminate with $x_s = 1$ at $\alpha = 8.15°$, and beyond this critical angle, the x_s''- curve takes over. Together with the C_L and Γ_1 behavior, these results show that the highest Γ_1 and C_L values at the end of the hysteresis loop is achieved when the reattachment location reaches the trailing edge. Further increase of C_L and Γ_1 beyond the second critical α (8.25°) is to be accompanied by the occurrence of a break-away point on the lower side. This implies that when Γ_1 is larger than the critical value, the recirculation region is so large that the reattaching streamline must reach the airfoil at the trailing edge from behind. [The latter thus acts as a leading edge (with a smooth entry) to the local flow.]

A helpful feature in the x_s' curve is its extremely small values at moderate α, meaning that the problem of flow separation at the apex may be less critical. [For the lower solution branch, the apex singularity in surface speed is hardly noticeable.] The curves x_s' and x_s'' meets at $\alpha \approx 13°$, beyond which the stagnation point leaves off the lower surface and the entire wing is then surrounded by a clockwise recirculating flow.

The flow patterns corresponding to the different ranges of x_s, x_s' and x_s'' are illustrated schematically in Fig. 5a-e. Two contrasting streamline patterns computed for $\beta = 45°$, $c_1/c_0 = 1/3$ near the (second) critical incidence (8.25°) are shown in Figs. 6a,b. The one in Fig. 6a is computed for $\alpha = 8.20°$ from the lower branch and that in Fig. 6b for $\alpha = 8.27°$. [As pointed out to me by Prof. H. Sobiesky, the angles formed by the flat plate and the reattaching and separating streamlines must all be equal to 120° at the second critical angle 8.25°.] The cusp-shaped reattachment shown in Fig. 6a appears to agree with the analysis of a closure model which stipulates a uniform eddy pressure.[36,37]

Boundary-Layer Separation as a Limitation

Apart from the hydrodynamic instability and the 3-D influence, the usefulness of the analysis is limited by boundary-layer separations which may occur at locations other than the two trailing edges, as stipulated by the present model. For example, separation will occur on the top wing surface (under the flap), on the bottom surface upstream of the trailing edge (at α beyond the second critical incidence) or at the apex. As noted earlier, a slight modification of the apex geometry with a finite nose radius may avert separation, which would be quite sizeable however for α beyond the second critical (cf. sketch in Figs. 7a,b). [Rounding the apex should also help to recover the leading-edge suction, hence, to reduce drag.] Apparently some work of boundary-layer control on the under wing surface is required, in order to achieve the higher C_L branch beyond the (second) critical α. Encouragingly, the break-away location x_s'' is found to be stationary with respect to α at the second critical α where $x_s'' = 1$ (see Fig. 4). Therefore, the length $(1-x_s'')$ is small for a moderate lift increase. This, together with the fact that the flow between the trailing edge and the break-away point is nearly stagnant for small $(1-x_s'')$ means that, power expenditure to keep the boundary layer from separating near the trailing edge at α beyond the second critical may not be excessive. A more serious flow separation in the supercritical case may result from the adverse pressure gradient associated with the strong surface velocity on the top surface induced by the trapped eddy. Figure 8 illustrates such a surface pressure distribution for the same flap design at $\alpha = 8.5°$. The need of analyzing and controlling the inviscid-viscous interaction in this aspect is apparent.

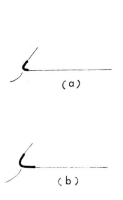

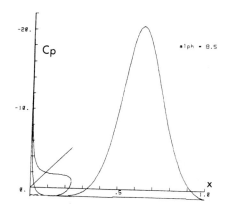

Fig. 7. Rounding off the apex to achieve the attached boundary-layer flow: (a) small incidence, (b) moderate incidence.

Fig. 8. Surface pressure for $\beta = 45°$, $c_1/c_0 = 1/3$ at $\alpha = 8.5°$.

3. APPROACH TO THE STEADY STATES: SEPARATION-VORTEX MODEL

The Separation Vortex Model

It is essential to examine if the steady state solutions are recoverable from a time-dependent analysis. In the following, this analysis will be made on the basis of the separation-vortex model which have
proven useful in the study of the leading-edge eddies.[1-4,11]
Early analyses of vortex separation from slender delta wings using
a discrete vortex pair encountered various problems with the potential
jump on the vortex-feeder sheet. (See Küchermann's book.[6])   The de-
finitive work which resolved these difficulties was given by R.H.
Edwards[1] who recognized that, as a physically sound representation of
the rolled-up free vortex sheet, the force on the discrete vortex
should not vanish but the sum of the forces on the vortex and on the
feeder sheet must be made zero. Edwards obtained results of the lift
and vortex position, simplifying the nonlinear effects as perturbations controlled by the angle of attack. With this version of the
rolled-up eddies, Brown and Michael[2] carried out a very thorough study
of the slender flat-plate conical wing; their much referred work also
includes a comparison with experiments. Meanwhile Cheng[3] adopted the
treatment to a time-dependent problem of a flat plate in transverse
motion, which is equivalent to a rectangular wing of low aspect ratio
at angle of attack; for very low aspect ratios, the predicted nonlinear
lift compares well with Winter's measurement. Independently, Rott[4]
made an extensive study of the evolution of the time-dependent eddy
behind a salient edge, using the same model; the loci of the vortex
centers are seen to compare consistently well with Schlieren photo
records of the rolled-up eddies in shock-tube experiments. Among subsequent works making use of this model for a rolled-up eddy is the more
recent study of Edwards and Cheng,[11] in which the separation-vortex
effect on an insect lift generation mechanism proposed by Weis-Fogh[38]
is studied; there, comparison with quantities deduced from photo record
by Maxworthy[10] again appears to be encouraging.

In passing, we observe that, in another well referred works on
slender delta wings by Mangler and Smith[5], where the
free vortex sheet is more completely treated, a simplifying description
near the center of the vortex spiral is used and is very similar to
that in Refs. (1-4,11). The model for the rolled-up eddy of Refs.
(1-4,11) will be referred to as the separation-vortex model.

Application to a Vortex Flap Problem

The time-dependent version of the separation vortex model is used below to study the transient behavior of the two systems of shed vortices associated with the edges B and C. The two discrete vortices of strengths Γ_1 and Γ_2 in the model, and their locations in the z and ζ-planes are sketched in Figs. 9a, 9b. In terms of the variable ζ, the complex velocity potential $W = \Phi + i\Psi$ in this case, normalized by Ua, is

$$W = e^{-i\alpha}(\zeta + \frac{1}{\zeta}) + \frac{i\Gamma_1}{2\pi} \ln\left(\frac{\zeta-\zeta_1}{\zeta-1/\bar{\zeta}_1}\right) + \frac{i\Gamma_2}{2\pi} \ln\left(\frac{\zeta-\zeta_2}{\zeta-1/\bar{\zeta}_2}\right) \tag{3.1}$$

where the subscripts "1" and "2" refer to the first and the second vortices of strength Γ_1 and Γ_2, respectively, and a bar signifies the complex conjugate. Essential are the cuts defining the angular ranges for the arguments under the four logarithms in (3.1), which are chosen to connect the vortices and their images to B and C, providing a feeder sheet for each vortex in the z-plane (cf. Figs. 9a,b).

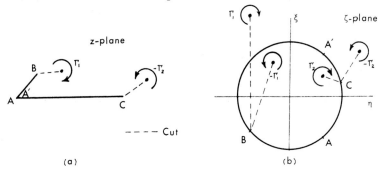

Fig. 9. Separation-vortex model and the cuts: (a) z-plane, (b) ζ-plane.

The force balance in the separation-vortex model, applied to the first vortex, yields

$$(z_1 - z_B)\frac{d\Gamma_1}{dt} = \Gamma_1(\bar{V}_1 - \frac{dz_1}{dt}) \tag{3.2a}$$

where $\bar{V}$ is the complex vortex velocity

$$V_1 \equiv \lim_{z \to z_1}\left[\frac{dW}{dz} - \frac{i}{2\pi}\frac{\Gamma_1}{z-z_1}\right]$$

and can be evaluated in the ζ-plane as

$$\bar{V}_1 = \left\{\lim_{\zeta \to \zeta_1}\left[\frac{dW}{d\zeta} - \frac{i\Gamma_1}{2\pi(\zeta-\zeta_1)}\right] - \frac{i\Gamma_1}{4\pi}\left[\frac{\zeta''}{(\zeta')^2}\right]_1\right\} \cdot (\zeta')_1 \tag{3.2b}$$

The same force-balance condition applies to the second vortex with "2" replacing "1" in the subscript of Γ_1 and ζ_1 in (3.2) above. The assumption of a bound velocity at the edges B and C (Kutta condition) calls for

$$\left|\frac{dW}{d\zeta}\right| = 0 \quad \text{at} \quad \zeta = \zeta_B, \ \zeta_C \qquad (3.3a, 3.3b)$$

These, together with the real and imaginary parts of the force-balance
requirement for the first and second vortices, suffice for the deter-
mination of the time rate of change for Γ_1, Γ_2, ζ_1 and ζ_2. The entire
history of the model flow from an initial data set may therefore be
analyzed. The pressure coefficient can be computed from

$$C_p \equiv \frac{p-p_\infty}{\frac{1}{2}\rho U^2} = 1 - \frac{2}{U^2}\left[\Phi_t + \frac{1}{2}\left|dW/dz\right|^2\right] \qquad (3.4)$$

The time t is normalized with a/U, except when indicated otherwise.

Recovering the Two Major C_L Branches

With this model, we solve (3.2) and (3.3) for the indicial (impulsively started) motion of an airfoil considered earlier in Section 2,
Figs. 3 and 4 ($\beta = 45°$, $c_1/c_0 = 1/3$), with $\Gamma_1 = \Gamma_2 = \zeta-\zeta_1 = \zeta-\zeta_2 = 0$
as initial data. For each angle of attack α, the indicial solution to
(3.2) and (3.3) approaches a single limit as $t \to \infty$ corresponding to
only one steady state. The limits inferred from the large-time solutions to (3.2) and (3.3) are shown as open circles in Figs. 3 and 4,
which compare closely with the equilibrium solutions on the lower
branches for α less than and up to $8.25°$; for $\alpha>8.25°$, the open circles all move up to the neighborhood of the upper branch. The segment
of the upper branch belonging to the hysteresis loop is not recoverable directly from the indicial solutions, they are nevertheless stable
and can be reached through other evolution processes. Those open circles marked with slashes on this segment are data recovered from the
time-dependent solutions by reducing, in succession, the angle of
attack from $\alpha = 8.5°$. We have experimented also with initial data
taken from the steady-state solutions on the third branch (dashes in
Figs. 3 and 4) and find the latter solutions unstable.

An Examination of the Indicial Response

It is instructive to examine closely the evolution behavior of the
indicial solutions themselves. Figures 10a,b,c, illustrate how Γ_1 and
Γ_2 change with time t for the configuration ($\beta = 45°$, $c_1/c_0 = 1/3$)
under study at three different incidences, $\alpha = 8.2°$, $8.5°$ and $12°$,
where, to be sure, t has been normalized by c_0/U, i.e., $3.65a/U$.
The Γ_1 and Γ_2 behave very similarly among solutions belonging to the
lower branch, of which the results shown in Fig. 10a for $8.20°$ may be
considered representative. Characteristically, the lower-branch solution for Γ_1 or Γ_2 features an early rapid rise with an overshoot,

returning to a level slightly below the equilibrium value, all taking place during the period $0 \leqslant t \leqslant 25$.

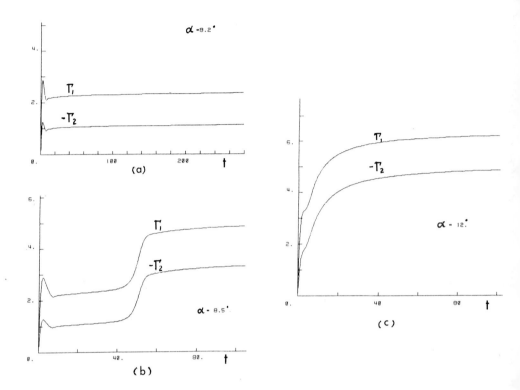

Fig. 10. Vortex strengths as function of time t (scaled by c_0/U) computed from separation-vortex model for $\beta = 45°$, $c_1/c_0 = 1/3$: (a) $\alpha = 8.2°$, (b) $= 8.5°$, (c) $= \alpha = 12°$. Note that the critical incidence for this geometry is $\alpha = 8.25°$.

While in Fig. 10b, the incidence $\alpha = 8.50°$ exceeds only <u>slightly</u> $8.20°$ of Fig. 10a, the drastic differences between Fig. 10a and 10b are obvious. Interestingly, the graphs of Γ_1 and Γ_2 in Fig. 10b can hardly be distinguised from the corresponding graphs in Fig. 10a until $t > 100$. Significant rises in Γ_1 and Γ_2 occur at about $t = 180$, reaching levels not far from their equilibrium levels on the higher branch at $t \gtrsim 250$. Apparently, the fluid dynamics during the earlier period ($t < 100$) cannot distinguish the slight difference between $\alpha = 8.50°$ and $\alpha = 8.20°$, and evolves along a solution trajectory towards the equilibrium point on the lower branch. A rapid rise to the higher branch must eventually take place, since only a single steady state on the higher branch is admissible beyond the (second) critical $\alpha(8.25°)$.

By contrast, the development for α exceeding $10°$ is generally very smooth and mono-tone, as is represented by the results for $\alpha = 12°$ shown in Fig. 10c. Values of Γ_1 and Γ_2 in the (supercritical) domain are seen to approach the equilibrium values on the higher branch much earlier than in the nearly critical case.

In passing, we point out that in contrast to the slow evolution of the indicial solution, the approach to equilibrium for the circles marked with slashes in Figs. 3 and 4 from the neighboring solutions is relatively fast, taking only 10 to 20 of the flow transit time.

The sharp reduction Γ_1 and in Γ_2 following an early overshoot in Figs. 10a and 10b signifies the temporary shedding of vortices with signs opposite to the prevailing ones. Questions may be raised as to whether the separation-vortex model at hand can be improved with the introduction of an additional vortex pair whenever $d\Gamma_1/dt$ and $d\Gamma_2/dt$ vanish. The questions are resolved (at least in part) in the subse-quent study with the point-vortex method, where large numbers of dis-crete vortices will be introduced. The stability found with the equi-librium solution belonging to the lower and the higher branches may be considered remarkable, since the approach to a steady-state cannot be assured by the time-dependent equations (3.2) a priori.

4. STUDY WITH THE POINT-VORTEX METHOD

Simulation study using the point-vortex method should provide dy-namical details of the evolving vortical structure far more complete than the separation vortex model does, and could be considered as a discretized Lagrangian equivalence of an Euler code. Two examples are studied below. The first example considers the evolution of the vor-tical field behind the leading-edge flap and in the wake from a flow which is (initially) dominated by a single, stationary, free vortex (as in Section 2). In the second example, we re-examine the indicial (impulsively-started) problem studied earlier in Section 3.

The studies presented below have been limited to flow developments in relatively early stages, owing to the limited computer resources (provided by a VAX/VMS shared with 200 potential users); more complete and conclusive study for t far beyond ten (10) is being made on a larger computer. Good agreement with the separation-vortex results (of Section 3) for the indicial problem is consistently found in t = 0(1). While computations carried out to t ~ 10 with two thousand vortices have furnished many more details, certain large-scale transient features found may still be explained by the separation-vortex solutions.

Only a primitive version of the point-vortex method (implemented by a new method to produce nascent vortices) has been used. Improvements via introducing vortex-core functions or cut-off,[16,19a] pivotal-point relocation[18a] amalgamation,[16] vortex-in-cell[18b] and other approaches have yet to be examined. In spite of the rudimentary nature of the program, surprisingly few instances of spurious vortex motion are detected (owing perhaps to the very small time step and large number of vortices used). Thus, pairing, merging and other familiar behaviors of vortex clusters[24-26] can be readily identified. Point vortices among a group of the same polarity (sign) are seen to behave extremely well, even in a tightly packed cluster.

The Nascent Vortices

In the application to a mixing-layer problem, the point-vortex method approximates the continuous shedding from a sharp edge by the release of discrete free vortices at successive time intervals.[17,19b,20] The procedure of determining the strength for each nascent (new-born) vortex varies among different works. Physically, it must comply with the Kutta condition which implies not only the boundedness of velocities and pressure at a sharp edge, but that the rate of vortex shedding is controlled by a convection velocity; the latter is the arithmatical mean of the velocities on the two sides of the edge.

In the present work, the strength and location of each of the nascent vortex pair are determined from the separation-vortex model. Denote the circulation and the z- coordinate of the jth vortex generated from the edge B and from edge C by

$$\Gamma_{1j}(t), \quad z_{1j}(t) \quad \text{and} \quad \Gamma_{2j}(t), \quad z_{2j}(t)$$

respectively. For simplicity, we shall take the interval between new births $\Delta_1 t$ to be a constant. During the period $n\Delta_1 t < t < (n+1)\Delta_1 t$, there will be a total of 2n vortices in the field (n from each edge). Refer to Fig. 11.

Fig. 11. Nascent and free vortices in the point-vortex method. The strength and movement of the nascent vortices are determined by separation-vortex model.

Within this period

$$\Gamma_{1n}(t), \quad z_{1n}(t) \quad \text{and} \quad \Gamma_{2n}(t), \quad z_{2n}(t)$$

of the growing nascent vortex pair must accordingly be determined by integrating forward in t the equivalent of (3.2) and (3.3) from t = $n\Delta_1 t$ when $\Gamma_{1n} = \Gamma_{2n} = z_{1n} = z_{2n} = 0$; in this application, the $\bar{V}$ in (3.2b) should of course be written as $\bar{V}_{1n}$ and $\bar{V}_{2n}$. For each of the other vortices (j≠n), the strength is frozen at the value when it was first set free, i.e.

$$\Gamma_{1j} = \Gamma_{1j}(j\Delta_1 t), \quad \Gamma_{2j} = \Gamma_{2j}(j\Delta_1 t) \quad . \tag{4.1}$$

The motion is determined by integrating

$$\zeta'_{1j} \frac{d}{dt} z_{ij} = \bar{V}_{ij}, \quad \zeta'_{21} \frac{d}{dt} z_{2j} = \bar{V}_{2j} \tag{4.2}$$

The vortex velocities

$$V_{ij} \quad \text{and} \quad V_{2j} \, ,$$

including those of the nascent vortices (j=n), are understood to be evaluated from (3.2b), with the subscript "1" therein replaced by "1j" and "2j", and the complex potential computed from

$$W = e^{-i\alpha}(\zeta + \frac{1}{\zeta}) + \frac{i}{2\pi} \sum_{j=1}^{n} [\Gamma_{ij} \ln (\frac{\zeta - \zeta_{1j}}{\zeta - 1/\bar{\zeta}_{1j}}) + \Gamma_{2j} \ln (\frac{\zeta - \zeta_{2j}}{\zeta - 1/\zeta_{2j}})] \tag{4.3}$$

Here, the need for appropriate cuts connecting the nascent vortices and their images to the edges B and C as in Section 2 is understood.

For the numerical integration in t, a first-order (Euler) scheme with a step Δt small compared to $\Delta_1 t$ is used; a criterion for choosing Δt based on increment of Γ_{1n} has been used, and, for most calculations, the ratio $(\Delta_1 t/\Delta t)$ is typically 5 to 10. The use of the separation-vortex model allows a nascent vortex to grow to a higher strength before being set free, and this, in principle, should reduce the total number of vortices, thus, computer work and storage required. More specifically, we note that computation work for each t step according the procedure outlined is proportional to n^2, and the total work up to $t = \Delta_1 t \cdot n$ is

$$\frac{\Delta_1 t}{\Delta t} \cdot n^3$$

If other procedures of determining the nascent-vortex strength, e.g.

Refs. 17,20a,b were used, the resulting time integration up to t,
would call for a larger number of vortices, say n' so that

$$n = \frac{\Delta t}{\Delta_1 t} \cdot n'$$

Therefore the foregoing estimate may become

$$\frac{\Delta_1 t}{\Delta t} \cdot n^3 = \left(\frac{\Delta t}{\Delta_1 t}\right)^2 \cdot (n')^3 \ll (n')^3 \quad,$$

suggesting a factor of $(\Delta t/\Delta_1 t)^2$ or 100 reduction of the work.

Stationary, Free Vortex as Initial Data

We shall investigate more critically the vortical-field development
by the procedure just described, using results of the stationary, free
vortex (of Section 2) as initial data. The hysteresis loops (Figs. 3,
4) and their stability according to the separation-vortex model suggest
that four cases should be considered, which correspond to four different sets of equilibrium free-vortex solutions as initial condition.
These four data sets represent the single free vortex in a (totally)
subcritical, a (totally) supercritical and the multiple steady-state
domains. The particular case to be examined below is $\beta = 45°$, $c_1/c_0 =$
1/3 at incidence $\alpha = 7°$, falling into the range $6.18°<\alpha<8.25°$ of the
multiple states. The initial data used for this particular calculation
was chosen from the lower branch (cf. Figs. 3,4). In the context of
point-vortex method, a strict stability of the potential-vortex solution appears to be very difficult to achieve, as long as vortex shedding is allowed at the sharp edges. It would be interesting to see how
far and fast the aerodynamic flow would depart from the equilibrium
solution, and whether the unsteady lift developed is higher or lower
than the equilibrium values.

Figure 12 shows results of the computation using $\Delta_1 t = 0.01$ for the
instantaneous positions of the point vortices and the original concentrated vortex sequentially at t = 0, 1.25, 2.50, 3.75, 5.0 and 9.87,
together with the surface distribution of the pressure coefficients at
the last stage. In terms of the flow-transit time c_0/U, the t sequence
reads 0, 0.342, 0.685, 1.027, 1.370, and 2.704, which are also included
parenthetically. The dominance of the concentrated potential vortex
makes the flow pattern quite smooth; representation of a continuous
vortex sheet by the point vortices appears to be adequate until t = 3.75
when a local instability of the sheet in the form of a (downstream-
pointing) finger occurs. Shortly thereafter, the regular recirculation
pattern around the original vortex resumes, while the vortices that

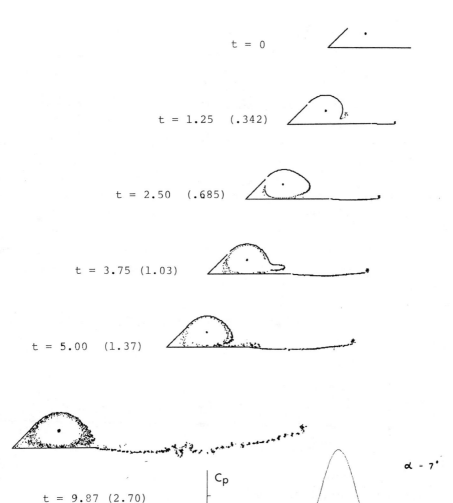

t = 0

t = 1.25 (.342)

t = 2.50 (.685)

t = 3.75 (1.03)

t = 5.00 (1.37)

t = 9.87 (2.70)

Fig. 12. Instantaneous vortex positions showing evolution from a field initially dominated by a single, stationary, free vortex, computed by the point-vortex method. $\beta = 45°$, $c_1/c_0 = 1/3$ and $\alpha = 7°$. The surface pressure at t = 9.87 is also shown. The time has been normalized by a/U, i.e., $c_0/3.65U$.

broke away from the recirculating path manage to join those shed from the trailing edge (see pattern at t = 5). Apparently, this break-up is the only occurrence of a large-scale instability observed during the entire period t = 0-9.87. In fact, the broken-away vortices (apparent-ly of the same polarity) remains in a cluster, becoming more and more distinct as a group among the array of the wake vortices as t increases. With only limited loss of vorticity to the wake, the recirculation re-gion identified by the ring-like vortex trace over the wing continues to grow, its downstream ends almost reach the trailing edge at t = 9.87, where the distinct figure of a slanted "S" may be noted amid the verti-cal wake, and traced back to the temporal break-up at t = 3.75.

The total circulations of vortices shed from edges B and C at t = 9.87 (2.704) are Γ_1 = 11.89 and Γ_2 =-5.43 which are 3.27 and -1.49, respectively, if normalized by Uc_0 (instead of Ua). Comparing these to the equilibrium single-vortex solution (Figs. 4), they are about mid-way between the upper- and lower-branch values. Aerodynamically, the surface-pressure coefficient giving the (inviscid) lift and drag co-efficients

$$C_L = 2.836, \qquad C_D = 0.0834$$

at t = 9.87 (2.704) is encouraging; future calculation beyond t = 10 should indicate when a maximum C_L will be reached and whether signifi-cant deduction in C_L, characteristic of an ordinary dynamic stall,[7,8] may occur. The value 2.836 is 40% higher than the lower-branch equi-librium C_L value (cf. Fig. 3).

From the viewpoint of point-vortex method, the rather smooth pres-sure distribution calculated from the central-difference formula pro-vides another sign of encouragement, indicating the absence of obser-vable spurious effects even near an impermeable surface. Notice that the pressure on the top and bottom surfaces are continuous at the two edges. The C_p value near the hinge line on the top surface is near zero (not unity), implying a poor pressure recovery owing to the transient nature of the large-scale flow. The peaky, yet smooth dis-tribution of the pressure on the top surface is very similar to that for an equilibrium single vortex shown in Fig. 8 for a larger incidence (α = 8.5°). Of interest also is a noticeable net force with upward and forward components on the flap.

The Impulsively Started Flow Problem

According to the result of the separation-vortex model, steady states of the higher C_L branch can be recovered from the indicial (impulsively started) solutions, provided the incidence α is higher than some critical value. For the particular configuration considered (β = 45°, c_1/c_0 = 1/3), this critical value is 8.25°, (cf. open circles in Fig. 3). The angle α in the specific case studied below is taken at α = 8.50 , which would promise a $C_L \sim 7$ according to Fig. 3. This choice overlooks, however, the excessively long rise time for Γ_1 and Γ_2 according to the separation-vortex solution, owing to the evolutionary behavior peculiar to α being close to 8.25° (cf. Fig. 10b). A meaningful comparison with the equilibrium single-vortex solution of Section 2 would require the computation to continue to a time t (based on c_0/U)

beyond 40. This is beyond our current computer resources unless the accuracy requirement could be relaxed. Nevertheless, results obtained thus far for t (based on a/U = $c_0/3.65U$) up to 9.12 provide many revealing features of the transient vortical structures, their comparison with the simpler model (of Section 3) as well as the point-vortex algorithm itself.

The program for this set of computations uses a $\Delta_1 t$ = 0.01 for the time interval between the new births. At an early stage up to t $\leq$ 1 (not shown), the instantaneous patterns of vortex positions are orderly

in a sense that a simulated vortex sheet of spiral shape, with sinuous roll-up patterns on a smaller scale, can be easily traced out. The centroid of all the vortices shed from the leading-edge flap and that from the trailing-edge agree well, in fact, with the prediction of the separation-vortex model even up to a time as large as t = 4. After t $\geq$ 1 the interaction among vortices of the local group turns most of the small-scale sinuous features into distinct clusters, and girations between some of the cluster pairs begin to be noticeable. Figure 13 shows vortex-location patterns during the period t = 3.63-5.62 in this case, during which a segment along the outer rim of vortex spiral from the leading-edge flap begins to break apart; an extended spiral arm is formed at t = 4.25 and tears away completely after t = 5.38. Note that up to t = 4.02, most vortices shed from the flap do not stray far from the wing (the arrow indicates the location of a vanguard of the run-away vortices from this group). Hence the agreement of the centroid of this group with that predicted by the separation-vortex model mentioned earlier is relevant up to t = 4.02. As is apparent, pairing girations of vortex clusters result mostly in their merging (cf. clusters 2,3,5 and 6 in the figure). There is evidently a continued self-growth of each cluster, whether paired or not, resulting from aggregation of the neighboring individual vortices of the <u>same</u>

216

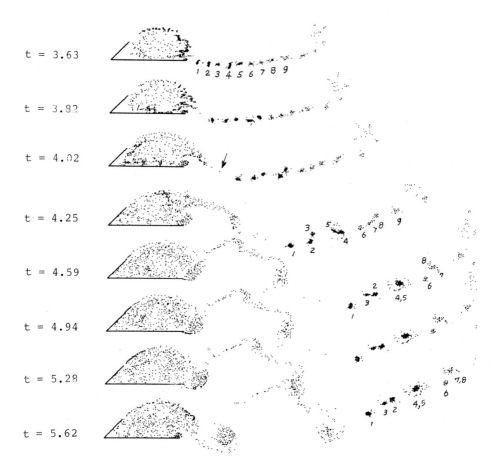

Fig. 13. Vortex-position patterns at successive stages after an impulsive start for $\beta = 45°$, $c_1/c_0 = 1/3$ and $\alpha = 8.5°$, computed by the point-vortex method. Here the time has been normalized by $c_0/3.65U$.

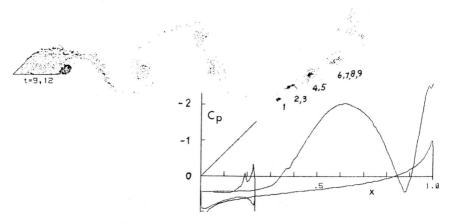

Fig. 14. Vortex-position pattern and surface pressure at $t = 9.12$. Other conditions are the same as in the preceding figure.

polarity. This appears to support the observation by Inoue.[39]

An important feature of this and subsequent periods associated with the spiral arm is a trailing-edge eddy which grows steadily in size (and population), and finally breaks loose at t ≈ 5 when the bulk of the broken spiral arm has moved far enough (more than two chord lengths) downstream. This is accompanied by an almost spontaneous appearance of another eddy at the trailing edge (cf. the trace at t = 5.62).

In the subsequent period, t = 5.63-9.12, the wake patterns resemble closely those of t = 5.62 except for a new, growing spiral arm and a noticeable larger, and more densely packed, trailing-edge eddy. Figure 14 shows the trace and the instantaneous pressure at t = 9.12 which, except for an increase in the wake length, differs little from those at t = 7.6-9.12.

The maximum C_p value for top surface ($C_p \approx 0.45$) occurs at the hinge line (x = 0) and at about the 87% chord, where the flow speed vanishes; the departure of C_p's from unity there indicates a strong unsteady aerodynamic influence. The relatively low lift coefficient C_L = 1.09 computed for t = 9.12 is not too surprising, as Fig. 10b would have suggested that a t of 9.12 is still far from the stage when more significant vortex shedding at the flap is to take place. This low C_L level is attributed also to the trailing-edge eddy last mentioned, which contributes to a negative circulation and results in a second stagnation point on the top surface.

Surprisingly large magnitudes of vortex-drift velocities are found during t = 3.6-4.6, corresponding to the period beginning with the formation of the spiral-arm pattern and terminating when the large trailing-edge eddy begins to break away. The large drift speed during this period (as high as seven times the free-stream velocity) accounts for the excessively long wake found in this example.

5. SUMMARY AND REMARKS

A stationary, free vortex behind a leading-edge flap can enhance the lift of an inclined flat plate. For a fixed flap angle β and flap size c_1/c_0, the (static) lift and other aerodynamic properties are found to develop hystersis loops as the incidence α varies, signifying that multiple steady states can exist. In the context of the separation-vortex model,[1-4] the flows corresponding to the lower- and the higher-lift branches are recoverable from unsteady solutions, which would indicate their dynamic stability.

The vortex dynamics of this problem is re-examined by a version of the point-vortex method, which adopts the separation-vortex model for generating the nascent vortices. Preliminary results of this evolution study has been obtained for two types of initial conditions, assuming a fixed β, c_1/c_0 and α: (i) the flow is initially given by that with a stationary, free vortex pertaining to the lower branch on the hysteresis loop, and (ii) the flow is initiated with an impulsive start (i.e. indicial motion) at an incidence which would lead to a unique solution on the higher-lift branch. Computation has been carried out to a stage corresponding to about three flow-transit times in both cases, generating two thousand free vortices at the end of each run.

In case (i), point vortices shed from the leading-edge flap evolve to a ring-like recirculating pattern around the original, slightly displaced, vortex, while the lift increases from $C_L \simeq 1.8$ at $t = 0$ to $C_L = 2.84$ at the end of the computation ($t = 9.87$). Thus the stability of a single stationary, free vortex pertaining to the lower-lift branch in the separation-vortex model is not supported by the solution of the point-vortex method, the evolution from it leads nevertheless to a substantially higher lift.

In case (ii), the shed vortices develop smooth spirals from both edges during the early phase $t = 0(1)$ after the impulsive start; their total circulation and centroid agree well with the separation-vortex solution. The subsequent stages computed up to $t \simeq 10$ furnishes details of the evolving structure on the global and the lesser scales, revealing the familiar pairing giration as well as self-growth of the vortex clusters prior to their merging. A recurrent scenario of the large-scale flow seems to be the appearance of an elongated vorticity patch in the form of a spiral arm and the build-up of a large eddy over the trailing edge, preceding their eventual detachment. The build-up of the trailing-edge eddy is believed responsible for a significant lowering of the C_L prior to its shedding. The exceptionally high wake velocites found during the short period prior to the breakaway of the large trailing-edge eddy furnishes a distinct dynamical feature to be identified in laboratory and other studies.

A major obstacle in realizing the lift as high as the present study may project is the high pressure gradient to be encountered by the boundary layer on the top wing surface (cf. Figs. 8,12). Unless wall suction or means of reducing local pressure gradients is effected, the resultant flow separation may reduce the flap's vortex-capturing capability. We note, in passing, that rounding-off the apex corner in the manner of Fig. 7 many also help to control the separation on both sides of the wing surfaces.

Much remains to be done concerning the continuation of the point-vortex computation to a larger time (on a larger computer) and studying the effects of starting the flow with an incidence smaller as well as larger than the present angle (8.5°), so that the possibility of achieving a high lift, as suggested by the separated-vortex model, can be fully ascertained. For the same reason, comparison with corresponding solutions by time-accurate Euler and Navier-Stokes codes, as well as inviscid-viscous iteration codes, are essential, even though whether these existing programs can describe adequately features such as a tightly wound spiral vortex sheet is open to question.

ACKNOWLEDGEMENT

The authors would like to thank Prof. Richard E. Kaplan for his generous advice and assistance on computer usage and resource, also Ms. Chelestine Holguin for her work on the manuscript. The research is sponsored in part by the U.S. Air Force Grant AFOSR-85-0318. The development has derived considerable benefit from earlier studies supported by the Fluid Mechanics Branch of the NSF Eng. Division and the ONR Fluid Dynamics Program.

REFERENCES

1. Edwards, R.H., J. Aero. Sci. 21, 134 (1954).

2. Brown, C. & Michael, W.H., J. Aero. Sci. 21, 690 (1954).

3. Cheng, H.K., J. Aero. Sci. 21, 212 (1954); also see J. Aero. Sci. 22, 217 (1955).

4. Rott, H., J. Fluid Mech. 1, 111 (1956).

5. Mangler, K.W. & Smith, J.H.B. (1956). Proc. Internat. Congr. in Appl. Mech., held at Brussels, 2, 137 (1956).

6. Kücherman, D., The Aerodynamic Design of Aircraft, Pergamon, (1978).

7. Metha, U.B., Unsteady Aerodynamics, AGARD CP-227, 23-1 (1978).

8. McCroskey, W.J., Ann. Rev. Fluid Mech. 14, 285 (1982).

9. Lighthill, M.J., Math. Biofluiddynamics, SIAM (1975).

10. Maxworthy, T., Ann. Rev. Fluid Mech. 13, 329 (1981).

11. Edwards, R.H. & Cheng, H.K., J. Fluid Mech. 120, 463 (1982).

12. Cox, J., Soaring, 37, 20 (1973).

13. Walton, D., Soaring 38, 26 (1974).

14. Saffman, P.G. & Saffield, J.S., Studies in Appl. Math. 57, 107 (1977).

15. Saffman, P.G., J. Fluid Mech. 106, 49 (1981).

16. Moore, D.W., J. Fluid Mech. 63, 225 (1974).

17a. Clements, R.R. & Moore, D.J., Prog. Aeron. Sci. 16, 129 (1975).

17b. Evans, R.A. & Bloor, M.I.G., J. Fluid Mech. 82, 115 (1977).

18a. Fin, P.T. & Soh, W.K., Proc. 10th Symp. Naval Hydrodyn., Cambridge, Mass. (1974).

18b. Christiansen, J.P., J. Comp. Phys. 13, 363 (1973).

19a. Kuwahara, K. & Takami, H., J. Phys. Soc. Japan 34, 247 (1973).

19b. Kuwahara, K., J. Phys. Soc. Japan 35, 1545 (1973).

20a. Sarpkaya, T., J. Fluid Mech. 68, 109 (1975).

20b. Kiya, M. & Arie, M., J. Fluid Mech. 82, 223 (1977).

21. Saffman, P.G. & Baker, G.R., Ann. Rev. Fluid Mech. 11, 95 (1979).

22a. Leonard, A., J. Computational Physics, 37, No. 3, 289 (1980).

22b. Aref, H., Ann. Rev. Fluid Mech. 15, 345 (1983).

23. Lamar, J.E. & Campbell, J.F., Aerospace America, January 1984.

24. Brown, G.L. & Roshko, A., J. Fluid Mech. 46, 775 (1974).

25. Winant, C.D. & Browand, F.K., J. Fluid Mech. 63, 237 (1974); also see Browand & Weidman, P.D., J. Fluid Mech. 76, 127 (1976).

26. Ho, C.M. & Huang, L.S., J. Fluid Mech. 119, 443 (1982).

27. Murman, E. & Powell, K., Paper presented in this Proceeding.

28. Marconi, F., Paper presented in this Proceeding.

29. Jameson, A., Schmidt, W. & Turkel, E., AIAA Paper 81-1259 (1981).

30. Huang, M.K. & Chow, C.Y., AIAA J. 20, 292 (1982).

31. Chow, C.Y., Huang, M.K. & Yan, C.Z., AIAA J. 23, 657 (1985).

32. Rossow, V.J., J. Aircraft 15, 618 (1978).

33. Saffman, P.G. & Tanveer, S., J. Fluid Mech. 154, 351 (1984).

34. Tanveer, S., Topics in 2-D Separated Vortex Flow, Ph.D. Thesis, Calif. Inst. Tech. (1983).

35. Batchelor, G.K., J. Fluid Mech. 1, 177 (1956) and 1, 388 (1956).

36. Cheng, H.K. & Smith, F.T., J. Appl. Math. & Physics (ZAMP) 33, No. 21, 1515 (1982).

37. Cheng, H.K. & Lee, C.J., in <u>Proc. 3rd Symp. Numerical & Physical Aspects of Aerodynamic Flows</u>, Springer Verlag (1985).

38. Weis-Fogh, T., <u>J. Exp. Biology</u> <u>59</u>, 169 (1973).

39. Inoue, O., <u>AIAA J.</u> <u>23</u>, 367 (1985).

NOTES ADDED AFTER PROOF

1. The values of t in Figs. 13 and 14 were incorrectly given; the t-values therein should be increased by 2.95 (e.g. t = 3.63 changed to t = 6.58). The same correction must also be made for each t-values given under the subsection entitled "The Impulsively Started Flow Problem." The t-value of 10 given in the 4th paragraph of Section 5 (SUMMARY AND REMARKS) should also be changed accordingly.

2. The instantaneous vortex positions at t = 5.62 (more correctly at t = 8.57) presented in Fig. 13 is incomplete, where vortex locations downstream of the clusters 1-8 at this stage was inadvertently deleted.

3. The straight-line cuts illustrated as dashes in Fig. 7 are introduced for the purpose of explaining how the total force on the vortex feeder sheet in the separation-vortex model can be simply computed. This calculation based on the straight-line cuts are correct, even though one of the lines cuts across the unit circle in the ζ-plane, since the force in question is not strictly dependent on the feeder-sheet geometry. For the pressure-field calculation, however, the cut lines would have to be so deformed as to avoid crossing the unit circle. The shape of the deformed cut will not affect the surface pressure, as long as the edge locations B and C are fixed.

4. The Γ_0 and Γ_1 presented in Fig. 4 has been normalized by $c_0 U$, i.e. 3.65 aU, not aU.

5. On the 5th line below Eq. (4.3), "is typically 5 to 10" should be changed to "varies from 10 or more for a small t to 1 for the larger t". On the 6th line from the top of the following page, delete "or 100".

6. The factor "$(\zeta + 1/\zeta)$" in Eq. (3.1) should be replaced by "$(\zeta + e^{i2\alpha}/\zeta)$".

7. More recent computations with the present version of the point-vortex method indicates that shedding interval ten (10) times larger than $\Delta_1 t = 0.01$ can be used without appreciable degradation of the large-scale description.

8. We appreciate the stimulating discussions and enthusiastic support of T. Maxworthy.

Methods for Numerical Simulation
of Leading Edge Vortex Flow

H. W. M. Hoeijmakers

SUMMARY

A review is presented of computational methods to simulate the aerodynamics of configurations with leading-edge vortex flow. The various methods in use at present are discussed in some detail, primarily with a view towards three-dimensional steady flow applications. The strengths and weaknesses of the methods are indicated and results of different methods are compared and discussed.

INTRODUCTION

In aerodynamics vortical type of flows usually arise from some form of flow separation (Ref. 1). The relatively large-aspect-ratio wings of subsonic and transonic transport aircraft are designed such that boundary-layer type of separation occurs in relatively small regions only. For this type of configurations vortical-type flows occur primarily in the tip region of wings, deployed flaps, extended slats and in the wake. In the rolling-up wake the trailing vorticity becomes concentrated into two or more vortex cores. Usually the details of the wake have only a minor influence on the pressure distribution on the generating airplane itself, leading to a simplified treatment in classical wing theory (e.g. Ref. 2), panel methods for linearized flow (e.g. Ref. 3) and current computational transonic potential flow methods (e.g. Ref. 4). The most important vortex flow problems for large-aspect-ratio wings are formed by the interaction of the rolling-up wake of an extended slat with the flow over the main wing component, the interaction of the wake of the wing with the flow about deployed flaps and to a lesser extent the flow near the wing tip and the interaction of the wake of the wing and the flow about the tail. In off-design conditions boundary-layer separation may result into a complicated and usually unsteady vortical flow pattern which may be related to buffet onset.

Vortex flow associated with flow separation from leading, side and trailing edges plays an important role in the high-angle-of-attack aerodynamics of various modern fighter aircraft (see Ref. 5 for an excellent overview). Examples are aircraft designed for high speed cruise employing the slender-wing concept with highly swept and aerodynamically sharp leading edges and aircraft designed for maneuvering employing the strake-wing concept or the close-coupled canard concept. For these configurations the introduction of sharp edges results in a vortical flow pattern that is "controlled", i.e. steady, stable and sustainable within a wide range of Reynolds number, Mach number, angle of attack, etc. The pressure distribution associated with the vortex flow is such that conventional type of ("uncontrolled") boundary-layer type of separation is delayed to a higher angle of attack. Moreover the lift increase more rapidly with angle of attack than without vortex flow, resulting in a substantial increase in lift capability. A limit to the favourable effects is reached when large-scale vortex breakdown occurs above the wing.

The attached-flow methods commonly used in present-day design procedures are inadequate to predict the characteristics of a configuration with vortical flow regions in the near proximity of its surface. Until now, the design of configurations in which vortical flow regions interact strongly with the flow about the configuration has been almost entirely empirical. In such design process the resulting geometry is inevitably the result of extensive and costly cut-and-try type of wind-tunnel tests, including measurement of detailed surface pressure distributions and often supplemented by surface-flow and flow-field visualisation studies. If the vortex-lift capabilities of slender-wing, strake-wing and canard- wing configurations are to be fully understood and utilized a detailed knowledge, adequate mathematical modelling and methods for the numerical simulation of rotational flow are required. The major contribution of computational methods for vortex flow, capable of simulating the details of the flow field, consists in guiding the designer towards a more optimal design and in defining the critical parameter region to be considered for extensive wind-tunnel testing.

It is the purpose of this paper to review computational methods for the simulation of the high-Reynolds number flow about configurations with, or designed to have, leading-edge vortex separation. The characteristic feature of this type of flow is its strong non-linearity; due to the vortical flow the airloads vary non-linearly with angle of attack and sideslip. As the Mach number is increased from low subsonic to transonic and supersonic also strong non-linear compressibility effects (shock waves) will occur and it is not uncommon that there is a strong coupling between both types of non-linearities (e.g. Ref. 6). While the development and use of computational methods for either one of these two non-linearities has been quite extensive, the development of methods for the coupled effects of both non-linearities has only recently attracted increased attention.

Computational methods for unsteady flow or for the flow about oscillating configurations have not yet been developed to the same extent as the methods for steady flow. Although some of the steady flow methods discussed here have been extended to the case of oscillating configurations, here we will consider steady flow only.

The various methods will be discussed primarily with a view toward three-dimensional flow applications. The present review is meant to give some insight into the weaknesses, strengths and prospects of the different methods. Recent reviews of various aspects of three-dimensional vortex flows are included in the proceedings of the AGARD FDP Symposium on Aerodynamics of Vortical Type Flows in Three Dimensions, AGARD CP 342, 1983.

In the next two sections we will first describe the main features of the physics of leading-edge vortex flow, followed by a discussion of mathematical models for the flow. In the subsequent sections we describe three classes of computational methods, termed the "rigid" vortex, the "fitted" vortex and the "captured" vortex methods, respectively. In the discussion we will follow a more or less chronological path through the abundant literature, starting with the most simplified methods and ending with the more sophisticated ones. Whenever appropriate we will compare results of different methods mutually as well as with experimental data.

DESCRIPTION OF THE FLOW

Delta-like wings

The topology of low-speed high-Reynolds number flow about slender delta-like wings with leading-edge vortex separation has been well established in primarily experimental investigations (e.g. Ref. 7-12).

At moderate to high angles of attack, the flow separates at the highly swept leading edge, with the boundary layers flowing off the upper and lower surface merging into a free shear layer (Fig. 1). Under influence of the vorticity contained in it, the free shear layer rolls up in a spiral fashion to form a relatively compact single-branched core with distributed vorticity, the so-called leading-edge vortex. The presence of this core in the proximity of the wing surface affects the pressure distribution to a large extent, the predominant effect being a low pressure region underneath the vortex core. It has been observed in experiments that the adverse pressure gradient in the region just outboard of the lateral position of the center of the leading-edge vortex core causes a so-called secondary separation. The free shear layer emanating at the line of secondary separation rolls up in a vortex core also, its vorticity being of opposite sign compared to that of the vorticity within the leading-edge vortex core. The extent of the effect of secondary separation on the surface pressure distribution has been observed to depend strongly on whether the boundary layer is laminar or turbulent. It appears that secondary separation effects are largest for the laminar case, and generally for both laminar and turbulent flow become more important with decreasing incidence and also with decreasing leading-edge sweep.

Viewed in the downstream direction the leading-edge vortex is continually being fed with vorticity shed at the leading edge which is transported through the free shear layer to the vortex core, so that the latter is increasing in strength as well as in cross-sectional dimension. Downstream of the trailing edge the free shear layer continues to feed vorticity into the leading-edge vortex core until in the far wake. In addition, it has been observed that the vorticity of sign opposite to that within the leading-edge vortex core, shed at the trailing edge roughly near the lateral position of the primary core, starts to roll up in a double-branched, so-called trailing-edge vortex core (Ref. 11).

A limit to the favourable effects induced by the vortex flow is reached when the angle of attack exceeds the value at which large-scale vortex breakdown occurs above the wing upper surface. Vortex breakdown manifests itself as an abrupt increase in the cross-sectional area and turbulence level of the core. It first occurs downstream of the wing, propagating upstream with increasing incidence. At the higher incidences vortex breakdown is often accompanied by an asymmetric vortex flow pattern, although for very slender delta wings also asymmetric vortex flow patterns occur without large scale vortex breakdown. For slender delta wings it has been observed that a stable, "controlled", vortex flow pattern persists for incidences up to 35 deg (see figure 2).

Strake-wing configurations

The vortex flow above strake-wing configurations is more complex. Investigations by, amongst others, Luckring (Ref. 13) and Brennensthul & Hummel (Ref. 14) have provided valuable insight into the vortex layer structure above strake-wing configurations. Verhaagen (Ref. 12) has acquired additional topological information employing a laserlight-sheet flow visualization technique. Manor & Wentz (Ref. 15) investigated the flow field using a total pressure scanning device. From these investigations it follows that above a strake-wing configuration the (single-branched) leading-edge vortex originating at the leading edge of the strake continues above the wing. The vortex modifies the wing flow field considerable, inducing additional favourable effects. In case of sufficiently (depending on leading-edge sharpness, angle of attack, etc.) swept wing leading edges the flow separates also at the wing leading edge and downstream of the kink the shear layer emanating from the strake leading edge is continued as the shear layer emanating from the wing leading edge (see figure 3). In this shear layer the so-called (double-branched) wing vortex develops.

This vortex starts at the kink and loops around the strake vortex. At some point downstream of the kink the two cores (with distributed vorticity of the same sign) merge into one core. At smaller wing sweep angles the flow picture is not completely clear. Most investigators assume that the strake shear layer detaches (tears) at the kink, while possibly a second shear layer emanates from the wing leading edge rolling up in a second single-branched vortex. In this case the "free end" of the strake shear layer will roll-up into another (single-branched) vortex core. As alternative for the latter it might possibly also be hypothesized that downstream of the kink the strake shear layer continues onto the wing upper surface, now emanating from some separation line, see also the discussion by Smith (Ref. 16). More detailed experimental investigation will be needed to clarify details of the flow structure for this case.

Reynolds number effects

It appears that for angles of attack not too close to the angle of attack for attached flow the flow pattern depends only weakly on Reynolds number. The latter with the exception of the location of the secondary separation which depends primarily on the state of the boundary layer. For edges that are rounded, i.e. not aerodynamically sharp, the flow does not separate all along the leading edge. At angles of attack near the attached flow condition the flow separates at the wing tip only. With increasing incidence the point on the leading edge where the vortex first forms progresses in upstream direction. The latter process of partial leading-edge vortex separation does appear to depend on Reynolds number.

Mach number effects

For thin slender wings at moderate incidences the influence of the Mach number on the characteristic features of the flow is small as long as $M_\infty \cos \Lambda$, the Mach number normal to the leading edge, is less than 0.7.

For higher Mach numbers, or at higher incidences or for thicker wings non-linear compressibility effects can become of dominating importance. This results in a wealth of possible flow patterns involving primary and secondary vortex cores (cross-flow and normal) shocks on the surface as well as in the flow field and associated regions with shock-induced distributed vorticity and all possible mutual interactions, sometimes resulting in an abrupt vortex breakdown. Some of these flow patterns are discussed by Szodruch & Ganzer (Ref. 6), Miller & Wood (Ref. 17), and Vorropoulos & Wendt (Ref. 18). Although it is recognized that vortex flow with non-linear compressibility effects are of increasing importance at transonic maneuvering and supersonic speeds, in the present paper the emphasis will be on the case of vortex flow without shock waves.

MODELS FOR LEADING-EDGE VORTEX FLOWS

In aerodynamics the Reynolds-averaged Navier-Stokes equations model essentially all flow details. However, a model is needed to model the turbulence in an appropriate manner, here for a rather complicated flow. Furthermore, the computer resources required for numerically solving the equations on a mesh that sufficiently resolves the boundary layers, free shear layers, vortex cores, etc. are quite excessive.

For most high-Reynolds-number leading-edge-vortex flows viscous effects are confined to thin boundary layers, thin free shear layers and centers of vortex cores, i.e. the global flow features are relatively independent of Reynolds number. This implies that Euler's equations, which allow for shock waves as well as rotational flow provide an appealing alternative. On a local scale, specifically at points where the flow separates and vorticity is shed into the flow field, some kind of model for viscous-flow dominated features will be required. Although the computer requirements of Euler codes can be met by the current generation of supercomputers, routine practical application of these codes is only just emerging.

In the absence of strong shocks the flow may be modelled as a potential flow with embedded free vortex sheets. Now the vortex sheets are "fitted" explicitly into the solution, rather than "captured" implicitly as part of the solution as is the case for the Navier-Stokes and Euler methods. This requires that one has to decide a priori on the presence of vortex sheets and that the topology of the flow must be well-defined and known in advance. In return the computer requirements of potential flow methods are relatively modest. In finite-difference/volume methods for solving the full non-linear potential equation the treatment of free vortex sheets, floating in a fixed spatial grid, poses considerable problems.

In case shockwaves are absent altogether the model is further simplified to linear potential flow, governed by the Prandtl-Glauert equation. Now the flow, including the position of the vortex sheets, can be solved for by a panel method. This panel method is non-linear because the position of the vortex sheet appears non-linearly in the boundary condition on the solid surface as well as in the boundary conditions on the vortex sheet itself. In the potential flow formulation a vortex core consists of a tightly wound spiraling vortex sheet of infinite extent. In order to simulate the flow field outside the core the sheet is necessarily cut at some angular extent, while the cut-off part is modelled by an isolated line vortex which is connected to the remainder of the sheet by a so-called feeding sheet. The latter feeds vorticity into the vortex allowing it to increase in strength when travelling in downstream direction.

In above mathematical models the aerodynamic characteristics of configurations with leading-edge vortex flow are determined through "capturing" or through "fitting" the details of the rotational flow regions. There are also methods that augment the conventional attached-flow model, with its "rigid" wake, through empirical concepts that account for the effects of the vortex flow interaction without modelling the rotational flow details. The major advantage offered by this latter class of methods is that they allow linear theory to be used for the prediction of non-linear flow phenomena, and, correspondingly, computer requirements are minimal.

RIGID VORTEX METHODS

In this section methods are discussed that rely on empirical concepts to account for the interaction of vortical flow without resolving the rotational flow details. The concepts are used to account for vortex flow effects in conventional attached (potential) flow methods, i.e. vortex-lattice and panel methods.

$\frac{1}{2}\alpha$-methods

Separation from the side edges of wings, flaps, etc. can often be modelled in an approximate sense by letting the trailing vortices leave the edge at an angle of $\frac{1}{2}\alpha$. This very simple method, found by Bollay (Ref. 19) and investigated further by Gersten (Ref. 20) provides reasonable agreement of computed and measured normal force without having to determine the position of the trailing vortex system. However, since the roll-up is neglected the local pressure distribution is not predicted accurately.

Leading-edge suction analogy

The most widely used method for predicting the characteristics of configurations employing vortex flow is the so-called "leading-edge suction analogy", conceived by Polhamus (Ref. 21). In classical lifting surface theory (e.g. Ref. 2) in general the velocity goes to infinity at all edges except the trailing edges, resulting in a finite suction force at leading and side-edges. This force acts normal to the edge and lies in the plane of the wing, partly counteracting the drag force. The magnitude of the suction force depends on the amplitude of the singularity in the loading at the edge. On a three-dimensional wing the normal force, found from integration of the net pressure distribution, combines with the suction force to yield the correct lift, side force and induced drag of the wing. When the flow separates at the edge the velocity remains finite and the suction force is lost. Now the analogy is based on the conjecture that in the case of the formation of a leading-edge vortex the suction force is recovered as a force rotated through 90 deg such that it acts in the direction of the normal force (Fig. 5). This implies that the aerodynamic forces and moments on configurations with vortex flow can be computed by utilizing any conventional linear attached flow method, without the necessity of having to solve a complex non-linear problem. The magnitude of the suction force follows then from the behaviour of the attached flow loading as the edge is approached. The suction analogy has been applied to a variety of cases in subsonic and in supersonic flow, as well as to oscillating wings. In general the correlation of computed forces, and to a lesser extent moments, with experimental data is surprisingly good, but no fundamental theoretical explanation appears to have been given. With a vortex-lattice method as attached-flow solver Lamar and his co-workers (e.g. Ref. 22) have developed the method towards a very useful engineering tool, both for analysis and design, of quite general configurations, including strake-wing configurations.

The major drawback of methods that use the suction analogy is that the pressure and velocity distribution cannot be predicted and that new situations, e.g. the encounter of a canard vortex with a wing, strake-wing vortex flow, vortex breakdown, etc. require the development of new empirical concepts.

FITTED VORTEX METHODS

The detailed flow field and surface pressure distribution can be provided by methods based on the potential flow model with embedded ("fitted") vortex sheets (Fig. 6). Although some progress has been made incorporating free vortex sheets in the full-potential flow formulation (Ref. 23, 24) here we will consider linear compressible flow only. This model then clearly only applies to flows without shock waves. In this linear potential flow model the Prandtl-Glauert equation (Laplace for zero Mach number) is solved subject to the following boundary conditions:

- the solid surface is a stream surface,
- the free vortex sheet is a stream surface and the static pressure is continuous across the sheet,
- the isolated vortex/feeding sheet model of the vortex core is force free,
- the flow leaves the surface at separation lines (edges) in a "smooth" manner (Kutta condition),
- at infinity upstream the flow is uniform.

The potential flow problem is solved by employing a singularity distribution on the solid surfaces as well as on the vortex sheets. Expressing the velocity field as an

integral over the singularity distribution and imposing the boundary conditions yields
a set of non-linear integral equations for the strength of the singularity distribution, the position of the vortex sheets and the position of the vortex cores.

Methods based on the slender-body approximation

Configurations that develop leading-edge vortex flow are mostly slender. This
suggests that the variation in streamwise direction of the flow field is much smaller
than the one in the cross-flow plane. In that case the full three-dimensional problem
can be reduced to a two-dimensional problem in successive cross-flow planes (Fig. 7).
The two-dimensional Laplace equation is the governing equation and the variation in
streamwise direction is accounted for through the boundary conditions. The streamwise
component of the perturbation velocity, required to compute the pressure, can be computed after that the cross-flow solution has been obtained. Moreover, observation
(surface oil, vapor screen, laser-light sheet, etc.) of the flow over highly swept
conical configurations suggests that the flow field is geometrically conical. In that
case the slender-body approximation requires only one cross-flow plane problem to be
solved.

Smith's conical-flow method

Methods employing the slender-body conical flow approximation were the first to
be applied with success to the computation of the flow about wings with leading-edge
vortex flow. In the early fifties Legendre (Ref. 7) and Brown & Michael (Ref. 25)
devised an approximate model in which all vorticity is concentrated in the isolated
vortex. Employing a conformal mapping technique the problem could be reduced to for
those days acceptable dimensions (three non-linear algebraic equations for as many
unknowns). Later Mangler & Smith (Ref. 26) and Smith (Ref. 27) improved upon this
early attempt by including a vortex sheet of finite extent, discretized by piecewise
continuous elements (panels). This greatly improved the treatment of the flow near
the leading edge and computed pressure distributions compare quite satisfactorily
with experimental data.
Clark (Ref. 28) has extended Smith's conical flow method to non-conical planforms. It
was found that for non-planar wings quite non-trivial vortex sheet geometries may be
expected, possibly involving multiple centers of roll-up.

Discrete-vortex methods

Discretizing vortex sheets by <u>discrete vortices</u> has a long history, primarily because of its simplicity and flexibility. In the present context Sacks et al (Ref. 29) computed the flow about wing-body configurations employing conformal mapping for the solid geometry and discrete vortices to represent the leading-edge vortex sheet. Although the vortex sheet appears quite chaotic the predicted forces compare reasonably well with experimental data. Recently Xieyuan (Ref. 30) applied Sacks' method to the flow about double-delta wings, also indicating disorderly motion of the vortices, through the position of the clusters of vortices agrees with the experimentally found vortex core positions. In the discrete-vortex method the onset of chaos can be delayed by a number of techniques, e.g. the rediscretization technique of Fink & Soh (Ref. 31); amalgamation of vortices in a core whenever an insufficient number of vortices is representing a loop of the sheet, as introduced by Moore (Ref. 32); regularisation of the velocity field by the sub-vortex technique of Maskew (Ref. 33), etc. Peace (Ref. 34) employed some of these techniques for an improved discrete-vortex method. The method was applied to various slender wings, including strake-wing configurations, yielding smooth vortex sheet geometries. For the strake-wing configurations two centers of roll-up were computed, demonstrating that discrete-vortex methods, have the flexibility to cope with more complex vortex flow topologies.

Realistic configurations may have a cross-section that is not easily handled by analytical conformal mapping techniques. To circumvent this it is often preferred to solve the problem in the physical plane. Maskew & Rao (Ref. 35) and Nathman (Ref. 36) use a low-order panel method to represent the solid surface in conjunction with an improved discrete-vortex representation of the vortex sheet to compute the flow about general slender planforms and wings with thickness.

The basic difficulty with the discrete-vortex representation is that at a large number of cross-flow planes a new discrete vortex must be shed to account for the increasing strength of the leading-edge vortex sheet. The strength and position of the newly shed vortex is obtained from Kutta-condition type of considerations, inevitably involving some empiricism. An additional difficulty poses the region near the apex where the vortex lines on the sheet are highly curved, requiring many cross-flow planes to follow the vortex trajectories in this area.

Higher-order panel method VORSBA

The method based on the slender-body approximation developed at NLR, solves the problem using a second-order panel method in the physical plane. It can treat arbitrarily shaped multiple-component cross-sections with vortex sheets attached to prescribed fixed positions. The method, designated VORSBA, is an extension of the VORCON (conical flow) and VORQCO (quasi-conical flow) methods described in Refs. 37 and 38. In the VORSBA method the cross-flow plane geometry is divided into curved panels each carrying a quadratically varying doublet distribution, resulting in an accurate approximation of the velocity field, with only very weak singular behaviour (ln) at panel edges. In the case of configurations with thickness a source distribution is included to account for the streamwise increase in cross-sectional area. The system of non-linear algebraic equations is solved by a full-Newton iterative procedure providing a rapid convergence to the solution for an initial guess sufficiently close to the solution. The latter is accomplished by continuation procedures, such as using the solution obtained earlier at a nearby angle of attack, nearby length of the vortex sheet(s), etc.

Applications of slender-body methods

Figure 8 shows the conical flow solution for a delta wing of unit aspect ratio (Λ = 76 deg) obtained by Smith (Ref. 27) for α = 13 deg and by the VORSBA code for α = 14 deg, both for a rather long vortex sheet. The geometry of the vortex sheets computed by the two methods differs somewhat, primarily by the different implementation of the Kutta condition at the leading edge. The computed pressure distributions show the characteristic suction peak underneath the rolled-up vortex sheet. There are also some differences in the location and height of the upper-surface suction peak, presumably due to slight differences in the formulation (approximation in Ref. 27 of tan α by α, etc.). On the lower surface the two results are nearly identical. At the leading edge the pressure on the upper and lower surface become equal to the pressure on the vortex sheet.

In figure 9 results of some of above described discrete-vortex methods, obtained for a delta wing of unit aspect ratio, are compared with results of panel methods. It shows that for this case the two types of method do yield similar results, though in general the vortex sheets computed by panel methods have a smoother and more regular appearance than the ones computed by the early discrete-vortex methods.

Figure 10 compares the computed shape of the vortex sheet with the total-pressure contours obtained from wind-tunnel measurements (Ref. 39). It demonstrates that the location of free shear layer and vortex sheet agree quite well, while also the position of the core is predicted satisfactorily.

That slender-body methods can be used for preliminary vortex-flow studies is demonstrated in the next selection of applications. In figure 11 we show the vortex sheet geometry and surface pressure distribution for a bi-convex delta wing of unit aspect ratio under side-slip conditions. The angle of side-slip is quite extreme, actually the port-side leading edge is a trailing edge and the port-side vortex is already swept off the wing surface. The starboard vortex is quite close to the upper wing surface resulting in a sharp suction peak on the starboardside of the wing. It must be expected that in the real flow secondary separation effects will preclude a sharp peak like this.

It is known from experimental investigations that secondary separation may have a large effect on the upper surface pressure distribution. Iterative inviscid-viscid procedures in which the inviscid flow is computed followed by a boundary layer calculation will require that the inviscid flow model includes the secondary sheet. Boundary-layer computations on delta wings have been reported by De Bruin (Ref. 39), Wai et al (Ref. 40) and DeJarnette et al (Ref. 41), using either measured pressure distributions or ones from single-sheet computations. Here we will consider the effect on the inviscid solution of the presence of a secondary vortex sheet, emanating from the wing upper surface at a fixed prescribed position. The results of this study are presented in figure 12. Four cases are considered, no secondary separation, separation at 70, 75 and 80 % semi-span. The 70 and 80 % location correspond approximately with laminar and turbulent separation, respectively. Including the secondary separation sheet has the expected effect on the position of the primary vortex core, it moves inboard and slightly upward. The effect on the upper wing surface pressure distribution is that the suction peak is reduced and that, just like in experimental data, a plateau of constant pressure is formed just outboard of the secondary separation point. However, between this plateau and the leading edge a second suction peak appears, which is more pronounced than found in experiments. The latter is directly associated with the concentrated vorticity in the secondary vortex core being located in the near proximity of the surface. Including tertiary separation would probably reduce the height of this peak, but it may also be argued that in the real flow viscous effects would have a relatively large effect on the (relatively weak) secondary vortex, spreading its vorticity, resulting in a less pronounced pressure signature on the surface. Considering the details of the computed vortex sheet geometry it is hard to envision that the classical picture of the topology, with the secondary attachment line located on the upper wing surface between the secondary separation point and the leading edge, applies here. Indeed, computing the surface velocity components along and normal to rays from the apex results in the situation indicated in figure 13. The most striking observation is that the secondary attachment line moves off the wing surface onto the vortex sheet, forming a half-sadlle point on the free sheet. The position of this point corresponds with a local maximum in the pressure, which is a prerequisite for its topological correctness (Ref. 43). Note that although on the sheet the mean cross-flow velocity is directed from the leading edge towards the primary vortex core, on the upper side of the sheet between the secondary attachment line and the leading edge the flow is towards the leading edge. It is difficult to assess whether this is a valid (unique) solution, but it is certainly an interesting outcome of the present study.

Without leading-edge vortex separation slender wings have a small lift capability, with leading-edge vortex separation the lift capability is greatly enhanced. However, the suction on the wing-leading edge is lost (or decreased considerably) and the drag is correspondingly increased. One device to alleviate the drag problem is the vortex flap (e.g. Ref. 44). When the vortex flap is deflected downward the vortex induced suction peak will contribute to a force component in forward direction, counteracting the drag force. Deflecting the flap at given incidence will result in the vortex becoming smaller and weaker and it will move towards the leading edge as well as closer to the wing upper surface. Methods based on the slender-body approximation can aid in preliminary studies of concepts like the vortex flap. This is illustrated in figure 14 which shows computed vortex sheet geometries and pressure distributions for a 65 deg swept delta wing at 15 deg incidence with (conical) vortex flap, hinged at the 70 % semi-span position. This configuration was also studied by Erickson (Ref. 45) using a three-dimensional free vortex sheet method. It is shown that at a flap deflection angle of 20 deg the leading-edge vortex is totally confined to the flap area (reattachment is very near the hinge line) as is the suction peak induced by its vorticity, indicating the vortex flap efficiency. The inviscid pressure distribution has a steep gradient and undoubtedly the real flow will involve secondary separation. The latter is clearly evident in the experimental data presented by for instance Hoffler & Rao (Ref. 46).

Figure 15 shows the application of the VORSBA panel method to the lengthwise cambered wing also treated by Clark (Ref. 28) and Peace (Ref. 34), the latter using a discrete vortex method. It is clearly indicated that for this case a second (double-branched) center of roll-up evolves. In order to obtain more accurate results this second vortex core will need to be modelled explicitly as a double-branched vortex. This is not yet possible with the present version of the code but will be pursued in the further development of the method.

In figure 16 we compare the solution obtained for a double-delta wing using the discrete-vortex method (Ref. 34) and using the panel method VORSBA. In Peace's solution two distinct centers of roll-up evolve and instabilities occur before the last station is reached. In the VORSBA panel method solution the change in sweep angle results in a dent (region of curvature with opposite sign) in the vortex sheet. Travelling in downstream direction the dent travels along the sheet towards the vortex core. At x = 15 the dent has been stretched out completely and there is no sign of any developing instability.

Methods for three-dimensional flow

In the slender-body approximation the upstream effect of the trailing edge, present in subsonic flow, is not accounted for. Furthermore not-so-slender configurations and large variations in streamwise direction on slender configurations as for example occur near the apex and near discontinuities in the leading-edge sweep angle are not allowed in the strict sense of the slender-body approximation. Also upstream interaction between the vortex flow from one component of a configuration with the (vortex) flow about another component of the same or another nearby configuration cannot be treated properly. Finally it must be realized that, strictly spoken, in subsonic flow a conical geometry cannot correspond with a conical flow. This in spite of the often very conical appearance of surface streamline patterns and vortical flow topologies observed in experiments. The usually not very conical measured (low-speed) pressure distribution on conical configurations is an indication of this.

To account for all these effects methods are required that, within the limitations of the mathematical model chosen, solve the full three-dimensional problem.

Non-linear vortex lattice methods

The extension of the discrete-vortex method to three dimensions is the vortex-lattice method. In the method the lifting surface is divided into elements. Each element carries a ("bound") line vortex along its ¼-chord line connected to ("trailing") line vortices along the side edges of the elements. The trailing vortices extend along the wing surface onto the wake of the wing to infinity downstream, so that a horse-shoe vortex is associated with each element.

In the case of separation from the leading or side edge discrete line vortices trail not only from the trailing edge but also from the edges at which the flow separates. Each of these vortices is still connected with a "bound" vortex on the wing. The trailing vortices form the discrete representation of the rolling-up vortex sheet. The boundary conditions on the vortex sheet are satisfied when the line vortices are aligned with the local flow direction. Their strength follows from satisfying the normal-velocity boundary condition at the midpoint of the ¾-chord line of the elements.

The non-linear vortex-lattice method has attracted considerable attention. Using the vortex-lattice method Belotserkovskii (Ref. 47) considered separation from side-edges of low-aspect ratio wings, relaxing the line vortices emanating from the side edges. Rehbach (Refs. 48, 49) applied the (non-linear) vortex-lattice method to the case of a slender delta wing. The solution was obtained by a geometry perturbation scheme, starting with the solution for a rectangular wing with side-edge separation and gradually decreasing the sweep of the side edge until the delta wing is obtained, the new position and strength of the vortex sheet being computed after each change. The same geometry perturbation procedure was also used to obtain solutions for a flat ogee wing, a flat gothic wing and two chordwise cambered delta wings. Correlation of computed vortex-sheet shapes with water-tunnel flow visualization is good, however, it is stated that although overall forces and moments are predicted satisfactorily the computed pressure distribution compares poorly with experimental data.

In Rehbach's method the vortex core is not modelled explicitly. More recently

Kandil (Ref. 50) improved upon this by including an amalgamation procedure for the core of the leading-edge vortex.

From a perusal of published results of non-linear vortex-lattice methods, including the related discrete-vortex methods for the case of the slender-body approximation, it must be concluded that employing a discrete-vortex representation for vortex sheets is apparently a too crude and singularly behaved approximation. In regions of strong interaction and large variations, where the vortex sheets stretch or contract correspondingly, considerable numerical problems are encountered which worsen when more vortices are employed in an attempt to improve the accuracy. The pressure distribution and vortex-sheet geometry are generally not well predicted. However, reasonable results for overall forces can be obtained if the number of free line vortices is kept small (e.g. see Ref. 51 for a recent account of applications).

Panel methods

The numerical problems associated with discrete-vortex methods can be overcome by employing a panel-method type of approximation. In panel methods the vortex sheets are modelled more accurately by piecewise continuous doublet distributions, eliminating the singular velocity field associated with discrete line vortices and approximating the velocity field due to a distributed surface vorticity distribution much more accurately.

Boeing's LEV method

The panel method developed at Boeing by Johnson and others (Refs. 52, 53) solves the Prandtl-Glauert equation for three-dimensional subsonic, subcritical flow about configurations with leading-edge vortex sheets (Fig. 17). The model consists of the wing surface, the leading-edge vortex sheet, the feeding ("fed") sheet and the wake. The wake is divided into two parts, the "near wake" immediately downstream of the trailing edge and the "trailing wake" extending (cylindrically) to infinity downstream. The geometry of the wake vortex sheet is not free but is specified as a smooth continuation of the wing and the free sheet. Consequently the geometry of the near wake is known and only one of the two boundary conditions can be enforced to determine the doublet distribution on the wake. This condition is the pressure-difference boundary condition (expanded to second-order accuracy), which furnishes the wake doublet distribution the freedom to adjust in lateral direction. On the far wake the first-order pressure-difference condition is applied. The latter implies a cylindrical geometry directed in free stream direction and carrying an in streamwise direction constant doublet distribution. In the latest version of the LEV method the panels are flat and carry a panelwise quadratically varying doublet distribution. The boundary conditions are applied at the panel midpoints. The main difficulty is that the boundary conditions on the sheet are of mixed analysis/design type. Constructing a numerically stable scheme for expressing the panelwise representation of the doublet distribution and geometry in terms of the unknown doublet and geometric parameters to be solved for has proven to be difficult. The scheme used presently involves a relatively costly least-squares procedure in which the doublet distribution on a panel is expressed in terms of 16 neighbouring doublet parameters (on a non-rectangular computational grid). Figure 17 also displays the kinematics of the geometry of the free and feeding sheet. The degrees of freedom are the panel inclination angles, θ, the length scale λ for the length of the sheet and the length scale ν for the feeding sheet. Note that the attitude of the feeding sheet is fixed with respect to the last panel on the vortex sheet.

The (square) system of non-linear equations for the singularity and geometric parameters is solved using a quasi-Newton procedure with controlled stepsize and recomputing (the computationally rather expensive) Jacobian matrix every three iterations. A solution derived from Smith's (Ref. 27) conical flow solution or from a solution for a similar case is used as initial guess for the iterative procedure. In cases where convergence problems are encountered a "no-panel-twist" condition is added and the now overdetermined system is solved in least square fashion. In general the published computed pressure distributions, forces and moments are in satisfactory agreement with experimental data (e.g. Luckring et al, Ref. 54, Erickson, Ref. 45). However, on the mostly rather coarse panel schemes pressure distributions are often insufficiently resolved to provide the details in region on the surface of the configuration underneath the leading-edge vortex.

NLR VORSEP method

The second-order panel method VORSEP developed at NLR (Refs. 37, 55) for simulating the three-dimensional incompressible flow about slender wings of general shape and planform with leading-edge vortex sheets is based on the same formulation as the LEV method. However, the numerical implementation differs on a number of points.

In VORSEP the potential-flow problem is solved in the rectangular computational domain depicted in figure 18. In the computational domain the wing is mapped onto the unit square, the leading-edge vortex sheet, the near wake and the far wake on the rectangular regions shown. In contrast with Boeing's method the wake is free to move to a position where both the zero-normal-velocity boundary condition and the zero-pressure-jump condition are satisfied.

The computational domain is divided into rectangular panels carrying panelwise quadratic representations for the geometry and doublet distribution. The panelwise quadratic representations involve function value, first and second derivatives at panel midpoints. The derivatives are approximated by second-order accurate finite difference type expressions, involving function values at neighbouring midpoints (Fig. 16). Because the computational domain is rectangular and globally defined the latter expressions are relatively simple.

It was found that for the boundary conditions on the vortex sheet numerical stability could be achieved by a central difference scheme for the second derivatives and a one-sided second-order accurate scheme for the first derivatives. The panel midpoints are taken as the collocation points and the number of unknowns equals the number of boundary conditions (Fig. 18).

The system of non-linear equations is solved using a quasi-Newton procedure. In this procedure the Jacobian matrix is stripped and the gradient matrix equation has, similar to the finite difference molecule, a four-block-diagonal form, which can be solved very efficiently. All the derivatives and in particular those of the induced velocity with respect to the geometric parameters are obtained from closed form expressions, avoiding the very expensive numerical differentiation. In the present quasi-Newton process the residue decreases by an order of magnitude or more in each iteration, no step-size control is necessary and the solution is usually obtained in 5 or 6 iterations, recomputing the Jacobian matrix during each iteration. The initial guess for starting the iterative procedure is obtained from the slender-body method VORSBA described earlier. Furthermore, continuation procedures are used to extend already obtained solutions to solutions with a longer or shorter vortex sheet or near wake. Also continuation to neighbouring angles of attack is often used. To aid the user is obtaining an initial guess a library of solutions is available. Implanting the solution for one specific case onto the the computational domain as initial guess for another case can be realized in a routine manner. Although originally set up for incompressible flow its extension to linear compressible flow is relatively straight-forward.

Applications of panel methods

(i) VORSEP solution of AR=1.0 delta wing at 20 deg incidence

In figure 19a the vortex sheet geometry as computed by VORSEP is shown in a three-dimensional view. Observe the evidence of the trailing-edge vortex as a wrink-ling of the near wake. Figure 19b shows the computed vortex sheet shapes in cross-flow planes. The geometry of the computed vortex sheet is conical, almost up to the trailing edge of the wing. Here also evident is the dent in the near wake where the trailing vorticity tries to create a second center of roll-up. However, in the numeri-cal solution this region is smoothed out. Figure 19c shows the computed doublet distribution in subsequent cross-flow planes. The doublet distribution is not as conical as the geometry, its gradient in chordwise-direction diminishes quite rapidly from a nearly constant value on the forward half of the wing to zero near the trailing edge. Observe that in the near wake the point with the largest positive gradient, which corresponds with the second center of roll-up, shifts in outboard direction.

The pressure distribution on the wing is shown in figure 19d. It shows that on the upper wing surface the pressure coefficient increases monotonically from apex forwards the trailing edge and is not conical at all. The pressure coefficient for this case as computed by the conical flow code approximately corresponds to the chordwise average value of the full three-dimensional solution. On the lower wing surface the three-dimensional pressure distribution agrees much better with the one computed by the slender-body method.

In the VORSEP method the near wake is free to attain the shape required to satisfy both the normal-velocity boundary condition and the pressure-difference boundary condition. For the panel scheme employed in figure 19 (3 near-wake strips of width 0.1 each) the second center of roll-up is evident as a dent. Increasing the number of near-wake strips from 3 to 6 halves the panel size in chordwise direction. The result for this case is shown in figure 20. Also shown is the result for a still finer panel scheme with 6 near-wake strips of width 0.025 each. In both cases the length of the sheet has been shortened for the sake of reducing the computational effort.

The results indicate that the details of the trailing vortex roll-up are better resolved as the discretization in the near wake is refined.

Fortunately, however, the direct influence of the near wake discretization on the leading-edge vortex sheet geometry, lift and pitching moment and pressure distribution is rather small.

(ii) Comparison of VORSEP and LEV results

In figure 21 we compare the solution for the AR=1.0 delta wing at α = 20 degrees obtained by the LEV code (Ref. 53) and the one obtained by the VORSEP code. The vortex sheet computed by the LEV code appears to be less curved near the leading edge. The vortex core computed by the LEV code is further inboard than the one computed by VORSEP. Correspondingly the suction peak is shifted slightly in inboard direction. This shift is probably more than might be expected from the difference in sheet length used in the computation. The upper wing surface suction peak and correspondingly the lift coefficient computed by the LEV code for the 60-wing-panel case is lower than the corresponding values computed by VORSEP for the 100-wing-panel case.

(iii) Comparison theory and experiment

Comparison of computed results with experimental data demonstrates that panel methods have the potential to predict detailed characteristics of the vortex flow. Figure 22a compares computed vortex locations with experimental data (Ref. 37). The position of the vortices is excellently predicted by VORCON/VOR2DT. The latter method computes the development of three-dimensonal wakes using the (approximate slender-body type of) two-dimensional time-dependent analogy (Ref. 59). The position of the leading-edge vortex core as predicted by VORSEP is satisfactory and it may be verified that the location of the dent in the vortex sheet in figure 19 and 20 corresponds with the position of the trailing-edge vortex.

In figure 22b the computed pressure distribution is compared with experimental data of references 56 and 57. Although the agreement of computed and measured pressure distributions is quite satisfactory, this figure also demonstrates one of the main difficulties in comparing computed and experimental vortex flow results. It concerns the secondary separation discussed earlier. In case of the experiment of reference 57 the boundary layer was fully turbulent, while in reference 56 transition from laminar to turbulent flow took place at 60 % root chord. This corresponds with the effects of secondary separation being more pronounced in the latter case as compared with the former one. Note further that VORSEP correctly predicts the upper surface pressure increase towards the trailing edge and also that the conical flow method VORCON gives a good prediction up to 50 % root chord. The effect of secondary separation is overly evident in the laminar flow case considered in figure 22c. It is the case of the 70 deg swept delta wing at 14 deg incidence. The suction peaks predicted by the VORSEP code are not attained in the experiment of Marsden et al. (Ref. 58). The computational result of the (early version of the) LEV code shows that the measured pressure distribution is predicted quite satisfactorily. However, clearly, the 30 wing panels used are insufficient to resolve the suction peaks. In general, for wings with turbulent boundary layers, as shown in figure 22b, correlation is better than for the laminar boundary layer case shown in figure 22c.

(iv) Other planforms

A variety of applications of the LEV code has been published, including configurations like arrow wings, cambered wings, leading-edge vortex flaps and double-delta wings. In the latter case it is assumed that the flow separates all along the leading edge, but the double-branched wing vortex (see Fig. 3) is not modelled explicitly. Results for a double-delta wing of the VORSEP code with its one-vortex system capability are presented in figure 23. Comparison of the vortex sheet shapes shown in figure 23a with the flow visualization results of reference 37 show that the shape of the sheet is correct, but the second vortex is not resolved for in the panel scheme employed. Note the strong tendency of the near wake to form a double-branched trailing vortex. The computed pressure distribution on the aft portion of the wing is shown in figure 23b.

CAPTURED VORTEX METHODS

In the fitted-vortex methods the topology of the vortical flow field must be known in advance, i.e. one must decide on the location of separation lines and on the

global structure of the vortex sheets and embedded cores. For most slender wing configurations at subcritical conditions, this is not much of a problem, one knows from
experiment that above the wing only one vortex develops. However, when non-linear
compressibility effects interact with the vortical flow structure, or for geometrically more complex configurations it is difficult to derive the flow field topology a
priori. In the methods that solve the Euler or Navier-Stokes equations the vortex
sheets and cores are captured, together with the shock waves, and it appears that an
a priori knowledge of the topology of the flow field is not required.

Euler methods

One of the first applications of a finite-difference or volume method for solving Euler's equations to slender delta wings was reported by Eriksson & Rizzi
(Ref. 60). It emerged that in this time-marching method (from rest to steady flow) a
vortical flow was generated without explicitly satisfying a Kutta type of condition
at the leading edge. Comparing the computed results with experimental data and also
with results from panel method computations suggest that for the relatively coarse
mesh used in these early applications, the amount of vorticity contained in the vortical flow region is too small and the vorticity is too much spread. As a result the
computed upper surface suction peaks are lower than computed by a panel method, even
lower than measured in experiment. A further anomaly is the appearance of a second,
spike-like, suction peak right at the leading edge, whether rounded or sharp. This
indicates that the flow turns around the edge before it separates at a location
slightly inboard of the leading edge. In later applications (Ref. 61, 62) the spike-
like suction peak is not so dominantly present, while studies on the influence of
mesh density (Ref. 62) reveal a rather large influence of mesh density and lay-out on
computed pressure distributions.

The treatment of flow separation at "aerodynamically sharp" edges in methods
solving Euler's equations is a major issue. When inviscid flow turns around the edge
it will expand to the limit of vacuum pressure. However, viscous flow will separate
at the edge before the inviscid expansion limit is reached. This suggests that truly
simulating high-Reynolds number flow using Euler's equations requires some kind of
modelling (Kutta conditions). This is common practise in Euler methods for supersonic
flow employing space marching (e.g. Ref. 63), but not so in the more recent Euler
methods that use time marching.

Two reasons have been put forward for the inviscid separation phenomenon occurring in the latter type of method: (i) During the time marching process, from rest to
steady state, shock waves appear at the leading-edge which produce vorticity due to
total pressure losses, resulting in the formation of the leading-edge vortex
(Ref. 60, 61). (ii) Separation occurs because of viscous-like terms due to (explicit

and/or implicit) dissipative terms in the numerical scheme (e.g. shock-capturing algorithm, terms or procedures added for stability, surface boundary conditions) (Ref. 64, 65).

Raj (Ref. 64) reports that for the low Mach number cases he studied (using FLO-57) the flow remained subcritical throughout the evolution to steady state and still a vortical type of flow emerged. In a careful study by Newsome (Ref. 65) for the case of supersonic flow about a delta wing with subsonic (not infinite sharp) leading edges two Euler solutions evolved, one for a relatively coarse mesh with and one for a fine mesh without leading-edge vortex separation. The latter solution, with the explicit dissipative terms turned off at the edge, involves a cross-flow shock on its downstream side accompanied by a small region of distributed vorticity (Fig. 24), similar to solutions found earlier by Marconi (Ref. 66) using shock-fitting. Following the time-accurate transient solutions it was observed that the coarse-grid solution evolved from one with a cross-flow shock to one with leading-edge vortex separation. Above investigations strongly suggest that in the time-marching Euler methods numerical and artificial dissipation are responsible for the flow separating at the leading edge.

The Euler methods used in above investigations are designed for transonic and supersonic flow and the convergence to steady state is degraded badly for decreasing Mach numbers. In the Euler equations for incompressible flow this difficulty is avoided by using the artificial compressibility concept (Ref. 67), also used for solving the Navier-Stokes equations for incompressible flow (e.g. Ref. 68). To shed some more light on the matter of computing flow separation using Euler methods and in particular how realistic is the leading edge vortex resulting from numerical and artificial dissipation, in Ref. 69 we undertook a comparison between incompressible flow results of the VORSEP panel method and the ones from Rizzi's incompressible Euler method. The test case chosen is a 70 deg swept flat-plate delta wing at 20 deg incidence, a case with a well-developed vortical flow above the upper surface of the wing.

The VORSEP solution has been obtained employing 13 spanwise strips of 36 panels each, 468 panels in total. The surface of the wing is represented by 10 strips of 20 panels each, 200 panels in total, providing sufficient resolution of the suction peak on the upper wing surface.
The grid used for solving Euler's equations is of 0-0 type with 80 cells around the half-span, 40 in chordwise direction and 24 in normal direction, 76800 cells in total.

In figure 25 the shape of the vortex sheet and the vorticity contours are superimposed for three cross-flow planes. Due to the artificial dissipation implied in the numerical method to solve Euler's equations the vortex sheet is spread over a number of cells. This means that in general the vortical flow region as computed by the Euler code occupies a larger region than that enclosed by the vortex sheet computed by the panel method. It follows from figure 25 that in this respect the two solutions agree quite well, except at the first station, $x/c = 0.3$, where the vortical flow region computed by the Euler code is somewhat larger and more inboard than indicated by the vortex sheet solution. In the vortex sheet solution the surface vorticity is largest near the leading edge where the vortex sheet is most curved. In the Euler code solution, a similar behaviour, although somewhat blurred in figure 25, is observed.

Figure 26 shows the spanwise pressure distribution at three stations x = constant. The upper wing surface suction peaks, so characteristic for leading-edge vortex flow, predicted by the two methods agree quite satisfactorily. At the first station shown, $x/c = 0.3$, there is a slight shift in the position of the suction peak. This appears to be associated with the differences in the vortical flow patterns already presented in figure 25. In general the height of the suction peak, its position and its width as predicted by both methods are in close agreement, much closer than has been achieved before (on a coarser and less tuned mesh, e.g. Ref. 38). On the lower wing surface the two pressure distributions are virtually identical. Note that there is no pressure spike at the leading edge, indicating that the flow separates correctly right at the edge.

The finite-difference/volume methods require a spatial grid that is sufficiently fine in regions of vorticity production and also, to avoid excessive spreading of shear layers, in regions with concentrated vorticity. An alternative method, due to Rehbach (Ref. 70) avoids excessive spreading and does not require the construction of a spatial mesh. In the (incompressible flow) method the velocity field is expressed as the contribution due to the "bound" wing-surface vorticity (approximated by the vortex-lattice method) and the contribution due to free floating "vorticity-carrying particles". The latter simulate the spatial vorticity distribution. The approach followed is a time-dependent one, starting from rest the vortical particles are traced. At each timestep the strength of the "bound" vortices is determined by satisfying, for given spatial vorticity distribution, the normal velocity boundary condition on the wing. Subsequently, from separation lines, here sharp edges, new vortical particles are shed into the flow field. The vortical particles already in the flow field are convected with the local velocity while the rate of change of the total amount of vorticity contained in any particle is obtained from an integral form of the vorticity transport equation. In this grid-free method for solving Euler's equations for incompressible flow vorticity is produced at predetermined locations only, employing a vortex shedding mechanism derived from Kutta condition type considerations. A drawback of the Rehbach's method is that, in general, just like the discrete-vortex methods discussed earlier, pressure distributions are not resolved very accurately.

Navier-Stokes methods

Krause, Shi & Hartwich (Ref. 71) solve the incompressible (low-Reynolds number) viscous as well as the inviscid flow about a delta wing of unit aspect ratio at 15 degrees incidence, using a finite-volume technique on a rectangular grid. It is found that the influence of the Reynolds number on the development of the leading-edge vortex is negligible. It is also indicated that in effect the truncation errors associated with the finite-difference scheme for inviscid flow introduce an artificial vorticity production term in otherwise irrotational flow. Fujii & Kutler (Ref. 72) using a finite difference method solve the Reynolds-averaged laminar thin-layer Navier-Stokes equations for compressible flow about delta wings and strake-wing configurations. At the leading edge similar pressure spikes occur as do show up in the results of early finite-difference methods for solving Euler's equations. These again indicate that the flow is attached at the leading edge and separates a short distance inboard of the leading edge. In general this results in a weaker vortex than in case the flow separates at the edge. Although laminar flow was assumed there is no sign of a large secondary separation so dominantly present in laminar-flow experiments.

Supersonic flow applications of the laminar Navier-Stokes equations have been reported by Newsome (Ref. 65), see figure 24, in his comparison of conical Euler and Navier-Stokes solutions. This study indicates a relatively large difference between the two solutions, due to differences in the way the flow separates at the leading edge (or does not separate at all) and due to secondary separation not accounted for in the Euler method. Rizetta & Chang (Ref. 73) obtained laminar and turbulent Navier-Stokes solutions for a sharp-edged delta wing in supersonic flow. As is the case in Ref. 65 some form of secondary separation appears to take place just underneath the primary vortex core while differences between laminar and turbulent flow appear to be small.

It must be emphasized here that it is not quite clear to what extent the above solutions are true solutions of the Navier-Stokes equations. This relates to questions like whether the residue has been reduced sufficiently so that viscous effects are no longer swamped by truncation error, how are non-linear instabilities suppressed, etc.

CONCLUDING REMARKS

-   In the last few years considerable progress has been made in the development of computational methods for the prediction of the aerodynamic characteristics of configurations with leading-edge vortex flow.

Three classes of computational methods had been distinguished:

(i) Rigid-vortex methods, specifically methods employing the leading-edge suctio
analogy and its modifications. These methods are widely used for preliminary analysis and design of configurations with leading-edge vortex flow. The methods predic
forces and moments on configurations with vortex flow, but do not provide the details of the flow. The next two classes do provide the detailed simulation of the
flow about configurations with leading-edge vortex flow.

(ii) Fitted-vortex methods, of which the discrete-vortex methods appear to have a
attractive flexibility but often require procedures to avoid or delay numerical di
ficulties due to the singular velocity field associated with an assembly of vortic
or line vortices. The panel methods explicitly use vortex sheets and vortex cores
during the computation. This renders panel methods a useful tool to generate accurate results for wings with a well-defined vortical flow pattern, but much less
practical for cases where a priori the vortex flow topology is not precisely known
(iii) Captured-vortex methods, methods solving the Euler or the Navier-Stokes equa
tions, represent vortex sheets and cores implicitly. Using these methods for computing the flow about a configuration with extensive vortex flow appears very promising. However, for these methods it appears that also here the topology of the
vortex flow has to be known in some detail to enable the construction of a computa
tional grid with sufficient resolution at vortex layers and cores. Also the sometimes excessive spreading of vortical-flow regions and the vorticity production pr
cess (Kutta conditions) for the Euler methods need further investigation. A furthe
important aspect of the captured-vortex methods is that they also account for non-
linear compressibility effects (shockwaves) and for the mutual interaction of vortical flow regions and shockwaves.

From the comparison of pressure distributions computed by panel and Euler methods
with measured values it is clear that secondary separation effects are often quite
important. It has been demonstrated that using a panel method some effects of
secondary separation can be simulated.

In Euler methods the secondary separation will have to be modelled explicitly by
specifying the separation line(s) and enforcing Kutta type of conditions. Ideally
they have to be found as part of the solution. The latter undoubtedly will involve
strongly coupled viscous-inviscid interaction of the external flow with the bounda
layer on the wing upper surface. This problem is closely related to the problem of
vortex flow separation from the smooth surface of slender bodies and has not yet
been considered extensively for wings or other three-dimensional configurations. I
Navier-Stokes methods the secondary separation is found as part of the solution.

The occurrence of vortex breakdown is of great importance in the design of slender
wings and especially strake-wing configurations. Although vortex breakdown has bee
investigated quite intensively, it is not yet fully understood. The prediction of
the breakdown (apparently turbulent and unsteady) of the leading edge vortex embedded within the full three-dimensional flow field has not been very succesful. C

the other hand, approximate (axisymmetric) theories for isolated or confined vortex cores have been proposed and have enjoyed some success in simulating some of the features of vortex breakdown. Luckring (Ref. 74) has set up a composite theory of matching an inner core solution with the outer LEV solution. The core flow was computed, a posteriori, using a quasi-cylindrical approximation to the laminar incompressible Navier-Stokes equations. However, breakdown of the quasi-cylindrical approximation could not be correlated with observed positions of vortex breakdown, amongst others due to the highly three-dimensionality of the flow field at the matching location. Luckring also derived from empirical analysis that observed location of vortex breakdown could be correlated with the occurrence (at the edge of the vortex) of a critical helix angle in conjunction with an adverse pressure gradient. Luckring's as well as other investigations quite strongly suggest that vortex breakdown might well be primarily inviscid in nature. This then suggests that Euler methods may exhibit some form of vortex breakdown.
Detailed comparison of results of different computational methods, say, for a sharp edged delta wing of unit-aspect ratio would be very valuable. Also, more exploratory flow field investigations that reveal the vortical flow field are indispensable, both for the fitted and the captured vortex methods.

REFERENCES

1. Peake, D.J.; Tobak, M.: Three-dimensional Interactions and Vortical Flows with Emphasis on High Speeds. NASA TM 81169, 1980, also AGARDograph 252, 1980.

2. Ashley, H.; and Landahl, M.T.: Aerodynamics of Wings and Bodies. Adison-Wesley Publ. Co. Inc., 1965.

3. Sytsma, H.A.; Hewitt, B.L.; Rubbert, P.E.: A Comparison of Panel Methods for Subsonic Flow Computation. AGARD AG-241, 1979.

4. Holst, T.L.; Slooff, J.W.; Yoshihara, H.; Ballhaus Jr., W.F.: Applied Computational Transonic Aerodynamics. AGARDograph 266, 1982.

5. Polhamus, E.C.: Applying Slender Wing Benefits to Military Aircraft. J. Aircraft, Vol. 21, No. 8, 1984, pp. 545-559.

6. Szodruch, J; Ganzer, U.: On the Lee-Side Flow for Slender Delta Wings at High Angle of Attack. AGARD-CP-247, 1979, Paper 21.

7. Legendre, R.: Écoulement au Voisinage de la Pointe Avant d'une Aile à Forte Flèche aux Incidences Moyennes. La Rech. Aéro. Vol. 30, 1952, pp. 3-8; Vol. 31, 1952, pp. 3-6 and Vol. 35, 1952, pp. 7-8.

8. Earnshaw, P.B.: An Experimental Investigation of the Structure of a Leading-Edge Vortex. ARC R&M No. 3281, 1962.

9. Wentz, W.H.; McMahon, M.C.: An Experimental Investigation of the Flow Fields about Delta and Double-Delta Wings at Low Speeds. NASA CR 521, 1966.

10. Fink, P.T.; Taylor, J.: Some Early Experiments on Vortex Separation. ARC R&M No. 3489, 1967.

11. Hummel, D.; On the Vortex formation over a Slender Wing at Large Incidence. AGARD-CP-247, 1979, Paper 15.

12. Verhaagen, N.G.: An Experimental Investigation of the Vortex Flow over Delta and Double-Delta Wings at Low Speeds. AGARD-CP-342, 1983, Paper 7.

13. Luckring, J.M.: Aerodynamics of Strake-Wing Interactions, J. Aircraft, Vol. 16, No. 11, 1979, pp. 756-762.

14. Brennenstuhl, U.; Hummel, D.: Vortex Formation over Double-Delta-Wings. ICAS Paper 82-6.6.3, 1982.

15. Manor, D.; Wentz, Jr. W.H.: Flow over Double-Delta Wing and Wing-Body at High J. Aircraft, Vol. 22, No. 1, 1985, pp. 78-82.

16. Smith, J.H.B.: Theoretical Modelling of Three-Dimensional Vortex Flows in Aerodynamics. AGARD-CP-342, 1983, Paper 17.

17. Miller, D.S.; Wood, R.M.: An Investigation of Wing Leading-Edge Vortices at Supersonic Speeds. AIAA Paper 83-1816, 1983.

18. Vorropoulos, G.; Wendt, J.F.: Laser Velocimetry Study of Compressibility Effects on the Flow Field of a Delta Wing. AGARD-CP-342, 1983, Paper 9.

19. Bollay, W.: A Nonlinear Wing Theory and Its Application to Rectangular Wings of Small Aspect Ratio, ZAMM, Vol. 19, pp. 21-35, 1939.

20. Gersten, K: Nichlineare Tragflächentheorie insbesondere für Tragflügel mit kleinem Seitenverhältnis. Ing. Arch. Vol. 30, 1961, pp. 431-452. Also AGARD Rept. 342, 1961.

21. Polhamus, E.C.: A Concept of the Vortex Lift of Sharp-Edged Delta Wings Based on Leading-Edge Suction Analogy. NASA TN D-3767, 1966.

22. Lamar, J.E.: Analysis and design of strake-wing configurations. J. Aircraft, Vol. 17, No. 1, 1980, pp. 20-27.

23. Murman, E.M.; Stremel, P.M.: A Vortex Wake Capturing Method for Potential Flow Calculations. AIAA Paper 82-0947, 1982.

24. Steinhof, J.; Suryanarayanan, K.: The Treatment of Vortex Sheets in Compressible Potential Flows, AIAA Paper 83-1881, 1983.

25. Brown, C.E.; Michael, W.H.: Effect of Leading-Edge Separation on the Lift of a Delta Wing. J. of Aeron. Sciences, Vol. 21, 1954, pp. 690-694.

26. Mangler, K.W.; Smith, J.H.B.: A Theory of the Flow past a Slender Delta Wing with Leading-Edge Separation. Proc. Roy. Soc. A, Vol. 251, 1959, pp. 200-217.

27. Smith, J.H.B.: Improved Calculations of Leading-Edge Separation from Slender, Thin, Delta Wings. Proc. Roy. Soc. London, A 306, 1968, pp. 67-90. Also RAE TR 66070, 1966.

28. Clark, R.W.: Non-Conical Flow past Slender Wings with Leading-Edge Vortex Sheets. RAE TR 76037, 1976.

29. Sacks, A.H.; Lundberg, R.E.; Hanson, Ch.W.: A Theoretical Investigation of the Aerodynamics of Slender Wing-Body Combinations Exhibiting Leading-Edge Separation. NASA CR-719, 1967.

30. Xieyuan, Y.: Roll-Up of Strake Leading/Trailing-Edge Vortex Sheets for Double-Delta Wings. J. Aircraft, Vol. 22, No. 1, 1985, pp. 87-89.

31. Fink, P.T.; Soh, W.K.: Calculation of Vortex Sheets in Unsteady Flow and Applications in Ship Hydrodynamics. Proc. 10th Symposium on Naval Hydrodynamics, Cambridge, Mass., 1974, pp. 463-491.

32. Moore, D.W.: A Numerical Study of the Roll-Up of a Finite Vortex Sheet. JFM, 63, 1974, pp. 225-235.

33. Maskew, B.: Subvortex Technique for the Close Approach to a Discretized Vortex Sheet. J. Aircraft, Vol. 14, No. 2, 1977, pp. 188-193.

34. Peace, A.J.: A Multi-Vortex Model of Leading-Edge Vortex Flows. Int. J. Num. Meth. in Fluids, Vol. 3, 1983, pp. 543-565.

35. Maskew, B.; Rao, B.M.: Calculation of Vortex Flows on Complex Configurations. ICAS Paper 82-6.2.3, 1982. See also Flows with Leading-Edge Vortex Separation. NASA CR 165858, 1982.

36. Nathman, J.K.: Estimation of Wake Roll-Up over Swept Wings, AIAA Paper 84-2174, 1984.

37. Hoeijmakers, H.W.M.; Vaatstra, W.; Verhaagen, N.G.: On the Vortex Flow over Delta and Double-Delta Wings. J. of Aircraft, Vol. 20, No. 9, 1983, pp. 825-832.

38. Hoeijmakers, H.W.M.: Numerical Computation of Vortical Flow about Wings. VKI-LS Computational Fluid Dynamics, 1984. Also NLR MP 83073 U, 1983.

39. Verhaagen, N.G.; Kruisbrink, A.C.H.: The Entrainment Effect of a Leading-Edge Vortex. AIAA Paper 85-1584, 1985.

40. De Bruin, A.C.: Laminar and Turbulent Boundary-Layer Calculations on the Leeward Surface of a Slender Delta Wing at Incidence. NLR MP 84040 U, 1984.

41. Wai, J.C.; Baillie, J.C.; Yoshihara, H.: Computation of Turbulent Separated Flows over Wings. Proc. 3rd Symp. on Num. and Physical Aspects of Aerod. Flows, Long Beach, Calif., Jan. 1985, Paper 11.

42. DeJarnette, F.R.; Woodson, S.H.: Numerical and Experimental Determination of Secondary Separation on Delta Wings in Subsonic Flow. AIAA Paper 84-2175, 1984.

43. Bakker, P.G.: Private communication, Delft University of Technology, Delft, The Netherlands, May 1985.

44. Frink, N.T.: Analytical Study of Vortex Flaps on Highly Swept Delta Wings. ICAS Paper 82-6.7.2, 1982.

45. Erickson, G.E.: Application of the Vortex Sheet Theory to Slender Wings with Leading-Edge Vortex Flaps. AIAA Paper 83-1813, 1983.

46. Hoffler, K.D.; Rao, D.M.: An Investigation of the Tabled Vortex Flap. AIAA Paper 84-2173, 1984.

47. Belotserkovskii, S.M.: Calculation of the Flow about Wings of Arbitrary Planform at a Wide Range of Angles of Attack. RAE Lib. Transl. 1433, 1968.

48. Rehbach, C.: Étude Numérique de Nappes Tourbillonnaires Issues d'une Ligne de Découllement près du Bord d'Attaque. Rech. Aérosp. No. 1973-6, 1973, pp. 325-330.

49. Rehbach, C.: Numerical Investigation of Leading-Edge Vortex for Low-Aspect Ratio Thin Wings. AIAA Journal, Vol. 14, No. 2, 1976, pp. 253-255.

50. Kandil, O.A.; Balakrishnan, L.: Recent Improvements in the Prediction of the Leading and Trailing Edge Vortex Cores of Delta Wings. AIAA Paper 81-1263, 1981.

51. Rusak, Z.; Wasserstrom, E.; Seginer, A.: Numerical Calculation of Nonlinear Aerodynamics of Wing-Body Configurations. AIAA Journal, Vol. 21, No. 7, 1983, pp. 929-936.

52. Johnson, F.T.; Tinoco, E.N.; Lu, P.; Epton, M.A.: Three-Dimensional Flow over Wings with Leading-Edge Vortex Separation, AIAA Journal, Vol. 18, No. 4, 1980, pp. 367-380. Also AIAA Paper 79-0282, 1979.

53. Johnson, F.T.; Lu, P.; Tinoco, P.; Epton, M.A.: An Improved Method for the Solution of Three-Dimensional Leading-Edge Vortex Flows. Volume I - Theory Document, NASA CR 3278. Volume II - User's Guide and Programmer's Document, 1980.

54. Luckring, J.M.; Schoonover Jr. W.E.; Frink, N.T.: Recent Advances in Applying Free Vortex Sheet Theory for the Estimation of Vortex Flow Aerodynamics. AIAA Paper 82-0095, 1982.

55. Hoeijmakers, H.W.M.; Bennekers, B.: A Computational Model for the Calculation of the Flow About Wings with Leading-Edge Vortices. AGARD CP-247, 1979, Paper 25.

56. Verhaagen, N.G.: Measurement of the Pressure Distribution on a Biconvex Delta Wing of Aspect Ratio 1, Unpubl. Rep. Dep. Aero. Eng. Delft University of Technology, 1979.

57. Hummel, D.; Redeker, G.: Experimentelle Bestimmung der gebundene Wirbellinien sowie des Strömungs-Verlaufs in der Umgebung der Hinterkante eines Schlanken Deltaflügels. Abh. Braunschweig, Wiss. Ges., Vol. 22, 1972, pp. 273-290.

58. Marsden, D.J.; Simpson, R.W.; Rainbird, W.J.: The Flow over Delta Wings at Low Speeds with Leading-Edge Separation. Cranfield College of Aeronautics. Rep. No. 114, 1957.

59. Hoeijmakers, H.W.M.; Vaatstra, W.: A Higher-Order Panel Method Applied to Vortex Sheet Roll-Up. AIAA Journal Vol. 21, No. 4, 1983, pp. 516-523.

60. Eriksson, L.E.; Rizzi, A.: Computation of Vortex Flow Around Wings Using the Euler Equations. Proc. 4th GAMM Conf. on Num. Meth. in Fl. Mech., ed. H. Viviand, Vieweg Verlag, Paris, 1981, pp. 87-105.

61. Hitzel, S.M.; Schmidt, W.: Slender Wings with Leading-Edge Vortex Separation: A Challenge for Panel Methods and Euler Solvers, J. of Aircraft, Vol. 21, No. 10, 1984, pp. 751-759.

62. Rizzi, A.: Euler Solutions of Transonic Flows around the Dillner Wing - Compared and Analyzed. AIAA Paper 84-2142, 1984.

63. Klopfer, G.H.; Nielsen, J.N.: Computational Fluid Dynamic Applications to Missile Aerodynamics. AGARD CP 336, 1982, Paper 3.

64. Raj, P.: Computational Simulation of Free Vortex Flows Using an Euler Code. ICAS Paper 84-1.3.1, 1984.

65. Newsome, R.W.: A Comparison of Euler and Navier-Stokes solutions for Supersonic Flow over a Conical Delta Wing. AIAA Paper 85-0111, 1985.

66. Marconi, F.: Shock Induced Vortices on Elliptic Cones in Supersonic Flow. AIAA Paper 85-0433, 1985.

67. Rizzi, A.: Computation of Inviscid Incompressible Flow with Rotation, JFM, Vol. 153, 1985, pp. 275-312.

68. Chang, J.L.C.; Kwak, D.: On the Method of Pseudo Compressibility for Numerically Solving Incompressible Flows. AIAA Paper 84-0252, 1984.

69. Hoeijmakers, H.W.M.; Rizzi, A.: Vortex-Fitted Potential Solution Compared with Vortex-Captured Euler Solution for Delta Wing with Leading-Edge Vortex Separation. AIAA Paper 84-2144, 1984.

70. Rehbach, C.: Numerical Calculation of Three-Dimensional Unsteady Flow with Vortex Sheets. AIAA Paper 78-111, 1978.

71. Krause, E.; Shi, X.G.; Hartwich, P.M.: Computation of Leading-Edge Vortices. AIAA Paper 83-1907, 1983.

72. Fujii, K.; Kutler, P.: Numerical Simulation of the Leading-Edge Separation Vortex for a Wing and Strake-Wing Configuration. AIAA Paper 83-1908, 1983, see also AIAA Paper 84-1550, 1984.

73. Rizetta, D.P.; Shang, J.S.: Numerical Simulation of Leading-Edge Vortex Flow. AIAA Paper 84-1544, 1984.

74. Luckring, J.M.: Theory for the Core of a Three-Dimensional Leading-Edge Vortex. AIAA Paper 85-0108, 1985.

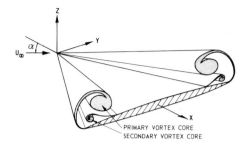

Fig. 1   Vortex flow about delta wing

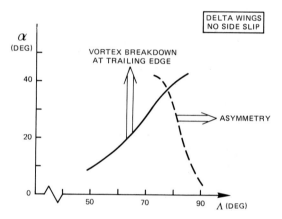

Fig. 2   Boundaries for onset of vortex breakdown and asymmetry
(low speed)

253

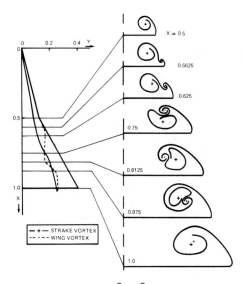

Fig. 3 Flow pattern above $76^{\circ}/60^{\circ}$ double-delta wing (sketch)

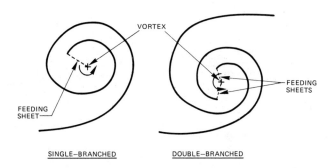

Fig. 4 Isolated vortex/feeding sheet(s) vortex core model

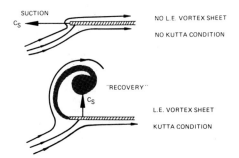

Fig. 5   Polhamus' leading-edge suction analogy

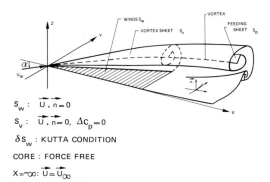

$S_w : \quad \overrightarrow{U} \cdot \overrightarrow{n} = 0$

$S_v : \quad \overrightarrow{U} \cdot \overrightarrow{n} = 0, \ \Delta c_p = 0$

$\delta s_w : \text{KUTTA CONDITION}$

CORE : FORCE FREE

$X = -\infty : \overrightarrow{U} = \overrightarrow{U_\infty}$

Fig. 6   Potential flow with free vortex sheets

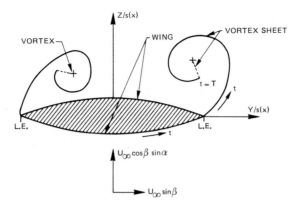

Fig. 7   Slender-body approximation, cross-flow plane

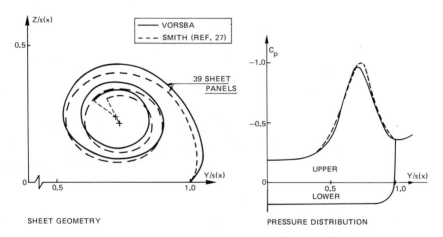

SHEET GEOMETRY

PRESSURE DISTRIBUTION

Fig. 8   Flat-plate delta wing, AR = 1.0, α = 14 deg computed conical
solution

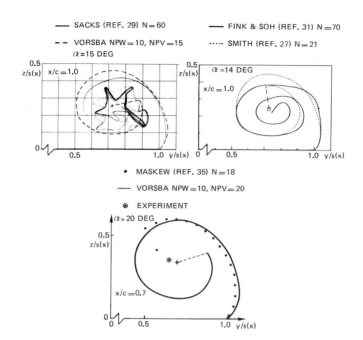

Fig. 9   Comparison of discrete-vortex methods with panel methods

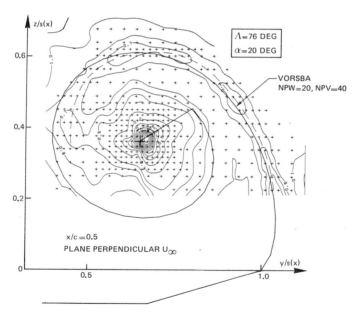

Fig. 10   Comparison of measured total pressure contours and computed
(VORSBA) vortex sheet geometry

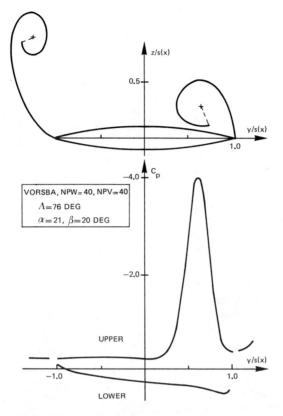

Fig. 11 Conical-flow solution thick delta wing

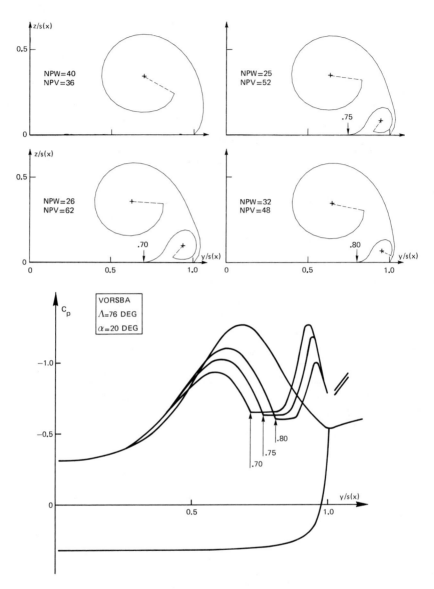

Fig. 12 Computed secondary separation effects

259

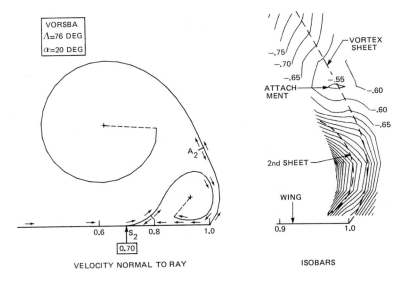

VELOCITY NORMAL TO RAY   ISOBARS

Fig. 13 Details of secondary separation solution

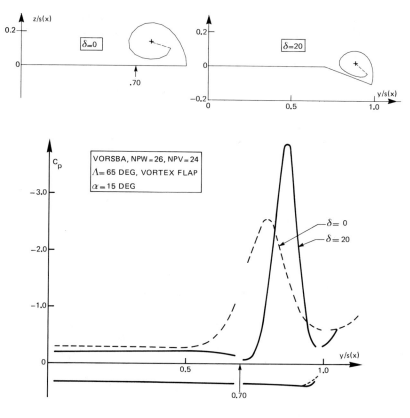

Fig. 14   Vortex-flap study

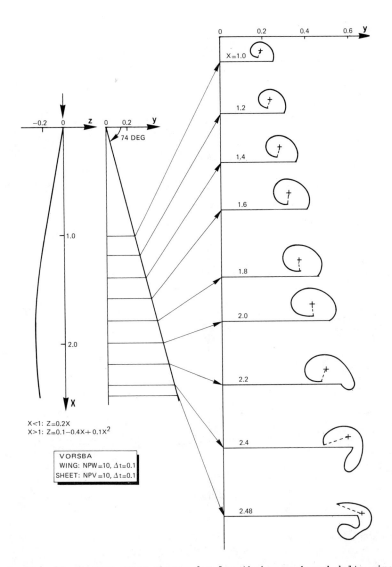

Fig. 15 Vortex sheet shapes for lengthwise cambered delta wing

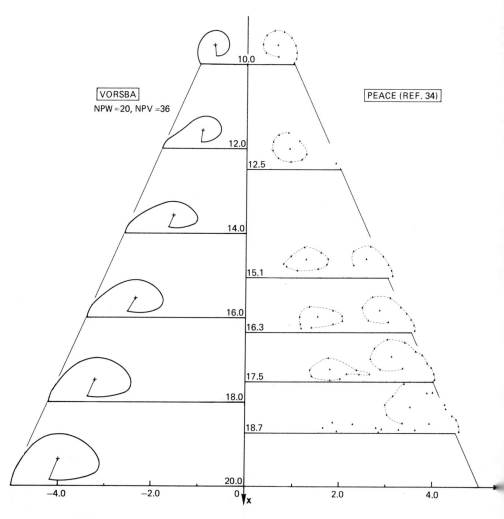

Fig. 16 Vortex sheet shapes for 84.3/68.2 double delta wing, α = 10.3 deg.
(Portions of Peace [Ref. 34] reprinted by permission of John Wiley & Sons, Ltd.)

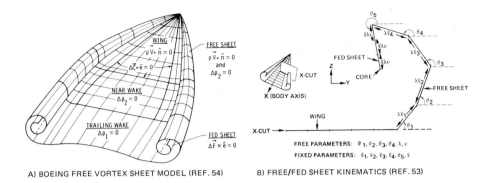

Fig. 17   Three-dimensional free-vortex-sheet method LEV.
(Research performed at the NASA Langley Research Center.)

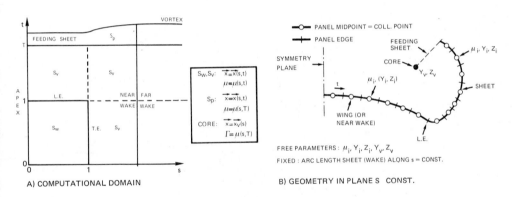

Fig. 18   Three-dimensional free-vortex-sheet method VORSEP (Ref. 55)

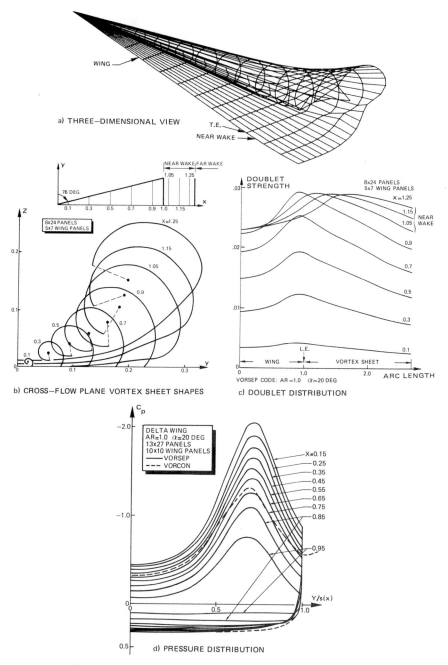

a) THREE—DIMENSIONAL VIEW

b) CROSS—FLOW PLANE VORTEX SHEET SHAPES

c) DOUBLET DISTRIBUTION

d) PRESSURE DISTRIBUTION

Fig. 19   Solution for AR = 1.0 delta wing at 20 deg incidence

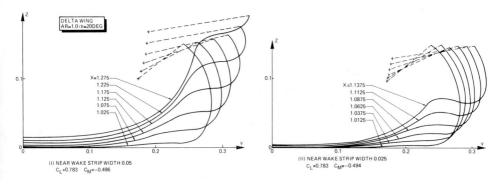

Fig. 20   VORSEP: Computed near wake shape, AR = 1.0 delta wing,
α = 20 deg

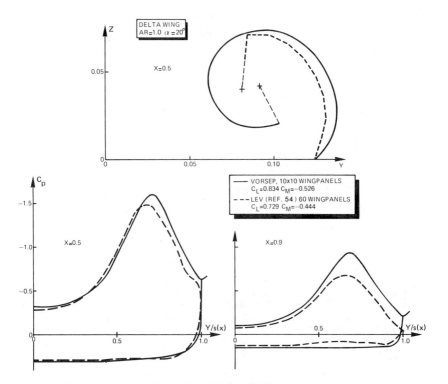

Fig. 21   Comparison of panel-method solutions

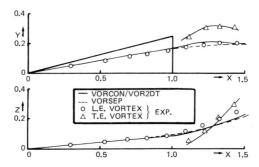

Fig. 22a   Position of vortex cores above AR = 1.0
delta wing at α = 20 deg. Comparison
with experiment (Ref. 37)

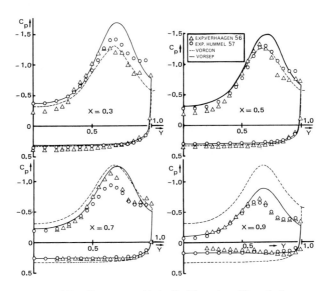

Fig. 22b   Pressure distribution for AR = 1.0
delta wing at α = 20 deg. Comparison
with turbulent flow experiment

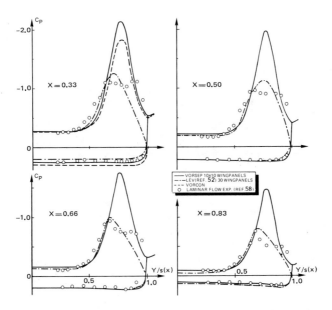

Fig. 22c Pressure distribution for AR = 1.46
delta wing at α = 14 deg. Comparison
with laminar flow experiment

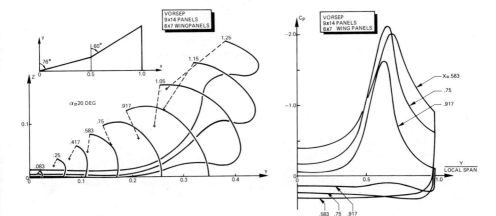

Fig. 23a Vortex sheet shape computed for
double-delta wing at α = 20 deg

Fig. 23b Computed pressure distribution for double-
delta wing at α = 20 deg

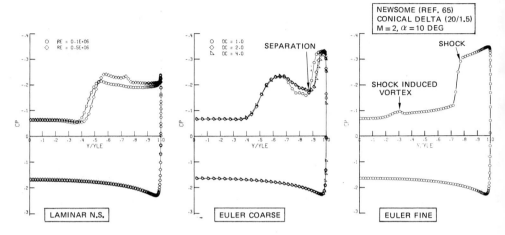

Fig. 24 Comparison of Euler and Navier-Stokes solutions (Ref. 65)

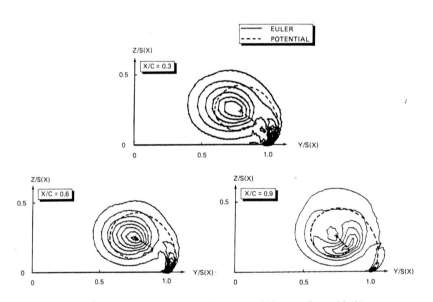

Fig. 25 Comparison of vortex sheet position and vorticity
magnitude contours ($\Lambda = 70$, $\alpha = 20$ deg)

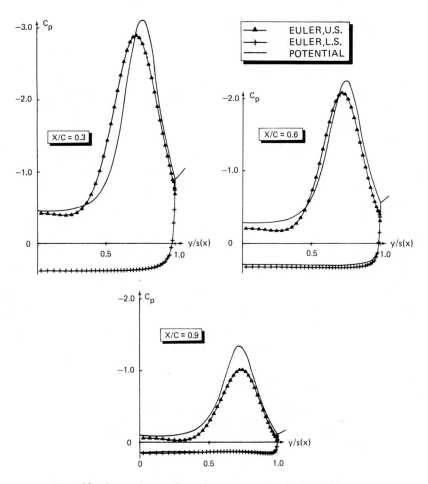

Fig. 26 Comparison of surface pressure distribution

Comparison of Measured and Computed Pitot Pressures In a Leading Edge Vortex From a Delta Wing

EARLL M. MURMAN AND KENNETH G. POWELL

Introduction

A number of papers[1-9] have appeared in recent years reporting numerical solutions of the Euler equations for flows past sharp edge delta wings with leading edge vortices. A schematic diagram of the problem under consideration is shown in Figure 1. Solutions have been compared with experiment [1,8], with incompressible panel method calculations [6], with other Euler equation solutions [3,5], and with Navier-Stokes calculations [9]. Although one cannot conclude that these comparisons show complete agreement, it is reasonable to say that they establish that finite volume calculations of the Euler equations appear to give reasonable predictions of the primary vortex generated by this configuration. A point of some concern in these papers [1,3,4] relates to the mechanism by which the Euler equations cause the flow to separate from the body and form a strong vortex which has associated with it a significant total pressure loss.

Powell *et al* [10] investigated the mechanism by which discrete solutions of the Euler equations can lead to the formation of a leading edge vortex. The particular focus of the study was the explanation of the large total pressure losses found in the vortex cores. The study concluded that the magnitude of these losses is insensitive to any of the computational parameters. A simple explanation was proposed based upon the kinematical argument that any vortical region with finite thickness must have associated with it a total pressure loss. Furthermore, the magnitude of the loss is established by the jump in tangential velocity across the vortical region, not by the distribution of velocity within it. Only a classical contact discontinuity which has no thickness permits a "uniform" total pressure field (although the total pressure is undefined at the contact surface location). The discrete form of the Euler equations admits a true contact discontinuity only under ideal circumstances which cannot be realized for calculations requiring realistic grids and added damping terms for the discrete equations. When discrete forms of the Euler equations are used to capture vortical regions, they only admit "sheets" with thickness, and, hence, there must be a total pressure loss.

From the analysis in the above paper, the following explanation was proposed for the observed results from finite volume solutions to the Euler equations for flows with leading edge vortices. For sharp edge geometries, the physical separation point for the flow is set by the geometry, not by magnitude of the viscosity (Reynolds number); i. e. there is a Kutta condition. Finite volume calculations using centered difference formulas appear to enforce the Kutta condition automatically through the added damping terms. Upwind methods appear to have somewhat different behavior[11]. Once the flow has separated, it forms a dominant leading edge vortex with a diffuse core in which there is a significant circumferential velocity variation and total pressure loss. The magnitude of the total pressure loss is established by the aerodynamic parameters (angle of attack, sweep angle, Mach number, etc.) and not the computational parameters. Grid spacing and levels of artificial viscosity do affect the size of the computed vortex core, with smaller values of each yielding tighter or less diffuse cores. Even with reasonably refined calculations [3,7], the core appears to be diffuse rather than a spiral vortex sheet type structure.

In this paper, calculations are presented for one flow condition for which experimental pitot pressure measurements through the vortex core were reported by Monnerie and Werlé [12]. The model is a 75 degree swept flat plate wing tested at a freestream Mach number of 1.95 and 10 degrees angle of attack.

The Reynolds number based on the wing chord is $.95 \times 10^6$. The computed solutions are compared with the measured data. They are also discussed in some detail to report interesting observed results.

Governing Equations

The three-dimensional unsteady Euler equations in conservation form are:

$$\frac{\partial}{\partial t}\begin{bmatrix}\rho \\ \rho u \\ \rho v \\ \rho w \\ \rho E\end{bmatrix} + \frac{\partial}{\partial x}\begin{bmatrix}\rho u \\ \rho u^2 + p \\ \rho uv \\ \rho uw \\ \rho u h_0\end{bmatrix} + \frac{\partial}{\partial y}\begin{bmatrix}\rho v \\ \rho uv \\ \rho v^2 + p \\ \rho vw \\ \rho v h_0\end{bmatrix} + \frac{\partial}{\partial z}\begin{bmatrix}\rho w \\ \rho uw \\ \rho vw \\ \rho w^2 + p \\ \rho w h_0\end{bmatrix} = 0. \tag{1}$$

The equation of state

$$\frac{p}{\rho} = (\gamma - 1)\left[E - \frac{u^2 + v^2 + w^2}{2}\right] \tag{2}$$

and the supplementary relation for the total enthalpy

$$h_0 = E + \frac{p}{\rho} \tag{3}$$

close the set of equations. Introducing the conical variables

$$\eta = \frac{y}{x} \qquad \xi = \frac{z}{x} \qquad r = \sqrt{x^2 + y^2 + z^2} \tag{4}$$

and assuming conical self-similarity (the solution is independent of r), the Euler equations become

$$\frac{r}{\sqrt{1 + \eta^2 + \xi^2}}\frac{\partial}{\partial t}\begin{bmatrix}\rho \\ \rho u \\ \rho v \\ \rho w \\ \rho E\end{bmatrix} + \frac{\partial}{\partial \eta}\begin{bmatrix}\rho\,(v - \eta u) \\ \rho u(v - \eta u) - \eta p \\ \rho v(v - \eta u) + p \\ \rho w(v - \eta u) \\ \rho h_0(v - \eta u)\end{bmatrix} + \frac{\partial}{\partial \xi}\begin{bmatrix}\rho\,(w - \xi u) \\ \rho u(w - \xi u) - \xi p \\ \rho v(w - \xi u) \\ \rho w(w - \xi u) + p \\ \rho h_0(w - \xi u)\end{bmatrix} + 2\begin{bmatrix}\rho u \\ \rho u^2 + p \\ \rho uv \\ \rho uw \\ \rho u h_0\end{bmatrix} = 0. \tag{5}$$

In the above, the unsteady terms have been retained for the iterative procedure used to reach the steady, conically self-similar solution. The equations are solved on the unit sphere by setting $r = 1$. The boundary conditions accompanying these equations are no flow through the body and freestream conditions ahead of the bow shock. The Kutta condition will be enforced implicitly by the numerical method.

Solution Procedure

The basic solution scheme is a finite volume spatial discretization with a multi-stage integration in time, as proposed by Jameson et al [13]. A set of partial differential equations of the form

$$\frac{\partial}{\partial t}\mathbf{U} + \frac{\partial}{\partial \eta}\mathbf{F} + \frac{\partial}{\partial \xi}\mathbf{G} + \mathbf{H} = 0 \tag{6}$$

is transformed to an integral equation by the divergence theorem and discretized as [3]

$$\frac{d\mathbf{U}_{ij}}{dt}A_{ij} + \sum_{\ell=1}^{4}[\mathbf{F}_\ell n_{\xi_\ell} + \mathbf{G}_\ell n_{\eta_\ell}] + \mathbf{H}_{ij}A_{ij} = 0. \tag{7}$$

A_{ij} is the area of a computational cell and the summation represents the fluxes through the four cell sides. The flux vectors $\mathbf{F}_\ell, \mathbf{G}_\ell$ on the cell sides are evaluated from simple averages of flux vectors at the cell centers $\mathbf{F}_{ij}, \mathbf{G}_{ij}$. A multistage integration scheme[13] is used to solve the equations in time to reach a steady state. In order to accelerate the iterative solution, the time step in each cell is determined by the linear stability criteria for that cell. Since the time steps are different for each cell, the solution is not time accurate.

This scheme requires added second and fourth order damping to capture shocks and damp unwanted high frequency modes of the discrete solutions. The damping formulation follows Rizzi and Eriksson [14]. The second order damping is pressure-weighted, and is of the form

$$\mathbf{D}_2(\mathbf{U}) = \nu_2 \left[\delta_i \left(\frac{\delta_i^2 p}{max|\delta_i^2 p|} \delta_i \mathbf{U} \right) + \delta_j \left(\frac{\delta_j^2 p}{max|\delta_j^2 p|} \delta_j \mathbf{U} \right) \right] \qquad (8)$$

where δ_i and δ_j are undivided central difference operators defined by

$$\delta_i(\mathbf{U}_{i,j}) = \mathbf{U}_{i+\frac{1}{2},j} - \mathbf{U}_{i-\frac{1}{2},j} \qquad \delta_j(\mathbf{U}_{i,j}) = \mathbf{U}_{i,j+\frac{1}{2}} - \mathbf{U}_{i,j-\frac{1}{2}}. \qquad (9)$$

The fourth-order damping is unweighted, and of the form

$$\mathbf{D}_4(\mathbf{U}) = \nu_4 \left[\delta_i^4 \mathbf{U} + \delta_j^4 \mathbf{U} \right]. \qquad (10)$$

The pressure switch in the second order damping is normalized to vary from 0 to 1 throughout the flow field. It is designed to turn on at embedded shock waves. In these calculations it was found that the second difference damping needed to be turned on at the wing tip. This was done by setting the pressure switch to 1 in a few cells in this region. In the neighborhood of the tip and embedded shock waves, the second difference damping is proportional to the first power of the mesh spacing. Away from these regions it is proportional to the cube of the mesh spacing. The fourth difference damping is unweighted and everywhere proportional to the cube of the mesh spacing. Values of $\nu_2 = .05$ and $\nu_4 = .01$ were used for the calculations.

The body boundary condition is a pressure extrapolation of the form

$$p_{j_{body}} = p_{j_{i=1}}. \qquad (11)$$

This is consistent since the radius of curvature is infinite on each side of the plate, and the j=constant grid lines are normal to the plate. The far-field boundary condition is enforced by fitting the bow shock as the outer boundary of the computational domain. This is preferable to capturing the bow shock, since it eliminates the need for high resolution far from the body.

The grid is generated by a conformal mapping. The (η, ξ) plane is transformed to a complex ς-plane in which the flat plate becomes a circle. This is achieved by a Joukowski transformation of the form

$$\xi + i\eta = \varsigma + \frac{tan^2 \left[\frac{\pi}{2} - \Lambda \right]}{2\varsigma} \qquad (12)$$

where Λ is the sweep angle. Grid points are generated along rays in the ς-plane by the formulas

$$r_i = -log \left[1 - (1 - e^{-\sigma}) \left(\frac{i-1}{i_{max}-1} \right) \right] \qquad (13)$$

$$\varsigma_{ij} = \varsigma_{body_j} + \left[\varsigma_{shock_j} - \varsigma_{body_j} \right] r_i \qquad (14)$$

in which σ is a stretching parameter. This radial distribution, proposed by Marconi [15], is linear near the body in the limit of small σ. Figure 2 shows the 32 x 32 grid used for the calculation. The horizontal line corresponds to the flat plate delta wing trace in the η, ς coordinates. Since shock fitting is used, the outer boundary shape is the bow shock geometry. In the region of the vortical flow the grid is nearly orthogonal and reasonably dense on a 128 x 128 mesh. A large number of points are concentrated at the tip with the mesh cells at the tip being extremely fine.

Computed Results

Solutions were obtained on a series of grids from 16 x 16 cells in the η, ξ directions to 128 x 128 cells. Results will be presented for the 128 x 128 and 64 x 64 calculations. The tested model [12] had a cross-sectional shape with a flat upper surface and a beveled geometry on the windward side. The calculations are for an idealized flat plate delta wing. Computations reported in Ref. 8 illustrate that the windward side shape has a negligible effect on the leeside flow patterns.

Pressure coefficient contours and wing surface values shown in Fig. 3 illustrate the overall nature of the flow field. The dominance of the vortical flow is clearly shown in the C_p contours as well as in the surface values. A rather strong cross-flow shock under the vortex is also evident. The possibility of such a shock was first proposed by Vorropoulus and Wendt [16] and has been confirmed by many calculations [3,5,7,8,15]. A plot of the total pressure loss across the bow shock is also shown in Fig. 3. The independent variable is the polar angle measured from the leeward symmetry plane ($\phi = 0°$) to the windward symmetry plane ($\phi = 180°$). The maximum total pressure loss is .00012 which is greater than 3 orders of magnitude smaller than the losses in the vortex core. Also note that for most of the leeward side the shock is effectively zero strength.

Figure 4 presents several plots which generally show the nature of the vortical flow. The cross-flow streamlines in Fig. 4a are traces which everywhere are tangent to the cross-flow velocity vectors on the unit sphere r = 1. The real three-dimensional streamlines are helical as shown in Fig. 1. By not including the radial velocity component, the cross-flow streamlines represent a projection (in a spherical sense) of the three-dimensional streamlines onto a flat surface. The cross-flow streamlines accentuate the topology of the flow but can exaggerate the magnitude of the cross-flow components in some cases. The primary vortex is hollow since the trajectory integration is terminated after a set number of cells have been traversed by the cross-flow streamline. It can be seen that there is an attachment line on the windward surface which divides the flow going inward towards the symmetry plane from that going outward around the tip. The latter separates forming a relatively thin "sheet" which feeds the primary vortex. There is also an attachment line on the leeward surface as schematically shown in Fig 1. A small secondary vortex is formed by the cross-flow shock under the primary vortex.

Contours of constant cross-flow Mach number are shown in Fig. 4b. Again, the veolcity magnitude in the numerator of this Mach number does not include the r component. The cross-flow velocity accelerates to supersonic speeds above the vortex and decelerates without a noticeable shock wave. Underneath the vortex it again accelerates through the sonic velocity before reaching 1.8 ahead of the cross-flow shock. The cross-flow streamlines indicate a nozzle-like structure in this region with sonic velocity being reached at the minimum area between the vortex and the wing. Figure 4c shows the total Mach number. Contours of total pressure loss $1 - \frac{P_t}{P_{t\infty}}$ are given in Fig. 4d. The "sheet" from the leading edge, the primary vortex, and the cross-flow shock are all clear features. As has already been mentioned, there is a substantial total pressure loss in both the vortex core and the sheet.

Figure 5 shows the same flow parameters as Fig. 4 but for the 64 x 64 grid. It can be seen that the vortex is considerably more diffuse and the flow structure not as well defined on the coarser grid. The shock under the vortex is diffuse enough that a secondary vortex is not realized. Note, however, that the magnitude of the total pressure loss is virtually the same as on the finer grid.

The experimental data reported by Monnerie and Werlé consists of pitot pressure contours on the leeside of the wing. With the assumption that the pitot probe is aligned with the local flow angle, the computed pitot pressures can be obtained from the Rayleigh pitot tube formula. However, the measured results utilized a probe which is aligned with the x-direction, and therefore contain an unknown angularity error. Generally, pitot pressure readings are accurate to one-percent if the flow angularity is less than 10° [17]. Figure 6 presents the computed pitot pressure contours for the 128 x 128 and 64 x 64 grids, the computed flow angles relative to the x-axis, and the experimental pitot pressure data. Note that the flow angles generally exceed 10° in the neighborhood of the vortex. However, in the middle of the vortex, the flow angularity is small and a direct comparison can be made. It is seen that the computed and measured values are in close agreement. The general features of the computed and measured pitot contours are in qualitative agreement outside of the region affected by the boundary layer. Inboard of the vortex the experiments show the losses in the boundary layer which are not included in the computations. Outboard of the primary vortex, the pitot pressures show a considerably different structure than the inviscid calculations. Clearly the interaction of the boundary layer with the cross-flow shock and the secondary separation of the boundary layer in the experiments will give quite different results than the inviscid theory. All in all, however, the comparison between calculations and experiments is remarkably good.

Figures 4-6 show that the 64 x 64 grid results are less well defined than the 128 x 128 grid results as would be expected, but that the overall features are similar. It is interesting to note that some key global parameters are insensitive to grid refinement as indicated in the following table.

Grid	32×32	64×64	128×128
Normal Force Coefficient	.2934	.2958	.2947
Max Pitot Pressure loss	.26	.23	.20
Max Total Pressure loss	.6171	.6103	.6104

The normal force and maximum total pressure loss are virtually unchanged with grid refinement. The pitot pressure varies more significantly due to the increase of the Mach number in the vortex core with decreasing mesh spacing (Figs 4c and 5c).

Additional information about the velocity field is given in Fig. 7. Contours of radial velocity (Fig 7a) show that in the core this velocity component is considerably greater than freestream. For this plot, the radial velocity has been normalized by the freestream velocity component projected along the wing leading edge. Surface velocity tufts shown in Fig. 7b have their lengths proportional to the local velocity magnitude. In the middle of the wing the flow is predominately streamwise, while outboard of the vortex it is almost aligned with the leading edge. The strong cross-flow is seen under the vortex. Qualitatively these compare well with the oil flow pictures in Ref 12.

Various vorticity contours are shown in Figure 8. The log of the magnitudes of the total vorticity vector, radial component, and cross-flow components are given in Figs. 8a-8c, respectively. The log of the magnitude of these components was selected for presentation to illustrate the core structure as well as the feeding sheet structure. The latter has significantly greater magnitude due to its relatively tighter structure. Using the log of the magnitude accentuates the computational noise, and care must be suggested in interpreting these plots. In particular we are not sure of the source of the contours outboard and below the tip. The total vorticity magnitude and radial component appear to have a similar structure. However, there is a significant structure in the cross-flow components in the feeding sheet region. Finally, in Fig. 8d we show the degree to which the solution satisfies the Crocco identity $T \ grad \ S + \mathbf{U} \times \omega - \ grad \ h_0 = 0$. The contours in this figure are the magnitude of the left hand side of this equation at each cell. They have been normalized by the maximum value over the field of the sum of the individual terms in this equation. The latter maximum is located near the tip. The smallest contour value shown is .5% of the maximum. The extent to which the Crocco relation is satisfied is a measure of the integrity of the inviscid nature of the discrete solution. It is clear that the strongest effects of the added damping terms are in the region of the tip. A post calculation check shows that the maximum error in the total enthalpy is less than 1%.

The final aspect of the computed solution we consider is the convergence history and "steady" state behavior as shown in Fig. 9. The global L_2 norm over all cells for all five equations is plotted versus iterations for all four grid densities. On the two coarsest grids, the solution reaches machine zero on a 32 bit computer in 500 iterations. However, on the finest grid it can be seen that a slow global convergence is accompanied by oscillations in the residual. Integral parameters as well as the general flow field are quite steady for the 128 x 128 grid. However, the detailed structure of the flow under the primary vortex outboard of the cross-flow shock has not reached a steady state. Figure 10 shows two "snapshots" of the cross-flow streamlines in this region at different points marked on Fig. 9. It can be seen that the secondary vortex is wobbling around. The shape of the primary vortex core in the region where the feeding sheet joins it is also changing slightly.

The present calculations are not time accurate, nor do they admit three-dimensional modes of instability, and little can be said except that they are not perfectly steady. It is interesting to note that in the experimental pitot pressure measurements in Fig. 6b cross-hatched areas in this same region of the flow. Perhaps the experiments also indicated unsteadiness. A possible source of unsteadiness is the interaction of the cross-flow shock and the vortical regions. As the shape of the primary vortex changes, the strength of the shock may change. This in turn could change the strength of the secondary vortex, and thereby affect the degree to which the regions of opposite signs of vorticity affect each other.

Concluding Remarks

The good comparison between measured and computed pitot pressures in the core of the vortex tends to support the argument that the total pressure losses predicted by the Euler equation solvers are not errors, but are realistic predictions. The mechanism which creates these losses must be attributed to the artificial viscosity in the discrete equations being solved. However, the magnitude of the losses is set by the aerodynamic parameters[18] and is insensitive to the computational parameters. Comparison with surface pressure data and flow visualization results for several different conditions are given in Ref. 8. Quite good agreement is obtained except for the effects of the secondary vortex.

A complete description must include realistic physical viscosity models such as the Navier-Stokes calculations in Refs. [9,19-21]. For blunt leading edges, there is no Kutta condition and the separation point is set by the true viscous conditions. The primary mechanism for generating the secondary vortex is boundary layer separation. The displacement effect of the secondary vortex affects the position of the primary vortex to some degree. However, the accuracy of Navier-Stokes solutions is no better than the accuracy of Euler solutions on the same grid, and it is important to understand the latter as well as the former.

An effect of some concern is the appearance of flow instabilities and the influence of real and artificial viscosity on such solutions. Inviscid rotational flows are subject to instabilities, and some indication of that may be illustrated in Figs. 9 and 10. Some thought needs to be given as to the ability of the current generation of Euler and Navier-Stokes solvers to accurately deal with these. With this in mind, we nevertheless close these remarks with some speculations on this topic.

The substantial total pressure losses which are predicted could be an explanation for the boundaries of vortex bursting. If a vortex such as analyzed in this paper is subjected to a rising pressure field, the fluid in the core which has low total pressure might stagnate. This in turn would cause an instability or bursting. The calculations of the present authors suggest that the level of total pressure loss is related to the circumferential velocity field through the vortex. Swirl angle is an important parameter in locating boundaries of vortex bursting. Our calculations also indicate that the magnitude of the total pressure loss increases with angle of attack [18] and changes with Mach number and sweep angle.

Euler equation solutions which capture the vortical regions produce diffuse vortex cores rather than a spiraling sheet ending in a small core. It is not known if finer grids will eventually produce a sheet structure, and this still needs to be investigated. However, flow visualization for this range of Reynolds numbers also shows a diffuse core [22,23]. A recent paper by Payne et al [24] clearly shows instabilities on the spiral sheet at much lower Reynolds numbers. It could be that the inviscid model of a leading edge vortex which best represents a high Reynolds number viscous flow is a diffuse core such as given by the discrete Euler equation solutions.

Acknowledgements

This research was supported by NASA Langley Research Center under NASA Grant NAG-1-358 monitored by Dr. Manny Salas and Mr. Duane Melson.

References

1. Rizzi, A., Eriksson, L.E. , Schmidt, W. and Hitzel, S. "Numerical Solutions of the Euler Equations Simulating Vortex Flows Around Wings," AGARD-CP-342, Paper 21, 1983.

2. Weiland, C., "Vortex Flow Simulations Past Wings Using the Euler Equations," AGARD-CP-342, Paper 19, 1983.

3. Murman, E., Rizzi, A. and Powell, K., "High Resolution Solutions of the Euler Equations for Vortex Flows," in Progress and Supercomputing in Computational Fluid Dynamics, pp 93-113, Birkhauser-Boston 1985.

4. Rizzi, A. and Eriksson, L., "Computation of Inviscid Incompressible Flow With Rotation," Journal of Fluid Mechanics, Vol 153, pp 275-312, April 1985.

5. Arlinger, B.G., "Computation of Supersonic Flow Including Leading-Edge Vortex Flows Using Marching Euler Technique," Presented at the International Symposium on Computational Fluid Dynaimics, Tokyo, Japan, Sept 1985.

6. Hoeijmakers, H.W.M. and Rizzi,A., "Vortex-Fitted Potential and Vortex- Captured Euler Solution for Leading-Edge Vortex Flow" AIAA Journal Vol 23, No 12, pp 1983-1985, Dec 1985.

7. Rizzi,A. "Three-Dimensional Solutions to the Euler Equations with One Million Grid Points", AIAA Journal, Vol 23, No 12, pp 1986, Dec 1985.

8. Murman, E.M., Powell, K.G., Miller,D.S., and Wood,R.M., "Comparison of Computations and Experimental Data for Leading Edge Vortices - Effects of Yaw and Vortex Flaps", AIAA Paper 86-0439, Jan 1986.

9. Rizetta, D.P. and Shang J.S. "Numerical Simulation of Leading-Edge Vortex Flows," AIAA Paper 84-1544, June 1984.

10. Powell, K., Murman, E., Perez, E., and Baron, J. "Total Pressure Loss in the Vortical Solutions of the Conical Euler Equations," AIAA Paper 85-1701, July 1985.

11. Newsome,R. and Thomas, J., "Computation of Leading-Edge Vortex Flows," Paper presented at NASA Langley Vortex Flow Aerodynamics Conference, Hampton Va. October 1985.

12. Monnerie, B. and Werlé, H., "Étude de l'Écoulement Supersonique & Hypersonique Autour d'Une Aile Elancée en Incidence," AGARD-CP-30, Paper 23, 1968.

13. Jameson, A., Schmidt, W., and Turkel, E., "Numerical Solution of the Euler Equations by a Finite Volume Method Using Runge-Kutta Time- Stepping Schemes," AIAA-81-1259, June 1981.

14. Rizzi, A. and Eriksson, L.E., "Computation of Flow Around Wings Based on the Euler Equations," Journal of Fluid Mechanics, November 1984.

15. Marconi, F., "The Spiral Singularity in the Supersonic Inviscid Flow over a Cone," AIAA-83-1665, July 1983.

16. Vorropoulos, G. and Wendt, J. F. "Laser Velocimetry Study of Compressibility Effects on the Flow Field of a Delta Wing," AGARD-CP-342, Paper 9, 1983.

17. Physical Measurements in Gas Dynamics and Combustion, Eds. R.W. Ladenburg, B.Lewis, R.N Pease, H.S. Taylor, Princeton Univ. Press 1964.

18. Murman, E.M., "Computational Studies of Compressible Vortex Flows" MIT CFDL TR 86-04, Jan 1986.

19. Fujii, K. and Kutler, P., "Numerical Simulation of the Viscous Flow Fields Over Three-Dimensional Complicated Geometries," AIAA Paper 84-1550, June 1984.

20. Buter, T.A. and Rizetta, D.P., "Steady Supersonic Navier-Stokes Solutions of a 75 degree Delta Wing at Angle of Attack," Paper presented at NASA Langley Vortex Flow Aerodynamics Conference, Hampton Va. October 1985.

21. Newsome, R.W. and Adams, M.S., "Numerical Simulation of Vortical-Flow Over an Elliptical-Body Missile at High Angles of Attack", AIAA Paper-86-0559, Jan 1986.

22. Miller, D.S. and Wood, R. M. "Lee-Side Flow Over Delta Wings at Supersonic Speeds," NASA TP 2430, June 1985.

23. Verhaagen, N.G., "An Experimental Investigation of the Vortex Flow Over Delta and Double-Delta Wings at Low Speeds", Paper 7, AGARD-CP-342, April 1983.

24. Payne,F.M., Ng,T.T., Nelson,R.C., and Schiff,L.B., "Visualization and Flow Surveys of the Leading Edge Vortex Structure on Delta Wing Planforms", AIAA Paper 86-0330, Jan 1986.

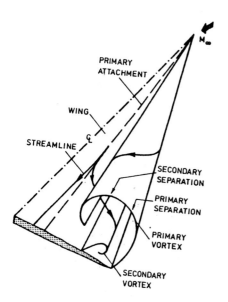

Fig. 1 Schematic illustration of leading edge vortex flows.

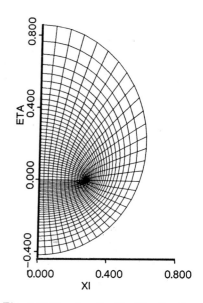

Fig. 2 Grid system for 32 x 32 cells. Outer boundary is bow shock. Horizontial line is wing.

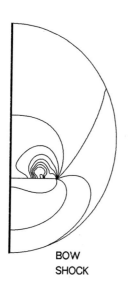

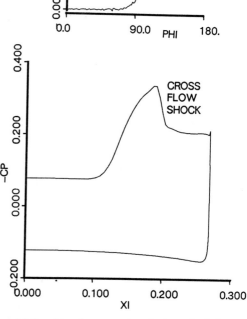

Fig. 3 C_p contours (left), wing surface C_p's (bottom right) and total pressure loss for bow shock (top right)

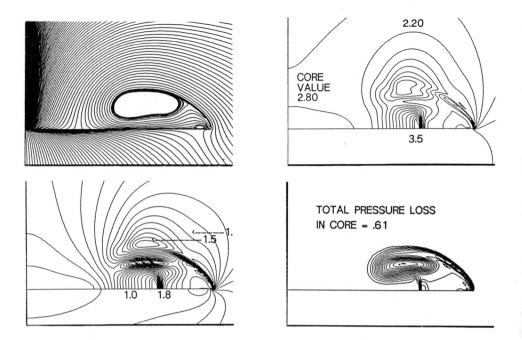

Fig. 4 Solution for 128 X 128 grid. Top left to bottom right: a.)Cross-flow streamlines, b.)Total Mach number, c.)Cross-flow Mach number, d.)Total pressure loss.

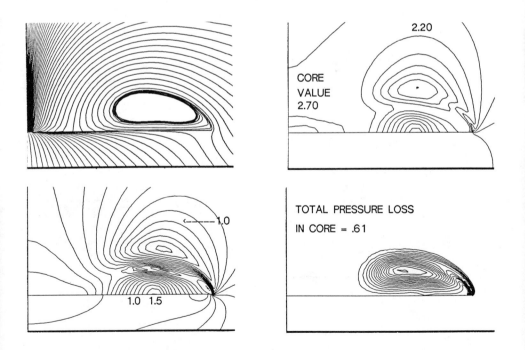

Fig. 5 Solution for 64 X 64 grid. Top left to bottom right: a.)Cross-flow streamlines, b.)Total Mach number, c.)Cross-flow Mach number, d.)Total pressure loss.

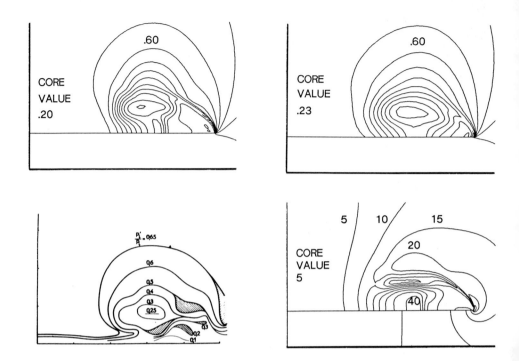

Fig. 6 Pitot pressure comparison. Top left to bottom right: a.) Computed pitot pressures on 128 X 128 grid, b.) Computed pitot pressures on 64 X 64 grid, c.) Measured pitot pressures, d.) Computed flow angles on 128 X 128 grid.

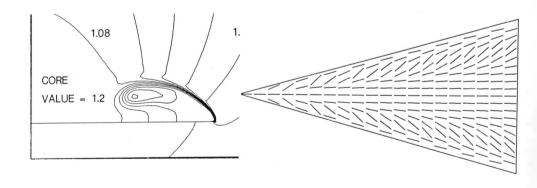

Fig. 7 Radial velocity contours (left) and surface velocity tufts (right).

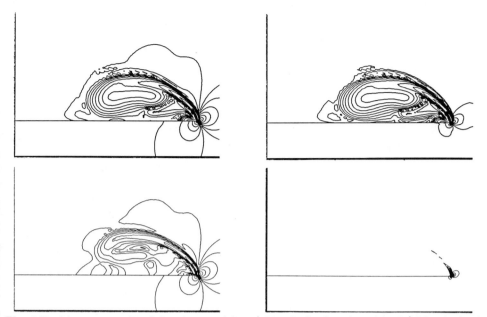

Fig. 8 Vorticity contours. Top left to bottom right: a.) Magnitude of total vorticity, b.) Magnitude of radial component of vorticity, c.) Magnitude of cross-flow components of vorticity, d.) Magnitude of Crocco's equation.

Fig. 9 Convergence history for all four grids.

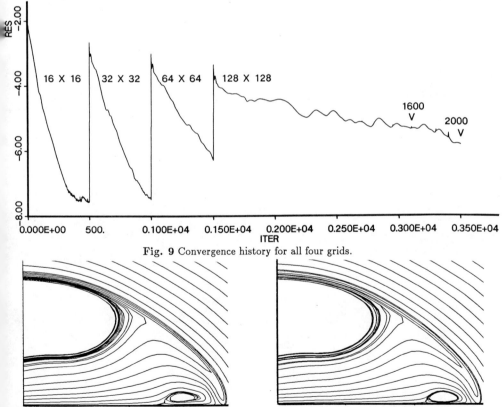

Fig. 10 Detail of cross-flow streamlines for 1600 iterations (left) and 2000 iterations (right) on 128 X 128 grid.

Section V

Conically Separated Flows

Separated Flow About Cones at Incidence—
Theory and Experiment

S. P. FIDDES

1   <u>INTRODUCTION</u>

The motivation for the work described here is the study, understanding
and ultimate prediction of separated flow past bodies at incidence.
The work has concentrated on modelling, mathematically, the vortical
flow about slender bodies at incidences where the separation starts at
the pointed apex of the body. Separation can, of course, occur further back along the body, particularly if the body is quite long, <u>ie</u>
of high fineness ratio. However, a major interest of the work has
been in understanding the development of lateral asymmetry in the flow
about laterally symmetric bodies at moderate to high incidences, for
cases where the separation has moved forward to the apex for typical
bodies, <u>eg</u> an ogive-cylinder combination. The importance of the nose
region in determining the development of separation asymmetry is emphasised by the fact that many of the 'fixes' proposed to reduce the degree of flow asymmetry involve modifying the nose, <u>eg</u> nose blunting,
strakes, blowing at the apex, adding small pieces of grit very near
the nose tip, etc.

With attention thus focussed on the nose of a body, a simplifying feature emerges in that a pointed nose is locally conical in shape and,
to borrow a concept from hypersonics, the flow may be regarded as
locally equivalent to that about a 'tangent cone'. In fact, the circular cone is representative of the noses of many aircraft and missiles,
and as such has been used as a prototype configuration for the experimental investigation of three-dimensional separated flows.

The results from experiments on cones, for example those described in
Refs 1 and 2, show that at high Reynolds number and moderate angles of
incidence, the boundary layers that grow from the windward attachment
line eventually separate to form free shear layers. The free shear
layers roll up to form concentrated vortices which lie on the leeward
side of the cone and have a considerable influence on the pressure
distribution over it. The vortices also persist downstream and

influence the flow over the rest of the configuration; for example if
the cone represented the nose of an aircraft, the nose vortices could
flow downstream and interfere with the flow over the wing or a fin.

At moderate angles of incidence - up to approximately twice the semi-
apex angle of the cone - the vortices are symmetrically disposed about
the incidence plane, but at larger angles of attack they become asymmetric. Associated with the development of asymmetry is the occurrence of large side forces ('phantom yaw') that can be of either hand.
This phenomenom is well known, and space does not permit a detailed
review of the various factors that have been found, experimentally, to
influence the development of asymmetry. Indeed, there is at present
no general agreement about the mechanisms involved in the creation of
side forces. However, the present paper describes work that has lead
to a quantitative theory for the development of these side forces on
pointed forebodies, a theory which has aided the understanding and
planning of experimental investigations of the phenomenon.

2   LINE VORTEX MODELS

One of the earliest mathematical models of the separated flow past a
cone at incidence was published by Bryson[3]. He assumes the flow is
symmetric about the incidence plane and represents the separations by
a pair of line-vortices embedded in a potential flow. Each vortex is
joined to the separation line, from which it is imagined to spring, by
a cut, as illustrated in Fig 1. The cut represents the feeding shear
layer in a rudimentary way, and serves the mathematical function of
rendering the velocity potential single valued. Both the vortex and
cut experience forces, and Bryson chooses to set to zero the crossflow components of the force acting on the combination. A form of
Kutta condition applied at the separation lines provides a third equation which completes the determination of the position and circulation
of the vortices.

Ref 4 describes an extension of Bryson's method in which the assumption of lateral symmetry is removed. Extra equations arise and
solutions are obtained by a Newton-Raphson iterative procedure.
Because of the relative simplicity of the line-vortex model it has been
possible to conduct an extensive survey of its solutions. During the
course of such a survey, a new type of solution was discovered. This
second family of solutions was distinguished from those discovered by

Bryson by the fact that they always have asymmetric vortex core
locations, even when they are fed from symmetric separation positions.
This is illustrated in Fig 2, which shows, for a pair of symmetrically
disposed separation lines, the loci of vortex positions as the incidence parameter (defined as the ratio of incidence of the cone (α) to
the semi-apex angle of the cone (δ)) varies from 3 to 10. Also shown
is the symmetric (first) family of solutions, discovered by Bryson.
It is not clear from Fig 2 how the two families of solution are
related. Fig 3 sheds some light on this, and shows the height of the
vortices above the cone centre-line as a function of the incidence
parameter. Solutions exist only to the right of the lower bound given
by Bryson. For a range of values above this lower incidence bound only
symmetric solutions are found. Eventually, as the incidence parameter
is increased, the two asymmetric positions bifurcate from the symmetric one, at an incidence parameter above a value of 2. The deter-
minant of the Jacobian matrix used in the Newton-Raphson scheme changes
sign at B as we progress along the symmetric branch. The value of the
incidence parameter at which bifurcation occurs, and thus above which
asymmetric solutions may be found, depends upon the position of the
(symmetric) separation lines. In the case of Fig 3, this is 56° above
the horizontal.

The nonlinear equations describing Bryson's line-vortex model for the
separated flow past a circular cone at incidence thus admit non-unique
solutions, arising via a classical bifurcation process. They show the
markedly asymmetric core locations found experimentally, and as Fig 4
shows, can produce large levels of side force, even with symmetric
separation positions.

Although these asymmetric solutions are striking, it must be admitted
that they are not very realistic, particularly for large values of the
incidence parameter. It is shown in Ref 4 that as the incidence parameter is increased, one vortex moves off to infinity, while the other
tends to a fixed position near the cone. These unreal solutions arise
because the form of the Kutta condition used in this model is satisfied at an attachment line as well as at a separation line. However,
at lower incidences, the solutions are able to represent qualitative
features of the flow with some degree of realism. For example, Fig 5
shows the incidence, as a function of (symmetric) separation position,
for the onset of asymmetric flow, ie the bifurcation boundary for second family solutions. For separation positions typically found in
experiments (about 40-50°), asymmetric flows are predicted to occur at

incidences about 2.5 times the nose half-angle, in accordance with observation. Also shown in Fig 5 is the lower limit found by Bryson for the existence of symmetric solutions.

3   VORTEX SHEET MODELS

Fiddes, in Ref 5, describes an alternative model for the symmetric separated flow about a circular cone. This uses the vortex sheet model to represent the free shear layers. This model was proposed by Legendre[6] implemented by Mangler and Smith[7], and further improved by Smith[8]. In this model, the vortex core is represented by a line vortex, as in the Bryson model, but the free shear layer connecting it to the separation line is replaced by a vortex sheet of finite extent (see Fig 6). The shape and strength of the vortex sheet is governed by the boundary conditions that it forms a stream surface of the three-dimensional flow across which the pressure is continuous, but the velocity is discontinuous. The specific form of these conditions in conical, slender flow are shown in Fig 6, along with the 'zero-force' condition on the line vortex and cut combination. A modified form of Kutta condition determines the circulation of the sheet and vortex. A local analysis of the solution in the vicinity of the separation line by Smith[9] revealed that the vortex sheet must leave the smooth surface of the body tangentially and with (generally) infinite curvature. Further analysis of the viscous problem for laminar flows in the vicinity of the separation line has been carried out by F.T. Smith[10]. He has shown that the inviscid behaviour discovered by Smith is consistent with a triple-deck description of the laminar viscous-inviscid interaction at separation. Fiddes[5] was able to incorporate these results in his method for calculating the separated flow past cones, and was able to predict the location of the separation lines for laminar flow.

More recently, the method has been extended to allow for asymmetric flows, and reported in Ref 11. This extension comprised a modification to the inviscid part of the method only, the triple-deck analysis being only a local one for the region near separation. Since a second, asymmetric family of solutions exists for the line-vortex model, it is natural to seek a second family of solutions for the vortex sheet model. A general technique for finding bifurcating solutions has been devised by Decker and Keller[12], but we choose a different approach, which avoids the numerical difficulties which can arise near the bifurcation point, and which exploits our knowledge of the line-vortex solutions. The essential step is to modify the boundary conditions for the vortex sheet model by the introduction of a parameter λ , as shown in Fig 7. When $\lambda = 1$, the equations are identical to those of Fig 6 for the conical vortex sheet. Further, when $\lambda = 1$, and we choose the cut to extend all the way to the separation position, so that Z_E is the separation point, we have the line-vortex model, described by the last equation in the figure. On the other hand, when $\lambda = 0$, the second equation is satisfied by $\Delta v_t = 0$, and the third equation is independent of Z_E . Consequently, when $\lambda = 0$, the solutions which correspond to the line-vortex model and the vortex-sheet model are identical. The procedure for obtaining asymmetric vortex-sheet solutions is first to reduce λ to 0 in the family of asymmetric solutions for the line-vortex model, then to construct a stream surface satisfying the first equation in Fig 7, starting from the separation line and ending at some convenient point E , and treat this surface as an embryonic vortex sheet. The last step is to increase λ again, until a solution of the modified vortex sheet equations with $\lambda = 1$ is obtained, ie a solution of the original conical vortex sheet equations. It is fortunate that the asymmetry did not disappear in the course of this construction. Fig 8 shows, on the right, one of the family of solutions originated in this way. Note that the separation positions are symmetric. The isolated points shown on the right-hand side show the solution to the asymmetric line-vortex model that acted as the 'seed' for the vortex-sheet solution, as described above, for the same separation positions and incidence. It is clear that the two models (line-vortex and vortex-sheet) have very different solutions. The left-hand side of the Figure shows the corresponding first family solution. As the separation lines are symmetric about the incidence plane, the first family solution is also symmetric.

4   EXPERIMENTAL RESULTS

Evidence that the second family of vortex sheet solutions so found is the appropriate one for describing the large side-force case comes from the experiments of Mundell[13]. In these low-speed experiments a cone-cylinder model was split (transversely) just beyond the nose and mounted on a balance so that only the forces on the conical portion of the nose, forward of a smooth fairing onto the cylindrical portion, were measured. Oil flow was used to record the separation positions, and by rolling the model at a given incidence, Mundell was able to find the condition of maximum side force. With the separation positions, and the direction of the side force, given from experiment, it was possible to construct two solutions of the vortex-sheet model to describe the flow. Typical examples are shown in Fig 9, along with a table showing experimental results for forces. The left-hand figure shows a first family solution, the right-hand a second family result. Note the separation positions, as recorded in the experiment, are not greatly asymmetric, although they correspond to a high side force condition.

Comparing the computed values of side force C_Y , with the experimental value, it is apparent that the second family is a more appropriate description of the flow than the first family solution. Unfortunately, pressure measurements or flow field surveys were not made, so the configuration of the vortices in the experiment are not known.

To investigate further the relevance of the second family of solutions to the high side force case, another series of experiments has been conducted to examine the circumferential pressure distribution associated with high side forces. Consider, first, the results from the vortex sheet model. Fig 10a shows the circumferential pressure distribution associated with the first family solution shown on the left-hand side of Fig 9. This pressure distribution shows that the largest contribution to the side force comes from the lateral differences in pressure ahead of separation. In effect, the separation reduces the amount of suction that was present in the attached flow by more than it introduces with suction under the vortex cores. Furthermore, the suction peaks associated with the vortex cores act on a part of the cone surface having little horizontal component, so further reducing the contribution to the side force. However, both suction peaks are well marked.

By contrast, Fig 10b shows the pressure distribution corresponding to the second family solution shown in Fig 9. Of note here is the very high suction peak under the left-hand core, which lies close to the body, while no suction peak can be detected on the right-hand side, as the core is too remote from the body. The lateral pressure difference ahead of separation also contributes greatly to the side force, as in the first family solution. The very high suction level under the near core is probably grossly inaccurate in quantitative terms, but no doubt qualitatively correct in its indication of locally high suctions.

To verify these predictions of the vortex sheet model, an experiment
has been conducted in the RAE 5m low-speed pressurised wind tunnel.
The model used in the experiment was geometrically similar to that
used by Mundell, being a cone-cylinder combination comprising a nose
cone of 10° semi-apex angle faired to a cylindrical afterbody. The
total length of the body was 2.26 m, with a diameter of 0.297 m. Further details of the model are given in Fig 11. The nose portion of
the model was pressure-tapped at six longitudinal stations, with each
station housing 36 holes equi-spaced around the circumference. The
design of the model allowed pressures to be recorded simultaneously at
all stations (216 pressure recordings), though results will be presented here for the front station only. The model was fixed to a
pitch sting via a roll box, and the model could be rolled through 360°.
Force measurement data for the complete model are presented in Ref 14.

During the experiment, the model was pitched to a particular incidence
(up to 35°) at zero roll, and the pressures recorded. The model was
then rolled, typically in 10° intervals, and the pressures recorded
again. The process was repeated until the entire roll range had been
covered. The pressures were integrated online to give side and normal
force coefficients in 'unrolled' body axes, ie normal force measured
in the incidence plane. (The definition of the force coefficients used
in the experiment differ from those used in the theoretical method in
the choice of reference area. However, this does not affect what
follows.) Results are presented here only for the front station, 0.15 m
from the apex. The experimental conditions corresponded to a Mach number of 0.15 and a Reynolds number of 3.7 million/m. An earlier investigation of the state of the boundary layer on the nose at these conditions, using the china-clay technique[15] revealed that the boundary
layer was laminar at the first pressure-measuring station for the
range of incidences used in the experiment.

Fig 12 shows the variation of side and normal force coefficients with
roll angle for the front station for 35° incidence. Note that there
is a large variation of side force with roll angle, including a change
of sign. The experimental pressure distributions corresponding to
maximum side force are now examined in detail. Fig 13 shows a typical
example, with some salient features annotated. Points to note are:

    (i)   The maximum pressure (attachment A) position is offset
            from the windward generator.

(ii) A small amount of pressure recovery takes place before separation. Separation is inferred to occur at the points labelled S1 and S2.

(iii) Only one suction peak (P), associated with a vortex core lying near the surface of the cone, is visible.

To illustrate just how typical of the maximum side force cases this pressure distribution is, Fig 14a&b show a compilation of pressure distributions that give maximum side force from the whole roll angle range. It is apparent that they have the same characteristic features as the pressure distribution shown in Fig 13. To complete the comparison of experimental pressure distributions, Fig 15 shows the pressure distributions giving maximum positive C_Y and maximum negative C_Y, with the negative side-force pressures plotted against the complementary angular position, $360-\theta°$. The two pressure distributions are seen to be identical. It thus appears that the high side force state is associated with a characteristic circumferential pressure distribution that is independent of roll angle, apart from a change of hand. Also of note is the large lateral pressure difference ahead of separation – this produces the bulk of the side force. The only visible suction peak associated with a vortex core acts on the upper surface of the cone and contributes little to the side force.

The maximum side force condition is not associated with a large asymmetry in separation position, confirming Mundell's observations. Fig 14a&b show little variation in separation position, and the difference in separation position between port and starboard sides does not exceed 15°. Furthermore, no mechanism related to a transitional boundary layer is possible – the flow near the apex was laminar, as described above.

These results indicate that the mechanism for generating large side forces in this experiment is inviscid in origin. A viscous mechanism, for example model surface imperfections leading to different boundary layer development on either side of the attachment line, would be expected to give a wider range of separation positions. Furthermore, a mere perturbation of the separation positions is not sufficient to produce large side forces – there must be a marked perturbation in the vortex core positions. This supports the conclusions of the vortex-sheet model described earlier.

5   CONCLUSIONS

Non-unique solutions have been found for both the line-vortex and
vortex-sheet models of the separated flow past a circular cone at
incidence. For the case of the line-vortex model, the non-uniqueness
has been shown to be the result of a classical bifurcation process
from the symmetric solutions originally found by Bryson[3]. The line-
vortex has been 'embedded' in the vortex sheet model, and used to seed
starting solutions that lead to asymmetric vortex sheets, even with
symmetric separation positions. Comparisons with forces and pressures
measured in experiment have confirmed that the non-unique solutions
correspond to the high side-force condition.

An important class of physically relevant, non-unique solutions for a
vortex dominated flow have now been identified. It is thus important
to consider the possibility of non-uniqueness when considering poten-
tially more accurate inviscid models of separated flow, eg solution of
the Euler equations by finite-difference methods.

REFERENCES

1   W.J. Rainbird. "The external flow field about yawed circular
     cones." Paper 19 in AGARD CP-30 (1968)

2   D.J. Peake, F.K. Owen, D.A. Johnson. "Control of forebody
     vortex orientation to alleviate side-forces." AIAA Paper
     80-0183 (1980)

3   A.E. Bryson. "Symmetrical vortex formation on circular
     cylinders and cones." J. Appl. Mech. (ASME), 26, 643-8 (1959)

4   D.E. Dyer, S.P. Fiddes, J.H.B. Smith. "Asymmetric vortex
     formation from cones at incidence - a simple inviscid model."
     Aeronautical Quarter, 33, 293-312 (1982)

5   S.P. Fiddes. "A theory of the separated flow past a slender
     elliptic cone at incidence." Paper 30 in AGARD CP-291 (1980)

6   R. Legendre. "Ecoulement au voisinage de la pointe avant d'une
     aile a forte fleche aux incidences moyennes." 8th Int. Congr.
     Th. Appl. Mech., Istanbul (1952)

7   K.W. Mangler, J.H.B. Smith. "A theory of the separated flow past a slender delta wing with leading-edge separation." *Proc. Roy. Soc. London A*, 251, 200-217 (1959)

8   J.H.B. Smith. "Improved calculations of leading-edge separation from slender, thin delta wings." *Proc. Roy. Soc. London A*, 306, 67-90 (1968)

9   J.H.B. Smith. "Behaviour of a vortex sheet separating from a smooth surface." RAE Technical Report 77058 (1978)

10 F.T. Smith. "Three-dimensional viscous and inviscid separation of a vortex sheet from a smooth non-slender body." RAE Technical Report 78095 (1979)

11 S.P. Fiddes, J.H.B. Smith. "Calculations of asymmetric separated flow past circular cones at large angles of incidence." Paper 14 in AGARD CP-336 (1982)

12 D.W. Decker, H.B. Keller. "Path following near bifurcation." *Comm. Pure Appl. Math.*, 34, 149-175 (1981)

13 A.R.G. Mundell. RAE unpublished (1982).

14 J.S. Smith. "Preliminary tests of slender bodies in the 5m pressurised low-speed tunnel." RAE Technical Memorandum Aero 1973 (1983)

15 I.R.M. Moir. "Recent experience in the RAE 5m wind tunnel of a china-clay technique for indicating boundary-layer transition." RAE Technical Memorandum Aero 2007 (1984)

16 E.R. Keener, G.T. Chapman, L. Cohen, J. Taleghani. "Side forces on a tangent ogive forebody with a fineness ratio of 3.5 at high angles of attack and Mach numbers from 0.1 to 0.7." NASA TM X-3437 (1977)

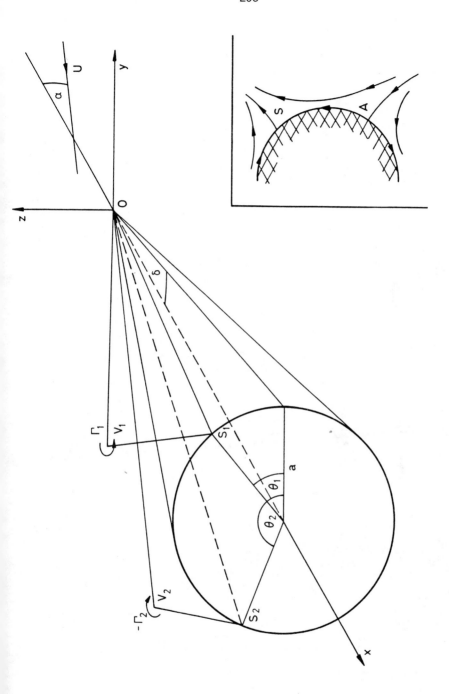

Fig 1 Configuration and coordinate system (Inset: flow structure)

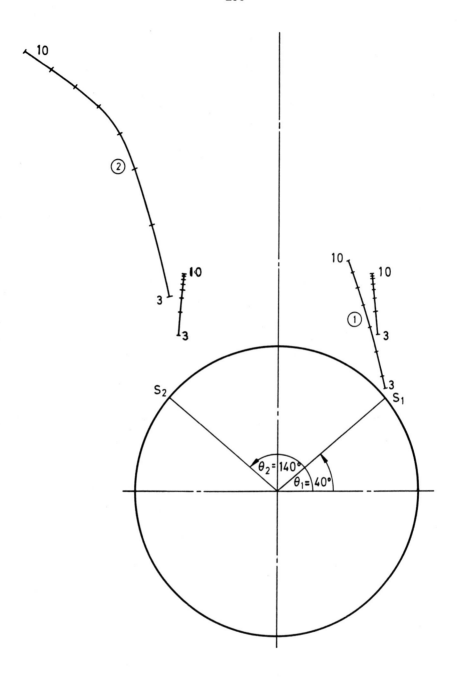

Fig 2 Symmetric and asymmetric vortex positions for symmetric separation
and $\alpha/\delta = 3(1)10$

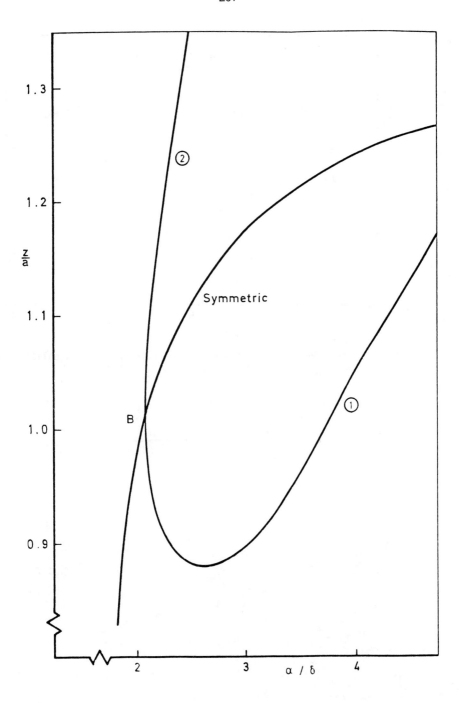

Fig 3 Bifurcation of asymmetric solution from symmetric solution, $\theta_1 = 56°$, $\theta_2 = 124°$; (b) vertical position of vortices

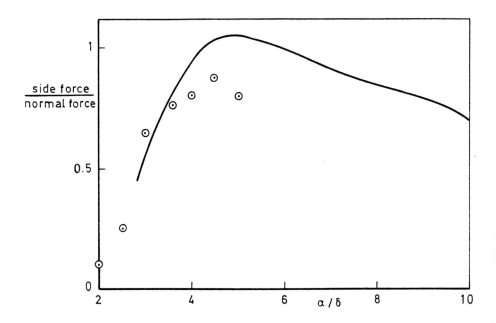

Fig 4   Ratio of side force to normal force on cone; present calculations for
$\theta_1 = 40°$, $\theta_2 = 140°$, and measurements of Keener *et al*[16]

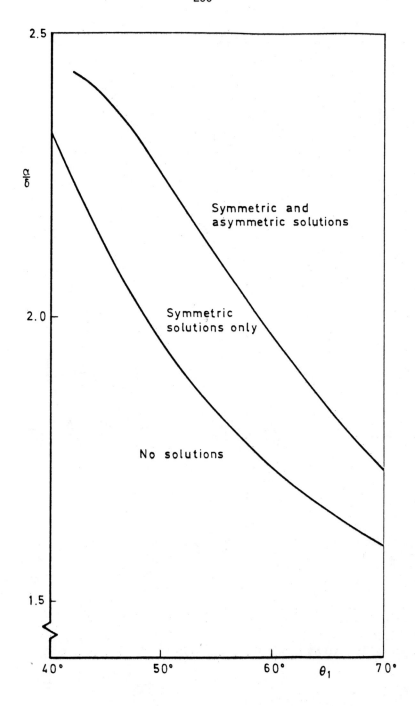

Fig 5 Regions of parameter plane in which solutions exist; $\theta_1 + \theta_2 = 180°$

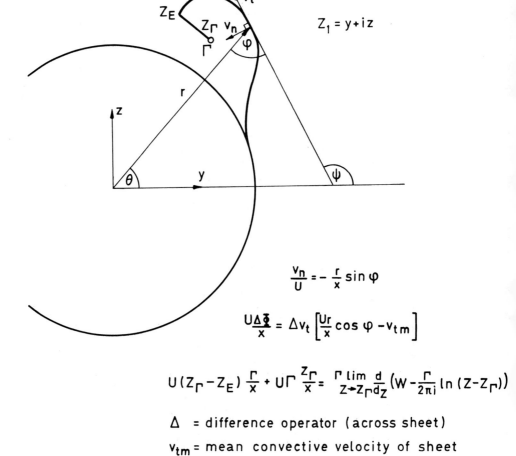

$$\frac{v_n}{U} = -\frac{r}{x}\sin\varphi$$

$$U\frac{\Delta\bar{\Phi}}{x} = \Delta v_t\left[\frac{Ur}{x}\cos\varphi - v_{tm}\right]$$

$$U(Z_\Gamma - Z_E)\frac{\Gamma}{x} + U\Gamma\frac{Z_\Gamma}{x} = \Gamma\lim_{Z\to Z_\Gamma}\frac{d}{dZ}\left(W - \frac{\Gamma}{2\pi i}\ln(Z-Z_\Gamma)\right)$$

Δ = difference operator (across sheet)

v_{tm} = mean convective velocity of sheet

W = complex potential

Fig 6 Boundary conditions for vortex sheet model

$$\frac{v_n}{U} = -\frac{r}{x}\sin\varphi$$

$$\lambda U\frac{\Delta\bar{\Phi}}{x} = \Delta v_t\left[\frac{Ur}{x}\cos\varphi - v_{tm}\right]$$

$$\lambda U(Z_\Gamma - Z_E)\frac{\Gamma}{x} + U\Gamma\frac{Z_\Gamma}{x} = \Gamma\lim_{Z\to Z_\Gamma}\frac{d}{dZ}\left(W - \frac{\Gamma}{2\pi i}\ln(Z-Z_\Gamma)\right)$$

Fig 7 Modified boundary conditions for vortex sheet

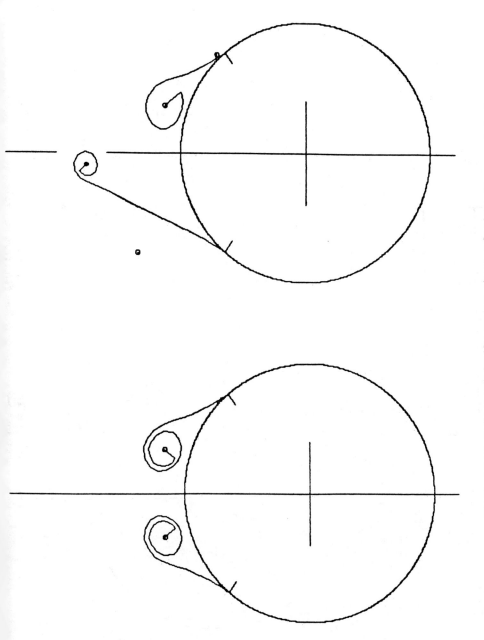

Fig 8   First and second family vortex sheet solutions for symmetrical separation position ($\alpha/\delta = 3.0$, $\theta_1 = 40°$, $\theta_2 = 140°$)

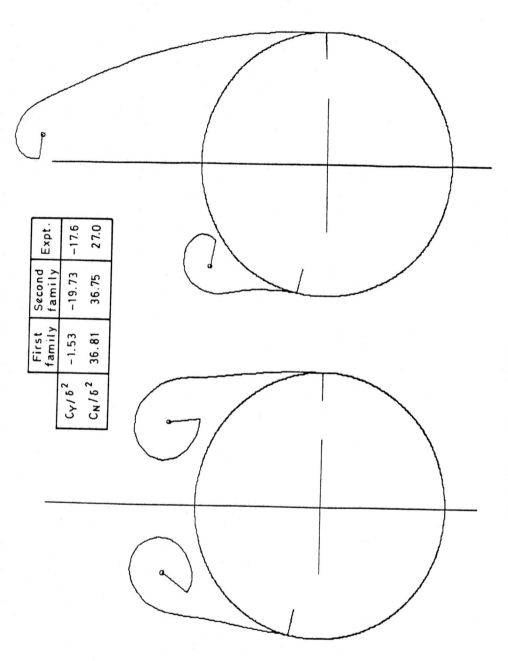

	First family	Second family	Expt.
C_Y/δ^2	-1.53	-19.73	-17.6
C_N/δ^2	36.81	36.75	27.0

Fig 9 First and second family vortex sheet solutions for separation positions as measured by Mundell (from Ref 13)

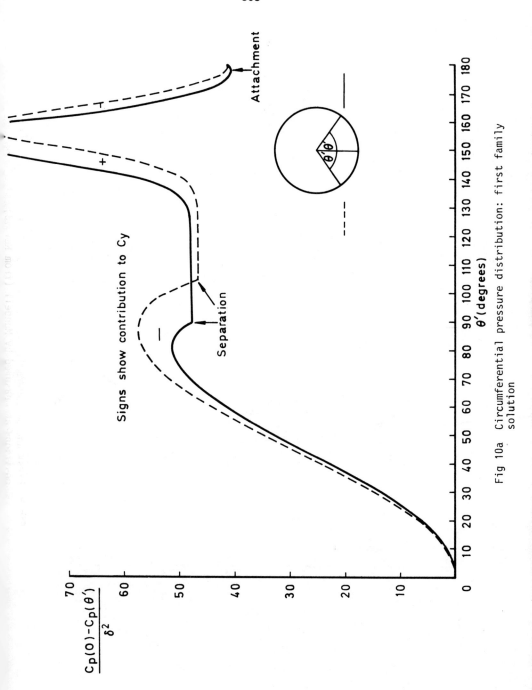

Fig 10a Circumferential pressure distribution: first family
solution

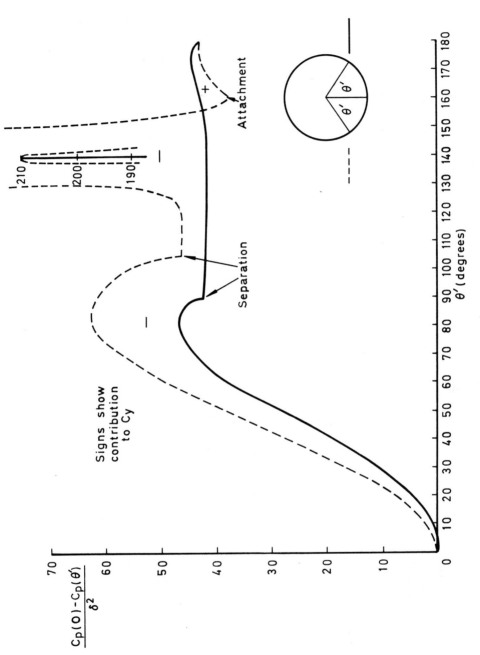

Fig 10b Circumferential pressure distribution: second family solution

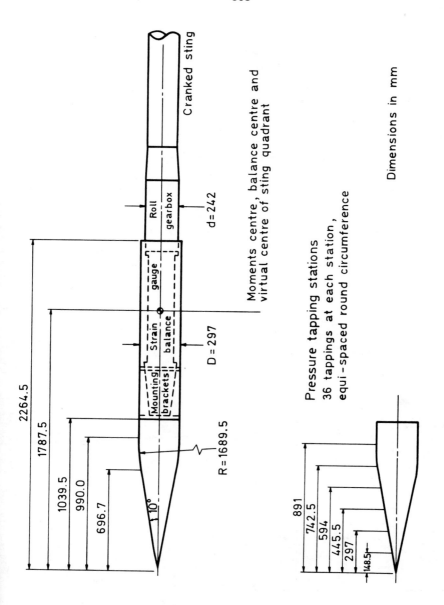

Cranked sting

Moments centre, balance centre and
virtual centre of sting quadrant

Dimensions in mm

Pressure tapping stations
36 tappings at each station,
equi-spaced round circumference

Fig 11 Details of cone-cylinder model

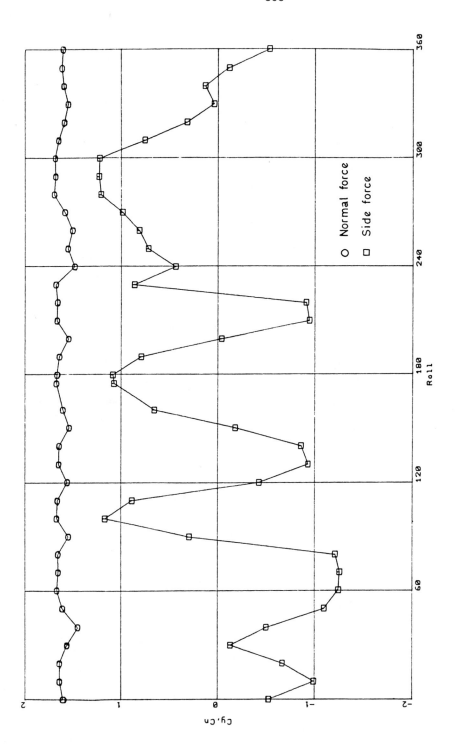

Fig 12 Variation of side and normal force with roll angle

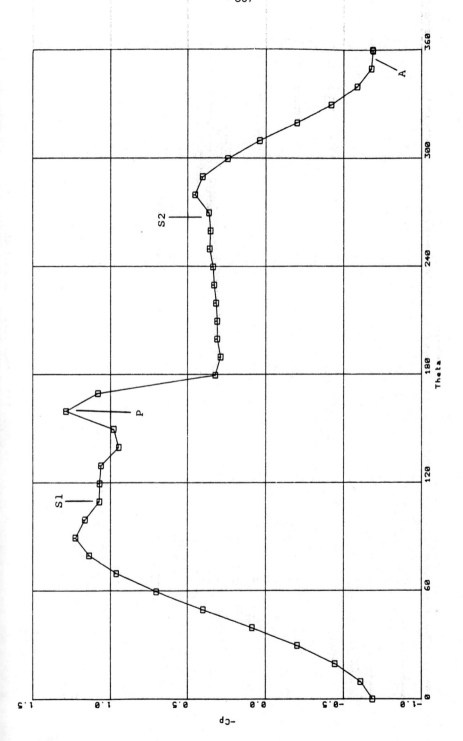

Fig 13 Circumferential pressure distribution giving large side-force
Incidence = 35 degrees, roll = 290 degrees

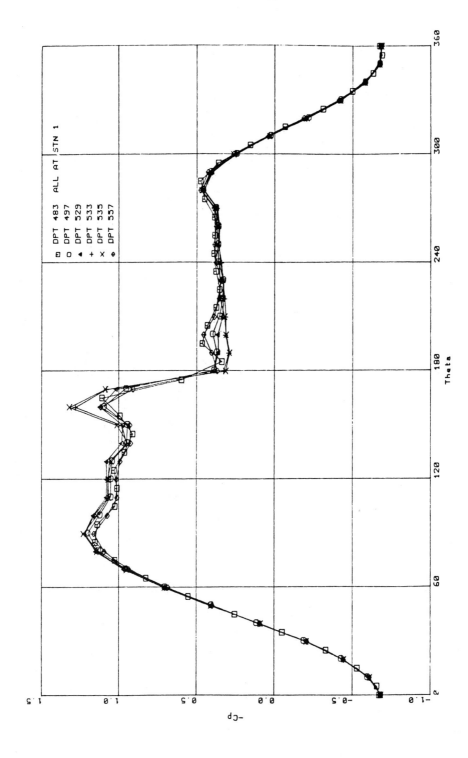

Fig 14a Circumferential pressure distributions giving largest +ve C_Y
Incidence = 35 degrees, roll = 100, 175, 180, 280, 290,

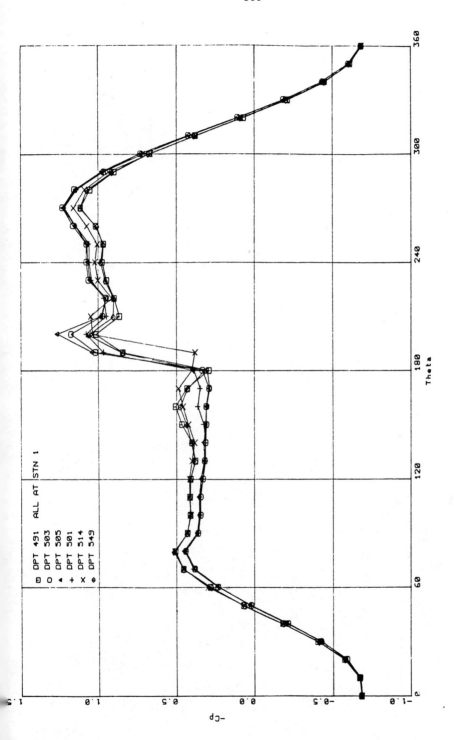

Fig 14b Circumferential pressure distributions giving largest -ve C_Y
Incidence = 35 degrees, roll = 10, 60, 70, 80, 130,
210 degrees

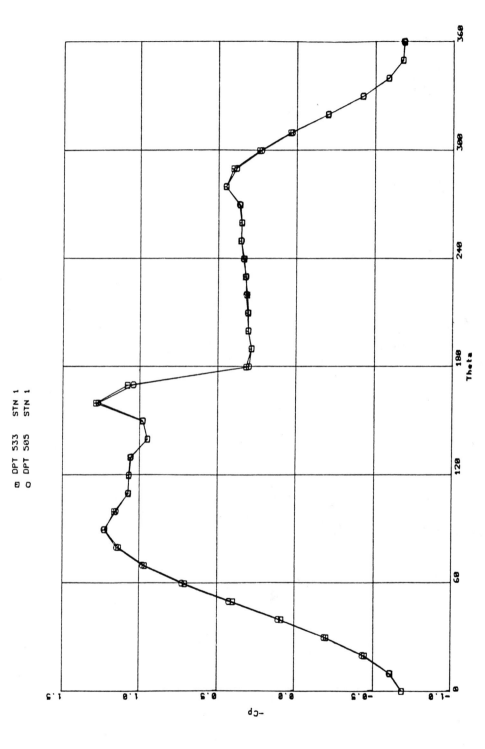

Fig 15 Overlay of pressure distributions for maximum -ve and +ve C_y

On the Prediction of Highly Vortical Flows Using an Euler Equation Model*

F. Marconi[†]

Abstract

An investigation of the power of the Euler equations in the prediction of conical separated flows is presented. These equations are solved numerically for the highly vortical supersonic flow about circular and elliptic cones. Two sources of vorticity are studied; the first is the flow field shock system and the second is the vorticity shed into the flow field from a separating boundary layer. Both sources of vorticity are found to produce separation and vortices. In the case of shed vorticity, the surface point from which the vorticity is shed (i.e., separation point) is determined empirically. Solutions obtained with both sources of vorticity are studied in detail, compared with each other, and with potential calculations and experimental data.

Introduction

Separation and the formation of vortices in the three-dimensional flow about wings and forebodies at high angles of attack is an important aspect of advanced vehicle design. The phenomenon will be especially important at maneuver conditions for the next generation fighter aircraft since the flight envelopes of these vehicles may extend to high angles of attack at supersonic speeds. An understanding of three-dimensional separation and accompanying vortex formation may help in adding to the aerodynamic efficiency of these aircraft. Design efforts which try to take advantage of vortex lift or utilize vortex flaps will surely benefit from basic studies of separated flows. Controllability problems associated with vortex interactions and asymmetric forebody separation will benefit from an understanding of flow separation. If separated flows can be studied accurately and efficiently using computational techniques, the development costs of these aircraft will be significantly reduced.

---

*This was supported by AFOSR under contract no. F49620-84-C-0056
[†]Senior Staff Scientist, Corporate Research Center

This paper presents a numerical study of separation and accompanying vortices in the flow about delta wings and circular forebodies at supersonic speeds. The governing equations considered in this investigation are inviscid (i.e., Euler's equation). It is well known that separation zones (i.e., spirals or closed separation bubbles in 2-D flow) are associated with the existence of vorticity in the flow[1,2,3,4]. There are two sources of vorticity in the high Reynolds number flows about supersonic aircraft: first, the vorticity in the boundary layer and second, the vorticity produced by shock waves. When boundary layer separation occurs, the vorticity of the boundary layer is shed into the flow field and may, to a good approximation, be considered as confined to an infinitesimal sheet which rolls into a vortex. This approximation has been used in conjunction with linearized flow field models for many years (see, for example, Ref. 5 and 6) with good results. The present author[7] showed preliminary Euler calculations which indicated that the more exact flow field model made the computation of vortex flows simpler and at the same time more accurate. At flight Reynolds numbers ($> 10^6$), viscous effects are confined to thin flow regimes (boundary layers and vortex sheets) with the majority of the flow inviscid in nature. This fact makes the solution of the Navier-Stokes equations with a unified procedure (no special treatment of sheets or boundary layers) very difficult. The Euler equations combined with special treatments for sheets and boundary layer may turn out to be a more reliable and accurate approach to the prediction of these flows. The power of the Euler equation model lies in its ability to transport vorticity while avoiding the evaluation of viscous terms. The extent of the capability of the Euler equations to predict separated flows is a major goal of this work.

In the past few years a number of investigators have studied numerical solutions to the Euler equations exhibiting separation. In Ref. 8, Salas showed how shock vorticity can cause separation at the base of a cylinder. Rizzi et al[9] and Murman[10] have shown solutions to Euler's equations with leading edge separation. In Ref. 9 and 10, the source of vorticity is unclear; there is no evidence of strong shock waves. Both authors indicate the possibility that separation is caused by numerical viscosity. In Ref. 10, the wing is a flat plate delta; therefore, there must be a singularity at the leading edge. The numerical scheme used could implicitly introduce a Kutta condition at the leading edge and shed vorticity into the flow field. R. Newsome[11] showed Euler solutions over an elliptic delta wing with

finite thickness. His crude grid results exhibited leading edge
separation (before any crossflow shock); his very fine grid results
exhibited separation after a crossflow shock. The numerical schemes
of Ref. 9, 10, and 11 require the addition of artificial viscosity for
stability. Their research has indicated a small sensitivity of
separation zones to reductions in artificial viscosity. It could be
that these separation zones are Reynolds number independent (even
numerical Reynolds number independent) and only zero artificial
viscosity will keep the flow attached. It is interesting to note that
the separated flows of Ref. 9, 10, and 11 compare well (at least
qualitatively) with experimental data even though the source of
vorticity is numerical (artificial viscosity or truncation error) in
nature.

In the Euler calculations presented here, separation occurs only
when a well defined source of vorticity is present. In addition to
vorticity shed from the boundary layer, shock waves in supersonic
conical flow can produce enough vorticity to induce separation. This
type of separation is qualitatively similar to that produced by
vorticity shed from a boundary layer. With boundary layer vorticity
excluded, any comparison with experimental data will be poor. It is
difficult to assess the relative magnitude of the vorticity introduced
into the flow from shocks and the boundary layer. It should be
pointed out that experimental data show large regions of separated
flow with no significant shock vorticity. This seems to indicate that
vorticity shed from the boundary layer is much larger than that
produced by shocks. In addition, boundary layer separation has a
tendency to reduce shock strengths and therefore shock vorticity.
Nevertheless, shock vorticity may still play an important role in the
separation process. Additionally, the investigation of shock
vorticity induced separation can shed some light on separated flow in
general. The author investigated the effects of shock vorticity on
circular cones[7,12] and on elliptic delta wings[13] and found that at
high angle of attack a crossflow shock (Fig. 1) can produce enough
vorticity to cause separation. This phenomenon will be discussed in
detail later in this paper.

All geometries considered in this work are either delta wings
which are elliptic in cross section or circular cones. In addition,
each section normal to the z axis (Fig. 1) is self similar so that the
geometry is conical. The flow is assumed supersonic everywhere and
separation lines are assumed to be on conical rays. With these
assumptions and that of a conical body, it follows that the flow is

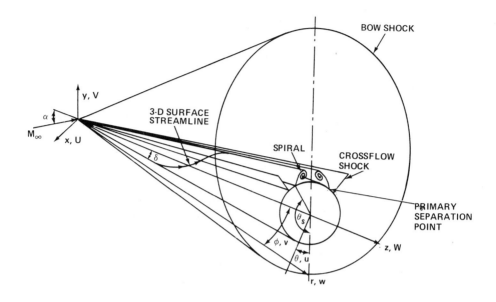

Fig. 1   Three-Dimensional Sketch of Flow Field & Coordinate System

conical (i.e., all flow variables are independent of the spherical radius r (Fig. 1)). However, the basic features of the flow, including separation and vortices, are three-dimensional in nature.

In this paper our previous work in highly vortical flows[7,12,13] will be reviewed after which our more recent findings will be presented. The overall computational procedure will be outlined and the model for shedding vorticity from specified separation points will be discussed. Flow fields with shock vorticity induced separation will be studied, in addition to flows with vorticity shed from both primary and secondary boundary layer separation points. A detailed comparison of these two different sources of vorticity will be presented. The investigation of the computed flow fields will be aided by comparisons with experimental data and computations of other researchers, including potential flow results. Finally, the findings of this work to date will be summarized.

Outline of Computational Procedure

The overall numerical procedure used in this study, although employing a number of new features, is essentially standard.[14] The fully three-dimensional Euler's equations are solved with an explicit marching technique. The marching direction, z (Fig. 1), is an iterative coordinate for the conical flow considered here. The scheme is restricted by the fact that the axial component of velocity, W, (Fig. 1), must be supersonic everywhere. The marching scheme is continued until the flow field is invariant with respect to the computational marching direction except for a scale factor. The finite difference scheme used is Moretti's characteristic based λ-scheme.[15] The bow shock and the primary crossflow shock are fit and are forced to satisfy the exact Rankine-Hugoniot jump conditions. The bow shock is fit as the outermost boundary of the flow field. On the low pressure side of the bow shock, freestream conditions exist. The crossflow shock is fit as an internal boundary of the flow field with its low pressure side being computed as the computation proceeds.[16]

The crossflow shock computation is a critical part of the overall procedure, particularly since it plays such a critical role in separation. As mentioned in the Introduction, it is the vorticity produced by the crossflow shock which is significant. Any scheme that captures the crossflow shock and introduces additional artificial viscosity to stabilize it runs the risk of distorting the separation and vortex. The bow shock also introduces vorticity into the flow field but not enough to produce separation. The crossflow shock is caused by the fact that at a large incidence the crossflow component of the velocity, $\sqrt{u^2+v^2}$ (Fig. 1), becomes supercritical. The flow expands in going around the body, and the crossflow can become supercritical if the incidence is high enough. This component of velocity must be small in the lee symmetry plane due to the boundary conditions. A supercritical crossflow generally passes through a relatively normal shock before stagnating in the lee plane on the surface. The crossflow shock exists to turn the three-dimensional flow parallel to the lee symmetry plane. The variation in strength of the crossflow shock can be quite large. The shock is strongest at the body where the crossflow is highly supersonic and approaches zero strength in the field where the crossflow becomes sonic. It is this variation in strength that produces a crossflow entropy gradient and thus radial vorticity. This vorticity in turn causes shock induced separation.

The low pressure side of the crossflow shock is computed with one-sided differences away from the shock. This is consistent with the fact that the crossflow is supersonic. The Rankine-Hugoniot conditions, together with the compatibility condition along a bicharacteristic reaching the shock on its high pressure side, supply enough information to compute the deviation of each shock point from a conical ray in addition to all the primitive variables on its high pressure side. The bicharacteristic used is the one in the plane containing the local normal to the shock and the marching direction. All shock points are computed with the post correction scheme proposed by Rudman[17] and independently by deNeef[18]. The shocks converge (i.e., each shock point becomes aligned with a conical ray) with the rest of the flow field. In all the computations presented here, the last crossflow shock point fit had a normal Mach number of approximately 1.05 (pressure ratio of 1.12). The finite difference scheme was able to capture weaker shock points. In addition, the finite difference scheme was used to capture all reverse crossflow shocks and oblique crossflow shocks in the flow field (to be discussed later). These shocks are usually weak and the scheme can capture them accurately.

The computational grid used in this work is developed in stages. In the first stage, the elliptic cross section in a constant z plane is mapped to a circle. This step is omitted in the case of a circular cone. The mapping is a simple Joukowski transformation and will not be detailed here. Next, a polar coordinate system is used in the mapped space. An exponential stretching is used in the radial direction to cluster grid points near the surface in order to resolve

the complex flow near separation. The Joukowski mapping, clusters
grid points in the circumferential direction automatically.
When thin ellipses (axis ratios > 10) were first considered, solutions
with large truncation error at the wing leading edge were found. One
result of this truncation error was the production of vorticity at the
leading edge. There was little error in entropy at the leading edge
(i.e., entropy was constant on streamlines), but the vorticity was
large, leading to a violation of Crocco's theorem. The error was
manifested by a layer in crossflow velocity at the surface. The
crossflow on the body was much lower than that one point away from the
body. While this vorticity was never large enough to cause leading
edge separation, as in the work of Ref. 9, 10, and 11, it did distort
the separation behind the crossflow shock. This error was eliminated
by increasing the resolution at the wing leading edge. An exponential
stretching was used in the circumferential direction.

Shock induced inviscid separation produces a contact surface
emanating from the separation point (Fig. 1). It is this contact that
ultimately spirals to form a vortex. The contact surface has a jump
in entropy and an accompanying jump in velocity. In conical flow,
entropy is constant on crossflow streamlines. Thus, one can see (Fig.
1) that the entropy on the windward side of the contact comes from the
windward stagnation point of a cone and the wing leading edge
stagnation point in the case of an ellipse. The streamline that wets
the body passes through the base of the crossflow shock to form the
high entropy side of the contact. There are a number of possibilities
for the entropy on the lee side of the contact. The crossflow
streamlines that wet the body are tracked in each step of the
iteration, and the proper entropy is imposed on the surface including
the jump in entropy at the separation point. The entropy
discontinuity off the body is captured with the finite difference
scheme. The computation of the vortex sheet in the shed vorticity
computation will be discussed in another section of this paper.

The finite difference scheme used in this work is an explicit
marching scheme which is notoriously inefficient for converging to a
conical or steady solution. One advantage of the explicit marching
scheme is that it is totally "vectorizable." The time consuming parts
of the code (i.e., interior point computation) utilized the vector
architecture of the Cray 1 computer. This made the computation about
20 times faster than on an IBM 3033. The computations shown here
typically took 45 CPU minutes on the Cray 1, and the grid used had 81
x 81 points in the cross sectional plane. In each case shown, the

maximum residual (i.e., the derivative of pressure in the marching direction) was reduced at least six orders of magnitude.

Study of Computational Results - Shock Induced Vorticity

In this section, computational results showing spiralling streamlines due to the vorticity produced by a crossflow shock will be analyzed. The first configuration considered is a 10° half angle cone at $M_\infty = 2$ and $\alpha = 25°$. The surface pressure distribution is shown in Fig. 2. This flow field has a shock induced separation at $\theta = 149.6°$. In Fig. 2, the computed surface pressures using two grids are shown. Both computations 73 x 73 and 89 x 89 give the same separation point and spiral location, indicating an independence of the results to numerical viscosity due to truncation error. As a matter of fact, this flow field was computed with an even coarser grid (37 x 37) that exhibits the same basic features (i.e., separation and spiral).

The crossflow streamlines are shown in Fig. 3, along with the computed bow and crossflow shocks. The streamlines show saddles in the wind and lee symmetry planes on the cone surface. In addition, there is a saddle at the separation point and a node at the center of the spiral. The separation point and spiral can be seen more clearly in the blowup of Fig. 4. If the nodes and saddles are summed using the procedure of Ref. 4, it can be shown that the proper number of nodes and saddles exist in the streamline pattern of Fig. 3. The separation point ($\theta_s = 149.6°$) corresponds to the plateau in pressure (Fig. 3) just after the shock. If one considers the momentum equation in the θ direction, it indicates that $\partial p / \partial \theta = 0$ at a crossflow stagnation point (u = v = 0). The separation point in this flow field is a real crossflow stagnation point (i.e., the crossflow passes continuously through zero). The streamlines (Fig. 4) show that the flow moves in the negative θ direction from the lee stagnation point toward the separation point. In doing so, the flow expands (i.e., there is a drop in pressure between $\theta = 180°$ and $\theta = 165°$). The flow then recompresses to the separation point. This recompression phenomenon is the cause of secondary separation. The inviscid separation point location is far from that of the viscous flow. This can be surmised by the fact that the shock is strong enough to separate a boundary layer at its base. The inviscid separation point is too far downstream of the shock. The streamlines of Fig. 4 clearly show how all the flow is ultimately swept up into the infinitely turning spiral. The apparent power of the Euler equations to describe the region near the center of the spiral should be noted. A detailed study of the streamline pattern near the center of the spiral revealed that the streamlines asymptote to an ellipse. This was first indicated by Smith (Ref. 5). It also should be pointed out that Fig. 4 shows that streamlines wrapping around the spiral very rapidly approach the lee side of the separation line (contact). Consider, for example, the third streamline off the cone in Fig. 4. It wraps around the top of the spiral, comes back and approaches the contact a short distance off the body, creating an entropy layer on the lee side of the contact. This entropy layer tends to weaken the contact.

Figure 4 also shows that the separation line leaves the cone at a finite angle. J.H.B. Smith's analysis (Ref. 5) concluded that at a forced separation the sheet comes off tangent. However, this difference is due to the fact that the separation in the shock induced case is highly nonisentropic. As such the flow can and does stagnate on both sides of the separating sheet. Figure 5 shows the crossflow velocity on the surface of the body (v = 0 from the body boundary condition). The point to be considered here is that the velocity passes through zero (separation) smoothly. The analysis of Smith indicates that separation occurs at a discontinuity in crossflow, the jump in velocity determining the sheet strength. This is a basic difference between forced and shock induced separation and is discussed in detail in Ref. 12. The jump in crossflow velocity in Fig. 5 is due to the shock. It is also interesting to note the maximum negative u that occurs under the spiral. This negative crossflow can become supersonic, causing a second reverse crossflow shock. A sample of this will be shown later.

Figure 6 shows the isobars for the same case ($M_\infty = 2$, $\delta = 10°$, $\alpha = 25°$). It should be noted how smoothly the crossflow shock transitions to zero strength in the field. The most interesting aspect of the figure is the closed isobar at the center of the spiral. It represents an absolute minimum in the flow field pressure. The component of vorticity in the spherical radial direction is given by:

$$\Omega_r = - (u \cos \phi + \frac{\partial u}{\partial \phi} \sin \phi - \frac{\partial v}{\partial \theta})/(r \sin \phi)$$

where u and v are the crossflow velocities defined in Fig. 1 and r, θ, ϕ are spherical coordinates (Fig. 1). Figure 7 shows lines of constant $r\Omega_r$. The figure shows how the vorticity is distributed and is produced by the crossflow shock. Note that Ω_r is small until the crossflow shock, indicating that the bow shock doesn't produce enough

vorticity to cause separation on its own. The vorticity produced by
the bow shock is an order of magnitude smaller than that produced by
the crossflow shock and two orders of magnitude smaller than that at
the center of the spiral. For this reason, there are no contours of
vorticity outside the region behind the crossflow shock in Fig. 7.
The vorticity is negative, causing the counterclockwise spiraling of
this flow, and its absolute magnitude is maximum at the center of the
vortex.

The existence of a crossflow shock does not necessarily imply
enough vorticity to cause separation. It is the variation in shock
strength which produces an entropy gradient and thus vorticity which
causes separation behind the shock. In the case of the circular cones
considered thus far, the shock strength decreases monotonically from
its base. This is not always the case, as will be discussed in the
case of elliptic cross sections. For a circular cross section the
shock strength at the surface is a good indication of the vorticity
produced by the shock. Figure 8 is a plot of the inviscid separation
point location vs the shock pressure ratio at the surface. The cone
has a 10° half angle, and the free stream Mach number is 2. The
crossflow shock strength was increased by increasing incidence. The
pressure ratio used is a good measure of the shock strength variation
and thus the vorticity produced by the shock. At each data point, the
corresponding α is noted. The plot shows that as the shock strength
decreases, the separation point moves to the lee symmetry plane. At
the same time, the spiral region is getting smaller. The highest
incidence computed was 25°, since above that value the axial Mach
number on the cone surface approaches sonic, making it impossible to
march. At the low end, $\alpha = 19°$, the spiral was so small that it was
difficult to resolve numerically, and no lower incidences are shown.
The interesting feature of this figure is that an extrapolation of the
curve would seem to indicate that inviscid separation moves to the lee
plane before the shock is eliminated ($p_2/p_1 = 1.$). An extrapolation
would indicate that the spiral is eliminated at $p_2/p_1 \approx 1.85$. This
corresponds to a normal Mach number slightly above 1.3, which is
approximately the region where the full potential approximation is
valid for these flow fields. It would seem that below a maximum
normal Mach number of 1.3, the crossflow shock may not produce enough
vorticity to cause separation.

Figures 9, 10 and 11 deal with an interesting case ($M_\infty = 3$, $\delta = 9.46°$, and $\alpha = 25°$). Figure 9 shows the streamlines and the crossflow
shock near the lee plane. The shock exhibits a kink as it passes from

the influence of the spiral. Near the cone, the shock must deflect the streamlines upward in order that they may pass over the spiral; beyond the top of the spiral, this is no longer true and the shock acts like a normal shock. The two regions are separated by the kink in the shock, and the shock slope in the crossflow plane had to be differenced away from this point. Another interesting feature of this flow is the fact that the expansion of the reverse flow from the lee stagnation point is so large that the negative crossflow becomes supersonic near the body. The smooth recompression shown in Fig. 2 is replaced by a reverse crossflow shock. This phenomenon has been noted experimentally in Ref. 19. The reverse crossflow shock, which was captured, can be seen in the isobars of Fig. 10. The second (reverse) shock is on the lee side of the primary crossflow shock. It is indicated by the clustering of the isobars between $\ln(p/p_\infty) = -1.2$ and $\ln(p/p_\infty) = -0.73$ on the cone surface. The shock is not strong enough to produce a secondary inviscid separation, whereas in the experiment of Ref. 19 it was strong enough to separate the boundary layer. The strength of the vortex is also indicated in Fig. 10 by the closed isobars representing a steep pressure minimum at the vortex center.

Figure 11 shows the surface pressure for the $M_\infty = 3$, $\delta = 9.46°$ and $\alpha = 25°$ case. The solid line is the present calculation with shock vorticity induced separation at $\theta = 155.3°$, again at the plateau in pressure just after the primary crossflow shock. The reverse crossflow shock can be seen in the pressure distribution at about $\theta = 167°$. The shock is captured, and so it is smeared over a mesh interval. The figure also shows the numerical results of Ref. 20. They are also a solution to Euler's equations but with the primary crossflow shock captured. In the results of Ref. 20, the separation point was forced to occur at $\theta = 120°$ in order to match the boundary layer separation point found experimentally. The forced separation model of Ref. 20 is in contradiction to the analytical work of J.H.B. Smith (Ref. 5). In addition, there are anomalies in the results of Ref. 20. Basically the comparison of Fig. 11 shows that the two results are very close, while the separation points are very different. It seems that forcing separation at $\theta = 120°$ simply inserted a wiggle in the surface pressure distribution of Ref. 20. The flow then came back to the shock induced flow field. The crossflow shock locations are very close. The shock of Ref. 20 is smeared over a few points but is close to the fit shock of the present calculations. If separation did occur at 120°, there must be a crossflow shock before $\theta = 120°$. However, there is no evidence of a shock before separation in the results of Ref. 20.

Figure 12 shows the pressure distribution on a 5° cone at $M_\infty = 4.25$ and $\alpha = 12.35°$. Compared are the results of the present calculation, those computed using the full potential equation supplied by M. Siclari[21], and the experimental results of Rainbird[22]. The potential results of Siclari have been corrected for nonisentropic bow shock effects while maintaining the irrotational assumption. Figure 12 indicates that the Euler and potential results are virtually identical until the shock. The potential result does not exhibit the minimum in pressure behind the shock typical of separation. Potential calculations cannot predict separation or spiraling without a shedding sheet. The comparison between the Euler and potential calculations affirms the fact that rotationality is important only after the crossflow shock. A comparison between the present calculation and the experimental results of Rainbird clearly shows that the vorticity produced by the shock does not separate the flow near the viscous separation point. The experimen-tal data show two separations, primary at $\theta = 120.3°$ and secondary at $160°$, while the inviscid separation is at $\theta = 151.3°$. The longer plateau in pressure in the experiment behind the primary separation point is due to the secondary separation; otherwise, the Euler and experimental pressure distribu-tions would be similar. It is the expansion and recompression of the reverse flow that causes secondary separation. It should be clear that while the vorticity produced by the crossflow shock is not the whole separation story, it may play an important role in the process. There are two sources of vorticity in this flow field: one is the shock and the other the boundary layer. Both these sources of vor-ticity play a role in separation and the resulting spiral or vortex. Boundary layer shed vorticity will be discussed later in this paper.

The eccentricity of the elliptic cross sections has a significant impact on the vorticity generated by the crossflow shock. In general, the thinner the cross section (higher axis ratio) the stronger the shock. There is an effect, discovered by the author[13], which seems to reduce the entropy gradient near the surface behind the shock as the section of elliptic delta wings gets thinner. Consider the isobar plots of Fig. 13 and 14; both show the leading edge of the wing, crossflow shock and isobars in the leading edge region. Figure 13 shows a 10:1 section and Fig. 14 a 6:1 section. The other parameters of the flow were chosen so that the shock was approximately the same strength at the surface. The two angles noted in the figure define

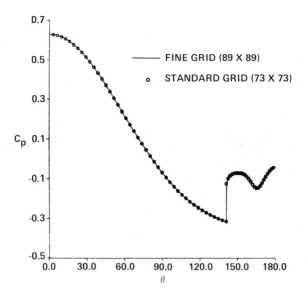

Fig. 2   Comparison of Computation with Two Grids, Surface Pressure ($M_\infty = 2$, $\delta = 10°$, $\alpha = 25°$)

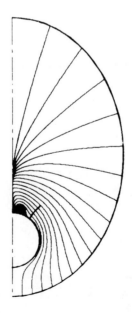

Fig. 3   Cross Flow Streamlines in Field ($M_\infty = 2$, $\delta = 10°$, $\alpha = 25°$)

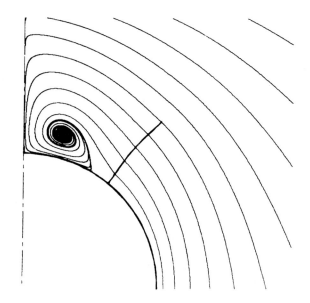

Fig. 4   Cross Flow Streamlines Near Lee Plane
          ($M_\infty = 2$, $\delta = 10°$, $\alpha = 25°$)

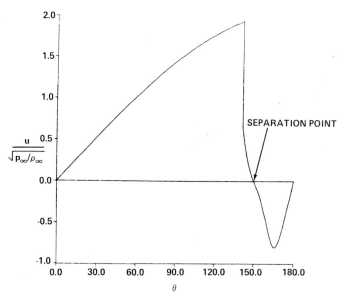

Fig. 5   Surface Cross Flow Velocity ($M_\infty = 2$,
          $\delta = 10°$, $\alpha = 25°$)

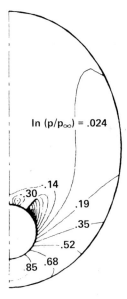

Fig. 6 Isobars in Cross Plane ($M_\infty = 2$,
$\delta = 10°$, $\alpha = 25°$)

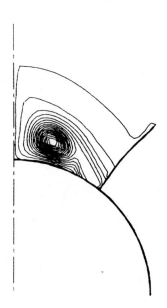

Fig. 7 Lines of Constant Radial Vorticity X Spherical
Radius ($M_\infty = 2$, $\delta = 10°$, $\alpha = 25°$)

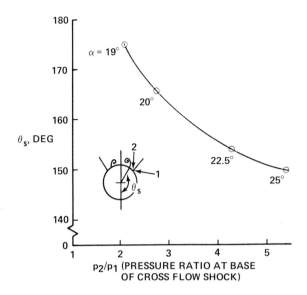

Fig. 8   Inviscid Shock Induced Separation Point Location
vs Shock Strength ($M_\infty = 2$, $\delta = 10°$)

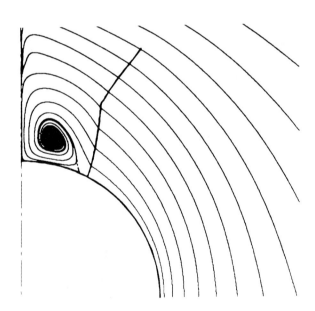

Fig. 9   Cross Flow Streamlines Near Lee Plane
($M_\infty = 3$, $\delta = 9.46°$, $\alpha = 25°$)

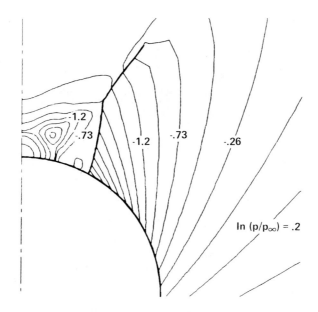

In $(p/p_\infty) = .2$

Fig. 10   Isobars Near Lee Plane ($M_\infty = 3$, $\delta = 9.46°$, $\alpha = 25°$)

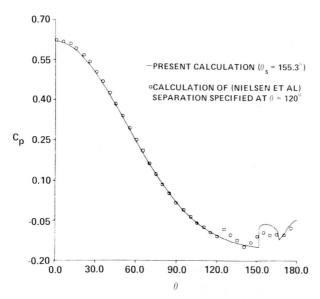

—PRESENT CALCULATION ($\theta_s = 155.3°$)

○CALCULATION OF (NIELSEN ET AL)
SEPARATION SPECIFIED AT $\theta = 120°$

Fig. 11   Surface Pressure Comparison ($M_\infty = 3$, $\delta = 9.46°$, $\alpha = 25°$)

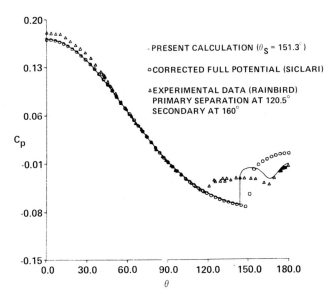

Fig. 12 Surface Pressure Comparison ($M_\infty = 4.25$, $\delta = 5°$, $\alpha = 12.35°$)

the wing geometry, λ is the sweep angle of the delta, and δ is the slope of the surface in the symmetry plane. In the thinner case (Fig. 13), there is a gap between the surface and the last wave of the expansion from the wing leading edge. This is indicated by the tendency of the isobars to become more vertical near the shock. The effect is similar to the uniform flow region downstream of a Prandtl-Meyer expansion below the last wave of the expansion. The result of this phenomenon is a region of uniform crossflow Mach number in front of the shock near the surface. This region is indicated in Fig. 13 by the portion of the shock near the surface before it begins to curve. An inspection of Fig. 14 shows that this effect is not present because the leading edge radius of curvature is too large in this case.

The result of this phenomenon is shown in Fig. 15, which is a plot of the entropy distribution along the shock on its high pressure side. Note the much smaller entropy gradient in the 10:1 case as compared with the 6:1 case. Also included in the figure is the entropy distribution along the crossflow shock for the 10° cone, $M_\infty = 2$, and $\alpha = 25°$ discussed previously. Although the shock is much stronger for the case of the cone, this curve is included for reference. Note that the entropy drops rapidly at the surface in the case of the cone. The vorticity just behind the shock at the surface is much larger in the 6:1 case than in the 10:1 case. The result is that the 6:1 case separates and the 10:1 does not. Figure 16 shows the crossflow Mach number distribution on the surface for the three cases. First note that the 6:1 and 10:1 cases have the same shock strengths as were indicated by their surface entropies at the shock (Fig. 15). The separation in the 6:1 case is indicated by the negative Mach number starting at $\bar\theta \approx 160°$ ($\bar\theta$ is the polar angle in the mapped space). While it is true that increasing the eccentricity of the section tends to decrease the entropy gradient behind the shock, this does not always eliminate separation. The vorticity just behind the shock at the wall may be reduced, but it can become large downstream and cause separation. In addition, once a spiral forms, the shock has a tendency to move toward the leading edge where the extent of the uniform flow region just discussed is reduced. The problem is complicated by a number of factors.

The next case considered is a very thin delta wing. The ellipse axis ratio is 14:1 ($\delta = 1.5$, $\lambda = 70°$); the freestream conditions are $M_\infty = 2$ and $\alpha = 10°$. The crossflow shock is somewhat stronger in this case than the ellipse discussed in the last paragraph. The shock crossflow Mach number at the surface is 1.9 in this case and was about

1.8 in both the 6:1 and 10:1 cases discussed previously. This is due to the higher eccentricity. The eccentricity of this section combined with a shock position relatively far from the leading edge results in a significant region of uniform flow near the base of the shock. This is indicated by the isobar plot of Fig. 17. The result is a very small derivative of entropy normal to the surface just behind the shock. The flow in this case does separate although the effect is mild. It seems that the eccentricity of the section has reduced the vorticity resulting in a weak vortex. Figure 18 shows the surface pressure distribution for this case compared with the computed results of R. Newsome.[11] The effect of the vortex is seen as a slight drop in the upper surface pressure, the local minimum is at about $X/X_{LE} =$ 0.3. The influence of the separation is negligible in this case, as demonstrated in the next paragraph. The comparison is very good, although the influence of the vortex in the result of Ref. 11 seems a little greater. In the work of Newsome, the shock was captured and is spread over four mesh intervals. This is good in light of the fact that no post or preshock overshoots are present. It seems that these are controlled with artificial damping (see Ref. 11 for details). No such damping was necessary in the present calculation because the shock is fit. It is interesting that the artificial damping can be controlled so that the shock vorticity induced separation is not distorted significantly.

In an effort to study a substantial shock vorticity induced vortex on a delta wing, a flow situation was developed in which the crossflow shock strength at the surface was large. It was shown by Siclari[23] that for potential flows, the shock strength is increased as the wing leading edge sweep is increased. A 10:1 ellipse whose leading edge was swept 72° (i.e., $\lambda = 72°$ and $\delta = 1.86°$) was considered. The freestream conditions were the same as those considered previously ($M_\infty = 2$, $\alpha = 10°$). The surface shock Mach number was 2.3 in this case while it was only 1.8 for the 10:1 ellipse considered previously. The flow did separate in this case, and the vortex was substantial (maximum reverse Mach number of -0.78). The crossflow streamlines are shown in Fig. 19. The dashed line that intersects the body near the leading edge (the stagnation point is a saddle) is the streamline which wets the surface. The entropy on this crossflow streamline wets the surface from the saddle to the wind plane node, and over the leading edge through the shock and separates onto the spiralling contact sheet. The dashed line that attaches to the surface near the lee plane carries the entropy which wets the surface from its saddle to the lee plane node and back in the reverse flow region to the low entropy side of separating contact. Figure 20 shows the region near the leading edge in more detail. The figure shows the crossflow shock whose shape is affected by the vortex. The dashed line near the vortex is the separating streamline, i.e., the contact sheet.

Figure 21 shows the isobar pattern (for the case discussed in the last paragraph) near the leading edge. Note that the separation has moved the shock toward the leading edge so that the region of uniform flow near the base of the shock does not exist in this case. The entropy gradient behind the shock is quite large so that the vorticity generated by the shock is significant. The closed isobar downstream of the shock represents a minimum in pressure which corresponds to the center of the vortex. The effect of the separation on all aspects of the flow field is significant. In particular, the surface pressure is affected significantly. Figure 22 compares the computed surface pressure with that calculated using the full potential equation. Again the two calculations give the same result up to the location of the shock. The shock position is moved from $X/X_{LE} \approx 0.73$ (potential) to $X/X_{LE} \approx 0.83$ (Euler) by the separation. In addition, all the typical separated flow features behind the shock do not exist in the potential result. The maximum in pressure just behind the shock represents the separation or stagnation point and the minimum in pressure at $X/X_{LE} \approx 0.6$ represents the maximum in reverse flow velocity just under the vortex.

The results shown thus far have indicated that shock vorticity can cause separation which is at least qualitatively similar to boundary layer separation. In addition, the computations have shown that shock vorticity can have a significant impact on the flow field. The rest of this paper will deal with the impact of vorticity shed from a separating boundary layer.

Shed Vorticity - Computational Model

In this section, the concept of shedding vorticity from a specified separation point (the intersection of a conical separation line and the cross plane) will be discussed. The conceptual process of shedding vorticity from a conical surface will force separation and an accompanying spiral. The model used here to force separation at a specified location follows the work of J.H.B. Smith[5]. Smith assumed irrotational flow outside a vortex sheet in order to analyze the local flow at separation. The present work uses only the basic concept,

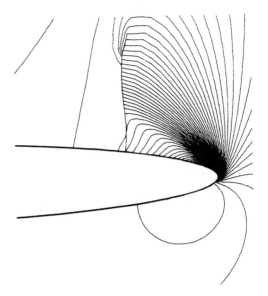

Fig. 13   Isobars Near Leading Edge & Shock ($M_\infty = 2$,
$\alpha = 10°$, Ellipse, $\delta = 2.08°$, $\lambda = 70°$)

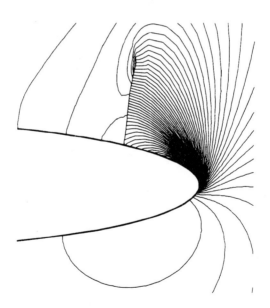

Fig. 14   Isobars Near Leading Edge & Shock ($M_\infty = 1.97$,
$\alpha = 10°$, 6:1 Ellipse, $\delta = 3.17°$, $\lambda = 71.61°$)

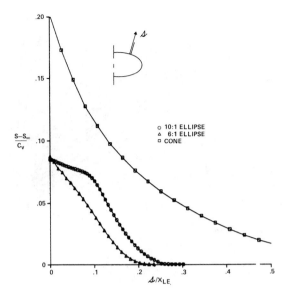

Fig. 15 Entropy Distribution Along the Shock
(10:1 Ellipse, $M_\infty = 2$, $\alpha = 10°$,
$\delta = 2.08°$, $\lambda = 70°$) (6:1 Ellipse,
$M_\infty = 1.97$, $\alpha = 10°$, $\delta = 3.17°$, $\lambda = 71.61°$)
(Cone, $M_\infty = 2$, $\alpha = 25°$, $\delta = 10°$)

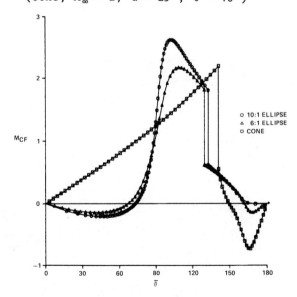

Fig. 16 Surface Crossflow Mach Number Distribution
(10:1 Ellipse, $M_\infty = 2$, $\alpha = 10°$, $\delta = 2.08°$,
$\lambda = 70°$) (6:1 Ellipse, $M_\infty = 1.97$, $\alpha = 10°$,
$\delta = 3.17°$, $\lambda = 71.61°$) (Cone, $M_\infty = 2$, $\alpha = 25°$,
$\delta = 10°$)

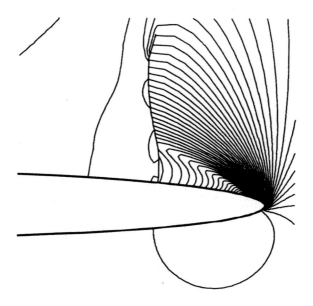

Fig. 17   Isobars Near Leading Edge & Shock ($M_\infty = 2$, $\alpha = 10°$, 14:1 Ellipse, $\delta = 1.5°$, $\lambda = 70°$)

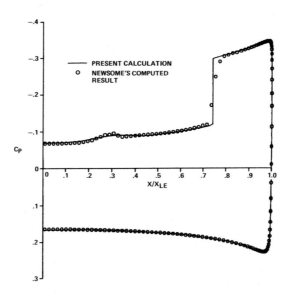

Fig. 18   Surface Pressure Comparison ($M_\infty = 2$, $\alpha = 10°$, 14:1 Ellipse, $\delta = 1.5°$, $\lambda = 70°$)

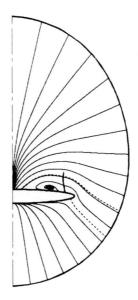

Fig. 19 Crossflow Streamlines (M_∞ = 2,
α = 10°, 10:1 Ellipse, δ = 1.86°,
λ = 72°)

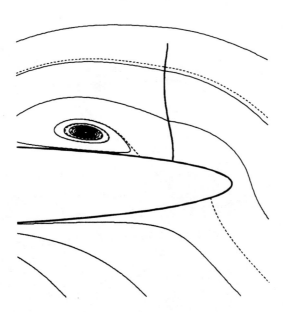

Fig. 20 Crossflow Streamlines Near Leading Edge (M_∞ = 2,
α = 10°, 10:1 Ellipse, δ = 1.86°, λ = 72°)

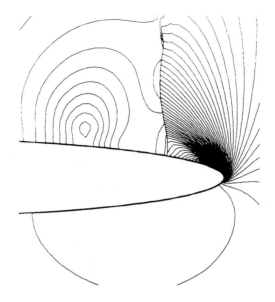

Fig. 21 Isobars Near Leading Edge ($M_\infty = 2$, $\alpha = 10°$, 10:1 Ellipse, $\delta = 1.86°$, $\lambda = 72°$)

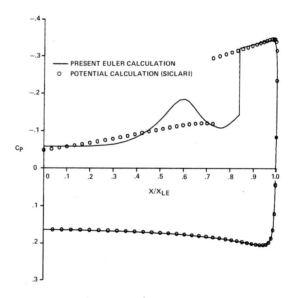

Fig. 22 Surface Pressure Comparison ($M_\infty = 2$, $\alpha = 10°$, 10:1 Ellipse, $\delta = 1.86°$, $\lambda = 72°$)

which doesn't depend on the irrotational assumption. In addition, any
model of inviscid separation should reduce to that of Smith as the
rotationality outside the vortex sheet gets small. The basic concept
acquired from Ref. 5 is that at a specified separation point, there is
a vortex sheet or contact surface which has a jump in velocity
direction. The condition at the intersection of the sheet and the
surface (i.e., the separation point) is that the crossflow velocity
stagnates on the lee side of the sheet (Fig. 23a). The crossflow
velocity on the wind side of the sheet is determined by the global
solution. In general, the crossflow velocity on the wind side of the
sheet is finite. The vortex sheet is a stream surface and the
pressure is continuous across it. In the case of isentropic
irrotational flow considered in Ref. 5, these conditions imply that
the modulus of velocity is continuous across the sheet. If the
crossflow stagnates on both sides of the sheet (u = v = 0) at
separation, the isentropic condition implies that the radial component
of velocity (w, Fig. 1) is continuous and the sheet doesn't exist
(i.e., no vorticity is shed into the flow field). These arguments
conclude that in isentropic flow, the separating sheet must leave the
cone surface tangentially if the wing side crossflow is subsonic. If
the surface crossflow is subsonic just upstream of separation and the
sheet leaves the surface at any angle relative to the surface, then
the flow on the windside of the sheet must stagnate and no vorticity
will be shed into the flow. On the other hand, if the crossflow is
supersonic the sheet can leave the surface at an angle, an oblique
crossflow shock will occur, and the flow on the wind side of sheet
will not stagnate so that vorticity is shed. This phenomenon will be
discussed further in the next section. None of these arguments hold
for the highly nonisentropic flow discussed in the previous section,
where the flow stagnates on both sides of the separating sheet. In
these cases vorticity is not shed from the surface, only shock
vorticity is present.

Separation was forced in the calculations presented here simply
by using a double point at separation in order to allow for a jump in
crossflow velocity. The crossflow velocity (u) on the wind side of
the sheet is determined by the governing crossflow momentum equation,
as at any other surface point. The crossflow component v must vanish
at every surface point from the boundary condition. The crossflow
velocity on the lee side of the sheet is set to zero (u = 0) in order
to force separation at a specified point.

The separation point in this forced separation model exhibits a
jump in crossflow velocity. The crossflow on the wind side (Fig. 23a)
of the separation is evaluated in each step of the iteration from the
momentum equation tangent to the body in the cross plane. The
remaining primitive variables are computed at this point with the same
governing equations as at any other body point, with the boundary
condition (v = 0) satisfied. All circumferential velocity derivatives
are taken one-sided in the negative θ direction in order to avoid
differencing across the sheet. The point just below the sheet (lee
side) is forced to be a crossflow stagnation point (u = v = 0). The
pressure is continuous across the sheet so that its value on the wind
side could theoretically be used on the lee side at separation.
Diffi-culties were encountered in computing the pressure in a small
region just after separation. The computation of the pressure in this
region will be discussed in the next paragraph. The entropy at the
lee side point is computed in the standard way (i.e., conserved along
the streamline that wets the lee side of the body). If the pressure
is known in addition to the entropy and total temperature (the flow is
assumed adiabatic), the last component of velocity w (Fig. 1) can be
evaluated from the energy equation. Again circumferential derivatives
of the velocities across the sheet at the body grid points near
separation are avoided. Differences in entropy across the sheet are
avoided naturally by the λ-scheme as in the shock induced separation
discussed in the previous section.

The difficulty encountered in the evaluation of the pressure on
the lee side of the separation point can be traced to the numerical
computation of the derivatives of velocities in a direction normal to
the surface. The derivatives appear in the continuity equation from
which the pressure is evaluated. These velocities are discontinuous
across the sheet, and since the sheet lies so close to the surface at
separation, there is no way to resolve the flow in the region between
the surface and the sheet near separation. The behavior of the
pressure in this region was evaluated analytically by Smith[5] and
numerically by Fiddes[6] under the assumption of slender body theory.
The results of Fiddes show a pressure plateau on the surface just
after separation. The full Euler equations predict that the pressure
derivative along the surface (p_θ) becomes zero at the lee side of the
separation point, where the crossflow stagnates. In the present
calculation it has been difficult to obtain this type of pressure
behavior. A number of procedures have been attempted with varying
degrees of success, but none has been totally satisfactory. In Ref. 7

the author showed preliminary results which exhibited a pressure plateau after separation, but the scheme used there proved unstable when the vortex sheet was moved during the iteration process. A reformulation of dependent variables has alleviated the problem. Currently, as long as the crossflow is subsonic (Fig. 23a) just before separation, no special treatment is used in evaluating the pressure except for circumferential and radial grid clusterings. Because of the difficulties associated with the pressure evaluation in the region on the surface just after separation the results of the next section should be considered somewhat preliminary.

In the case of supersonic crossflow (Fig. 23b) differencing of pressure across the separation point proved to be unstable. The results of a number of numerical experiments indicate that the model for the inviscid flow in this case should be such that the sheet leaves the surface at an angle relative to it, and an associated shock occurs at separation. The flow structure is sketched in Fig. 23b. This model does not violate the concepts of vorticity shedding proposed by Smith[5] since the flow on the wind side of the separation point need not stagnate and thus a jump in velocity does exist across the sheet. In order to capture the shock at separation in one mesh interval no pressure derivatives were taken across the separation point on the cone surface. In addition a small pressure plateau is imposed just after separation, the level being taken from just downstream of separation.

It should be pointed out that the discussion presented in this section thus far has assumed a circular cross section. The velocity u is tangent to the body and v is normal. For an elliptic cone the procedure is the same with u and v being replaced in the discussion by conical velocities tangent and normal to the body.

Secondary separation is forced by shedding vorticity from the body in the reverse crossflow region produced by primary separation. The model is the same as was just outlined for primary separation. Both separation point locations are obtained from experimental data. The only free parameters in the problem as posed here are the locations of the separation points. In the work of Ref. 6 primary separation points were found iteratively by matching an inviscid solution (slender body theory) to a boundary layer solution. A full viscid/inviscid interaction procedure using the current Euler calculation will be the subject of future work.

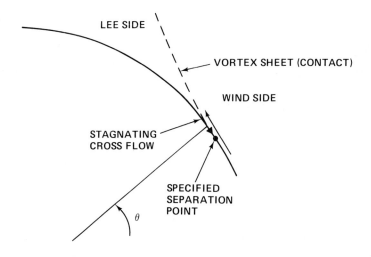

a) Subcritical

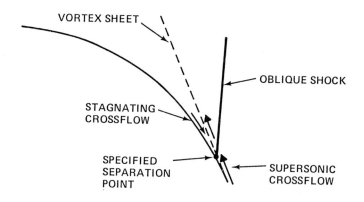

b) Supercritical

Fig. 23   Sketch of Forced Separation Model

Study of Computational Results - Shed Vorticity

The procedure for shedding vorticity in an inviscid flow was tested by comparisons with the experimental data of Rainbird[22]. A 5° half angle cone was tested at M_∞ = 1.79 (Re = 34 x 10^6) and M_∞ = 4.25 (Re = 68 x 10^6). The detailed surface pressure distributions presented in Ref. 22 were digitized for comparison. Both primary and secondary separation point locations were determined from surface shear stresses in the measurements and used in the Euler computations. In addition, pitot pressure surveys were taken in the vicinity of separation, which were used to compare the experimental and computed vortex core locations. All the computations which follow were performed on an 89 x 89 cross sectional grid with residuals reduced at least five orders of magnitude.

The first case considered is the 5° cone at M_∞ = 1.79 and α = 12.65°. The crossflow is subsonic in this case and primary separation occured at θ = 132°. Figure 24 shows the computed crossflow streamline pattern in the vicinity of separation. The vortex is clearly shown above the cone surface in Fig. 24. The dashed line leaving the surface tangentially is the vortex sheet. The dashed line off the surface is the crossflow streamline which stagnates at a saddle in the lee plane and partitions the flow which goes into a node in the lee plane (not shown in Fig. 24) from that which goes into the spiral node at the center of the vortex. The vortex center location compares reasonably well with that found experimentally. The experiment shows the location at θ = 165° and h = 0.2 (h is the radial distance from the surface normalized by the cone radius) and the computation predicts the vortex location at θ = 162° and h = 0.18. All the streamlines are well behaved including the separating streamline and those which spiral into the vortex. The power of the Euler equations to capture the flow features once vorticity is shed from the surface is clear from Fig. 24. Potential methods, linearized or fully nonlinear, require the inclusion of discrete vortices to model the sheet. This requirement usually precludes a description of the entire sheet, only a portion of the sheet is computed with the remainder being lumped into a single vortex. The only special treatment in the present work is that of shedding vorticity from one point on the surface as described in the last section.

Figure 25 shows the streamlines for the same case with both primary and secondary separation included. Secondary separation was imposed at θ = 156° (from the data of Ref. 22) and is indicated by the lifting off of the surface streamline in the reverse flow region. The secondary vortex was too small to resolve with the grid used. Figure 26 shows the surface crossflow velocity distribution for this case, the reversals in velocity at the two separation points can be seen clearly. The inclusion of secondary separation affects the flow behavior substantially, in particular the location of the primary vortex and the surface pressure distribution. Unfortunately, it moves the vortex center to $\theta = 165°$ and $h = 0.14$. With secondary separation included the computed radial location of the vortex center is further from the experimental data than with primary separation alone. It is obvious from the pitot pressure survey of Ref. 22 that secondary separation occurs in a region which has stronger viscous effects than primary separation. It may be that the prediction of flows with secondary separation with a purely inviscid model is impossible. The possibility that the computational difficulties in computing the flow in the region just after separation is affecting these results also exists.

The surface crossflow velocity distribution shown in Fig. 26 should be compared with that of Fig. 5 which only included shock vorticity. Figure 26 shows a discontinuity at primary separation from $u \approx .6$ to $u = 0$ and at secondary separation from $u \approx -.5$ to $u = 0$. Of course, it is these jumps in velocity which determine the vorticity which is shed from the separation points. In Fig. 5 the velocity passes through zero smoothly so that no vorticity is shed from the surface and only shock vorticity exists.

Figure 27 shows the surface pressure distribution computed assuming no separation, only primary separation and both primary and secondary separation also included are the experimental results. Primary separation occurs in the middle of the adverse pressure gradient of the attached flow ($\theta = 132°$). Primary separation forces the flow to compress more rapidly upstream of separation which is consistent with the findings of Ref. 6 and the experimental data. In addition a reverse flow region is developed behind the separation point. This reverse flow expands from the lee plane to a local pressure minimum just under the vortex ($\theta \approx 162°$) and recompresses to somewhat of a plateau (beginning at $\theta \approx 150°$). This region exhibits pressure variation because of the difficulties already mentioned. It is the recompression between $\theta = 162°$ and $\theta = 150°$ which causes secondary separation. With secondary separation included the expansion/recompression of the reverse flow is reduced and the computed results in this region approach the experimental data. The pressure plateau between primary separation ($\theta = 132°$) and $\theta = 150°$ becomes flatter and its level is very close to the experimental data. The inclusion of secondary separation moves the pressure distribution before primary separation away from the experimental data. This is due to the flattening of the vortex sheet discussed previously. Some of the differences between the experimental data and the computed results are surely due to viscous effects (i.e., boundary layer thickening before separation). This is true for both primary and secondary separation. It should be pointed out that the experimental separation is not at the beginning of the plateau in the experimental pressure. This is not typical of purely inviscid separation indicating a significant boundary layer thickening in this case.

The next case considered was at a lower angle of attack 10.6° (5° cone at $M_\infty = 1.79$). The surface pressures are shown in Fig. 28. The separations are at $\theta = 139°$ and 157°. The behavior is similar to that shown in Fig. 27. In this case the inclusion of secondary separation brings the surface pressure compression upstream of separation back to essentially the attached results. The reason for this becomes obvious after a comparison of crossflow streamlines with and without secondary separation (Fig. 29 & 30). The pressure plateau in this case is very flat. A large discrepancy between calculation and experimental data in the reverse flow region is eliminated with the inclusion of secondary separation. While the inclusion of secondary separation does bring the calculated results closer to the experimental data in the reserve crossflow region, it seems obvious that viscous effects are important there; the experimental data show this. In the subsonic crossflow cases considered thus far, any viscous effects which modify the vortex core location can effect the global solution. The discrepancies between calculated results and experimental data before primary separation are totally eliminated when supersonic crossflow is considered.

The last case to be considered involves supercritical crossflow. It is the high speed ($M_\infty = 4.25$) flow over the same 5° cone at $\alpha = 12.35°$. The streamlines for this case are very interesting. Figure 31 shows the streamlines with primary separation forced at $\theta = 120°$ and no secondary separation forced. The vortex very close to the body is a result of a shock induced separation at $\theta \approx 151°$. With only primary separation specified (i.e., shedding vorticity) the reverse crossflow becomes supersonic causing a reverse crossflow shock. This shock can be seen from the isobars of Fig. 32 ($\theta \approx 154°$). It is the vorticity generated by this shock which causes the secondary separation in Fig. 31. An oblique crossflow shock is apparent from the isobars of Fig. 32 at the primary separation point. This is due to the fact that the sheet comes off at an angle relative to the body as discussed previously. The dashed line off the surface at $\theta = 90°$ (Fig. 31) wraps around the vortex and stagnates at a saddle on the surface. This streamline partitions the flow which goes into the main vortex node from that which spirals into the secondary vortex node.

Figure 33 shows the streamlines for this case with both primary and secondary ($\theta = 160°$) separations specified. The figure shows a third vortex near the primary vortex sheet. The streamlines passing over the secondary vortex pinch together to form a saddle just above the secondary vortex. The existence of this saddle implies the formation of an additional node which is the spiral node close to the vortex sheet just after separation. The shock system in this case is quite complex and can be deduced from the isobars of Fig. 34. An oblique crossflow shock can be seen leaving the surface at the primary separation point, this shock becomes normal to the flow off the body. Visualization of this flow is aided by a look at the crossflow sonic lines (Fig. 35). The sonic line leaving the surface at the separation point coincides with the vortex sheet. The normal portion of the oblique shock formed at the primary separation points can to seem as the supersonic to subsonic transition off the body just after separation ($\theta \approx 132°$). There is another transition further downstream $\theta \approx 146°$) , a corresponding shock can be seen in the isobars of Fig. 34. It seems that the flow then re-expands to supercritical, and this region is terminated by yet another shock ($\theta \approx 165°$). As usual the reverse flow expands as it moves from the lee plane. The reverse crossflow becomes supercritical just beneath the primary vortex ($\theta \approx 165°$) . There is then the possibility of an oblique crossflow shock at the secondary separation point. However, the reverse crossflow Mach number is too low to make the deflection required by secondary separation. A detached normal shock can be seen before the secondary separation point in both the isobars (Fig. 34) and the sonic line (Fig. 35).

The comparison with the experimental data in this case is much better than those of the two subsonic cases studied previously. The primary vortex core location is computed to be $\theta = 164°$ and h = 0.22 which compares very well with Rainbird's data $\theta = 165°$ and h = 0.23. More importantly the surface pressure distributions (Fig. 36) compare very well. Figure 36 shows no upstream influence before the primary

separation point, a comparison with the attached flow (no forced
separation) of Fig. 12 shows no change before $\theta = 120°$. Of course,
this is not surprising for the inviscid calculation since the
crossflow is supersonic. The surprising result is that the
experimental data shows no influence of boundary layer thickening.
The pressure rise due to the shock at primary separation compares very
well with data, note how sharply the experimental pressure raises. In
the region of secondary separation the comparison is somewhat poor.
It is becoming quite clear that the flow in this region is influenced
significantly by viscous effects. The inclusion of secondary
separation moves the compression from $\theta \approx 152°$ to $\theta = 160°$ but the
supersonic reverse crossflow is not eliminated. It should be pointed
out that the expansion/recompression just after the secondary
separation ($\theta \approx 150°$) is consistent with the magnitude of the
secondary vortex.

The procedures used to shed vorticity from the surface of a
circular cone were applied to the flow about an elliptic delta wing
tested by Squire[24]. The grid used had 89 x 89 points in cross section
and the residual was reduced five orders of magnitude. While Squire
didn't determine precisely the primary separation point location, his
experimental data indicated leading edge separation for the ellipse
($a/b \approx 14/1$) at $M_\infty = 2$ and $\alpha = 10°$. The computed attached flow
surface pressure is shown in Fig. 18. There is a small vortex
($X/X_{LE} \approx .3$) due solely to shock vorticity in the result of Fig.
18. Figure 18 shows a rapid expansion around the wing leading edge
followed by a weak recompression before the shock. In the present
study it was found that no solution could be obtained with vorticity
being shed in the region of favorable pressure gradient near the wing
leading edge. Once the weak recompression was reached, separation
could be forced to occur. Figure 37 shows the computed surface
pressure distribution on the elliptic delta wing with separation
forced at $X/X_{LE} = .99$. (This is as close to the leading edge as
possible that separation could be forced.) Figure 37 shows that the
flow expands around the leading edge before the specified separation
point. In fact this separation is of the supercritical type discussed
previously. An oblique shock causes the recompression just after the
leading edge compression. Figure 37 also shows the experimental data
of Squire[24]. The comparison is good except for the pressure plateau
between $X/X_{LE} = .6$ and $.9$, where secondary separation was detached
experimentally. Figures 38 and 39 show the cross-sectional streamlines and isobars, respectively. A comparison of Fig. 37 and 38 shows

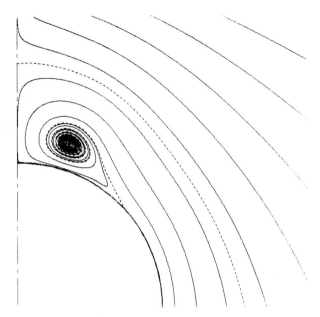

Fig. 24   Cross Flow Streamlines, Primary Separation
($M_\infty = 1.79$, $\delta = 5°$, $\alpha = 12.65°$)

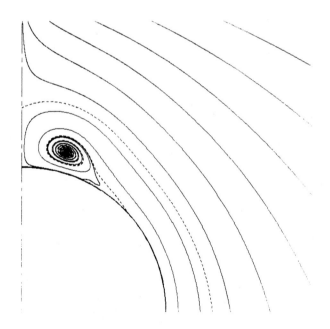

Fig. 25   Cross Flow Streamlines, Primary & Secondary
Separations ($M_\infty = 1.79$, $\delta = 5°$, $\alpha = 12.65°$)

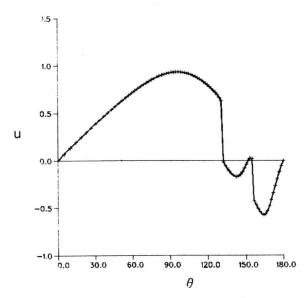

Fig. 26   Surface Crossflow Velocity Distribution,
Primary and Secondary Separations
($M_\infty = 1.79$, $\delta = 5°$, $\alpha = 12.65°$)

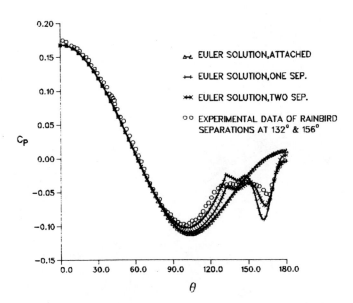

Fig. 27   Surface Pressure Comparison ($M_\infty = 1.79$,
$\delta = 5°$, $\alpha = 12.65°$)

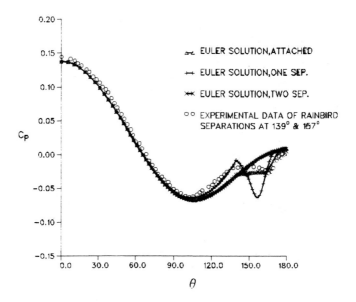

Fig. 28   Surface Pressure Comparison ($M_\infty = 1.79$, $\delta = 5°$, $\alpha = 10.6°$)

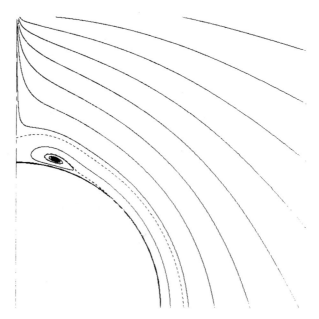

Fig. 29   Crossflow Streamlines, Primary & Secondary Separations ($M_\infty = 1.79$, $\delta = 5°$, $\alpha = 10.6°$)

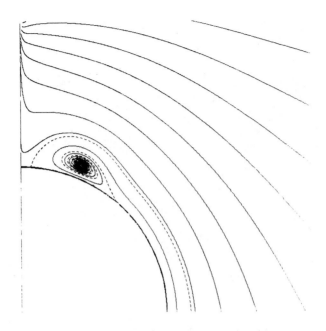

Fig. 30   Crossflow Streamlines, Primary Separation
($M_\infty = 1.79$, $\delta = 5°$, $\alpha = 10.6°$)

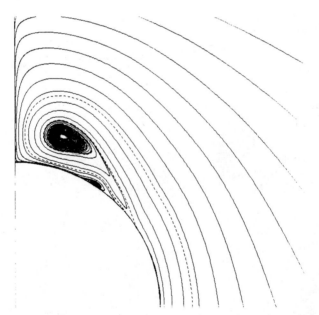

Fig. 31   Crossflow Streamlines, Forced Primary
Separation Shock Induced Secondary Separation
($M_\infty = 4.25$, $\delta = 5°$, $\alpha = 12.35°$)

350

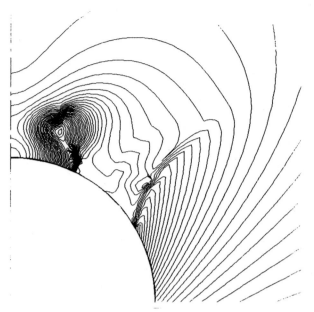

Fig. 32   Isobars, Forced Primary Separation, Shock
Induced Secondary Separation
$(M_\infty = 4.25, \quad \delta = 5°, \quad \alpha = 12.35°)$

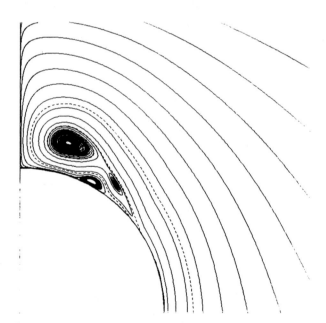

Fig. 33   Crossflow Streamlines, Forced Primary and
Secondary Separations $(M_\infty = 4.25,$
$\delta = 5°, \quad \alpha = 12.35°)$

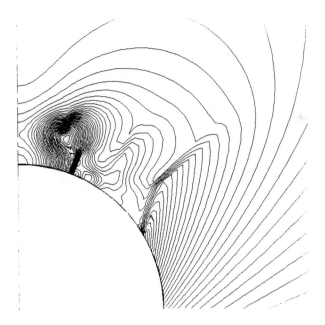

Fig. 34   Isobars, Forced Primary and Secondary
           Separations ($M_\infty = 4.25$, $\delta = 5°$, $\alpha = 12.35°$)

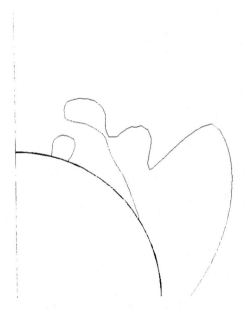

Fig. 35   Crossflow Sonic Lines, Forced Primary
           and Secondary Separations ($M_\infty = 4.25$,
           $\delta = 5°$, $\alpha = 12.35°$)

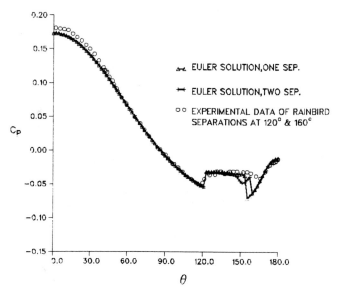

Fig. 36   Surface Pressure Comparison
($M_\infty = 4.25$, $\delta = 5°$, $\alpha = 12.35°$)

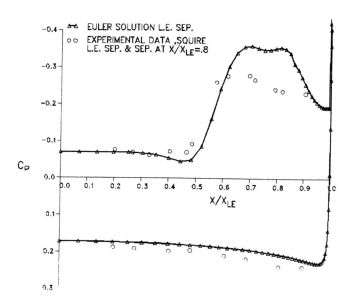

Fig. 37   Surface Pressure Comparison ($M_\infty = 2$, $\alpha = 10°$,
14:1 Ellipse, $\delta = 1.5°$, $\lambda = 70°$) Separation
Forced at $X/X_{LE} = .99$

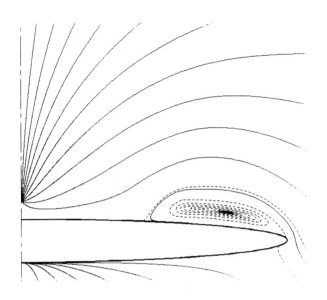

Fig. 38   Crossflow Streamlines ($M_\infty = 2$, $\alpha = 10°$, 14:1 Ellipse, $\delta = 1.5°$, $\lambda = 70°$) Separation Forced at $X/X_{LE} = .99$

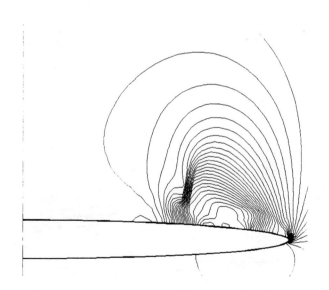

Fig. 39   Isobars ($M_\infty = 2$, $\alpha = 10°$, 14:1 Ellipse, $\delta = 1.5°$, $\lambda = 70°$) Separation Forced at $X/X_{LE} = .99$

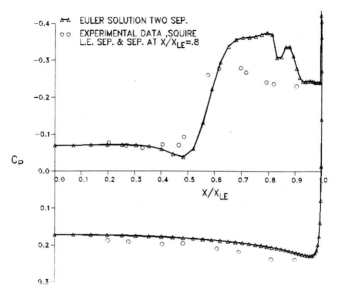

Fig. 40   Surface Pressure Comparison ($M_\infty = 2$, $\alpha = 10°$, 14:1 Ellipse, $\delta = 1.5°$, $\lambda = 70°$) Separation Forced at $X/X_{LE} = .99$ and $X/X_{LE} = .8$

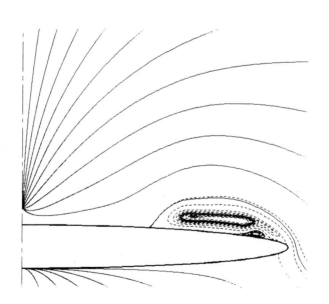

Fig. 41   Crossflow Streamlines ($M_\infty = 2$, $\alpha = 10°$, 14:1 Ellipse, $\delta = 1.5°$, $\lambda = 70°$) Separation Forced at $X/X_{LE} = .99$ and $X/X_{LE} = .8$

that the apparent pressure raise at X/X_{LE} = .55 is in the reverse crossflow region, so that it is in reality an expansion on the wing surface. A comparison of Fig. 38 and 39 shows a crossflow shock located above the reverse flow region at X/X_{LE} ≈ .6 . It is this shock's interaction with the reverse crossflow that causes the expansion at X/X_{LE} ≈ .55. The experimental results indicate secondary separation at X/X_{LE} = .8. Figure 40 shows the computed surface pressure distribution with both primary and secondary separation specified (secondary separation is specified at the experimental location). The inclusion of secondary separation brings the pressure level just after primary separation (X/X_{LE} = .9 to X/X_{LE} = 1.) to the experimental value. The reverse cross is supersonic after the expansion at X/X_{LE} = .55, so that the inclusion of secondary separation cannot affect the pressure level between X/X_{LE} = .6 and .8 and so there is a significant difference between the computed and experimental pressure level in this region. There is shocking of the reverse crossflow at the secondary separation point X/X_{LE} = .8 followed by a local minimum in pressure (X/X_{LE} ≈ .85) under the core of the secondary vortex. The streamline patter is shown in Fig. 41.

The results of the investigation of the delta wing flow field are the same as those for the circular cone. In particular, it was found that secondary separation is dominated by viscous effects, and an inviscid model (as the one used here) will have difficulties predicting details of the flow in this region. The region of the flow between X/X_{LE} = .55 and .8 in Fig. 39 is similar to the reverse crossflow region θ ≈ 170° to 140° in Fig. 37 (supercritical cone). In the case of the ellipse this region is longer because the vortex is elongated.

Comparison of Shock Vorticity and Shed Vorticity

In an effort to gain a better understanding of the relationship between shock vorticity and vorticity shed from the surface of a smooth body, the flow about the 5° cone tested by Rainbird[22] was computed with a number of different separation point locations specified. The case considered was M_∞ = 4.25 and α = 12.35°. As indicated previously, this flow is supercritical and, with no vorticity shed from the cone surface, the crossflow shock produces enough vorticity to cause separation. Figures 42 and 43 show the crossflow streamlines and isobars, respectively, for this flow with no vorticity shed from the body. Separation for this case is computed to be at $\theta = 151.3°$. Figure 42 indicates that the separating streamlines leave the surface at a large angle (57°) relative to it. As shown earlier in this paper (Fig. 5), when only shock vorticity is present there is no jump in crossflow velocity at the separation point, which is consistent with the fact that no vorticity is being shed from the surface. It should be pointed out that in the computational results that follow, all crossflow shocks are captured. Figure 43 indicates that the shock is captured very sharply (see the closely spaced isobars). Additionally, these captured shock results compare very well with the shock fit results for this case (Fig. 12).

Figures 44 and 45 show results for the other extremes of separation point location studied ($\theta = 115°$). Figure 44 shows the crossflow streamlines. The secondary separation shown is due to a strong reverse crossflow shock (see Fig. 45), and the third vortex off the surface is similar to the one discussed previously. In Fig. 45, the isobars are shown, and they indicate an oblique shock at the specified primary separation point. The jump in velocity at the separation point is significant with separation specified at $\theta = 115°$, indicating significant vorticity being shed from the surface. A comparison of Fig. 42 and 44 shows that the extent of the vortical regions are comparable, while the two sources of vorticity are very different.

The relationship between shock vorticity and shed vorticity is made clearer by considering Fig. 46. The figure shows the jump in crossflow velocity vs separation point location. The jump in crossflow velocity is directly related to the vorticity shed into the flow field from the separation point. The shock configuration transition from an oblique crossflow shock to a detached normal crossflow shock occurs at about $\theta_s \approx 128°$ (indicated by the shaded area in Fig. 46). It should be pointed out that the jump in velocity at the separation point in the oblique shock cases was computed by subtracting the oblique shock velocity jump from the numerical results. Thus, the jumps in velocity in Fig. 46 represent the jumps across the vortex sheet at separation. The figure shows that this velocity jump goes to zero smoothly as the separation point location due to shock vorticity alone is approached ($\theta_s = 151.3°$). This indicates that separation due to shock vorticity alone and that due to shed vorticity are related. In fact, it would seem that separation due to shock vorticity alone is a particular solution of the set of solutions in which vorticity is shed from the surface. In this particular solution the value of the vorticity shed is zero.

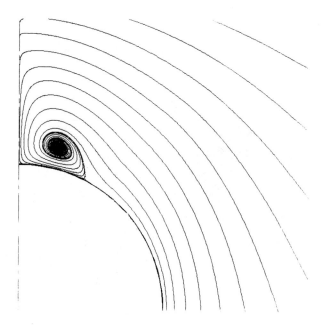

Fig. 42   Crossflow Streamlines on the 5° Circular
          Cone (M_∞ = 4.25, α = 12.35°) Separation Due
          to Shock Vorticity Alone at θ = 151.3°

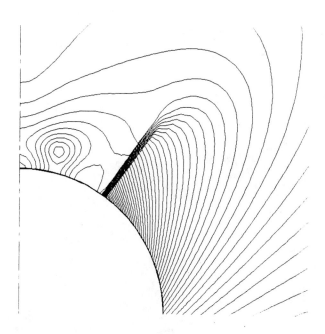

Fig. 43   Isobars on the 5° Circular Cone (M_∞ = 4.25,
          α = 12.35°) Separation due to Shock Vorticity
          Alone at θ = 151.3°

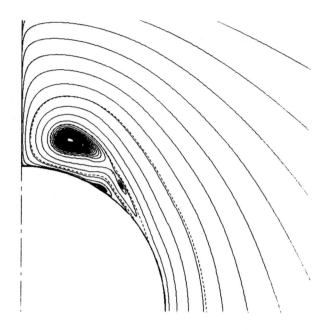

Fig. 44   Crossflow Streamlines on the 5° Circular Cone
(M$_\infty$ = 4.25, α = 12.35°) Separation Forced
at θ = 115°

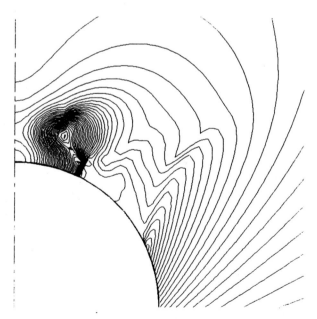

Fig. 45   Isobars on the 5° Circular Cone (M$_\infty$ = 4.25,
α = 12.35°) Separation Forced at θ = 115°

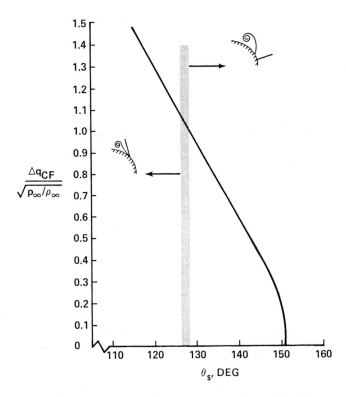

Fig. 46   Vorticity Shed into the Flow Field as a
Function of Separation Point Location

Summary of Findings

The research effort in this area is ongoing, and while some of the results presented here should be considered preliminary, a number of conclusions have been reached:

o   The vorticity produced by the shock system in supersonic conical flow can cause separation on its own and may add significantly to the vorticity shed from a separating boundary layer.

o   As the shock system becomes weak and approaches a potential (i.e., irrotational) shock, its vorticity no longer causes separation (Fig. 8).

o   The reverse crossflow can become supersonic beneath a vortex core and a reverse crossflow shock may form (Fig. 10 & 11). This shock can cause secondary separation on its own (Fig. 31).

o   Increasing eccentricity on elliptic cross sections has a tendency to reduce shock entropy gradients and thus vorticity. Yet, the separation caused by shock vorticity can have a significant impact on the flow field (Fig. 22).

o   The artificial damping required to stabilize captured shocks does not necessarily significantly distort shock vorticity (Fig. 18).

o   Both primary and secondary separation can be forced at specified locations by shedding vorticity from a smooth surface. With the vorticity shedding model of Smith[5], the basic features of the separated flow can be reproduced.

o   Euler calculations including vorticity shedding are more accurate and in some sense simpler than those using irrotational flow models.

o   There are computational difficulties in a small region after separation that may significantly affect the global results.

o   In the case of supercritical crossflow, the vortex sheet leaves the surface at an angle relative to it causing an oblique crossflow shock (Fig. 31 to 35).

o   The viscous effects (boundary layer thickening) upstream of separation are much more significant in the case of subsonic crossflow than in the case of supercritical crossflow (Fig. 27 & 36).

o   Secondary separation is influenced more by viscous effects than primary separation.

o   The vorticity shed into the flow field is reduced smoothly as the separation point is moved to its shock induced location (Fig. 46).

References

1. Küchemann , D. "Inviscid Shear Flow Near the Trailing Edge of an Aerofoil," R.A.E. TR 67068 (1967).

2. Fraenkel, L.E., "On Corner Eddies in Plane Inviscid Shear Flow," J Fluid Mechanics, Vol 11, pp 400-406 (1961).

3. Smith, P.D., "A Note on the Computation of the Inviscid Rotational Flow Past the Trailing Edge of an Airfoil," R.A.E. TM 1217 (1970).

4. Smith, J.H.B., "Remarks on the Structure of Conical Flow," Progress in Aeronautical Sciences, Vol 12, pp 241-271 (1972).

5. Smith, J.H.B., "Behavior of a Vortex Sheet Separating from a Smooth Surface," R.A.E. TR 77058 (1977).

6. Fiddes, S.F., "A Theory of the Separated Flow Past a Slender Cone at Incidence," AGARD CP 291, pp 30-1 - 30-14 (1981).

7. Marconi, F., "The Spiral Singularity in the Supersonic Inviscid Flow Over a Cone," AIAA Paper 83-1665 (1983).

8. Salas, M.D., "Recent Developments in Transonic Euler Flow Over a Circular Cylinder," NASA TM 83282 (1982).

9. Rizzi, A., Erickson, L., Schmidt, W. and Hitzel, S., "Numerical Solutions of the Euler Equations Simulating Vortex Flows Around Wings," AGARD CP-342, pp 21-1 - 21-14 (1983).

10. Murman, E., "Solutions of the Conical Euler Equations for Flat Plate Geometries - Preliminary Results," MIT CFDL-TR-84-4 (1984).

11. Newsome, R., "A Comparison of Euler and Navier-Stokes Solutions for Supersonic Flow Over a Conical Delta Wing," AIAA Paper 85-0111 (1985).

12. Marconi, F., "Supersonic Conical Separation Due to Shock Vorticity," AIAA J, Vol 22, No. 8, pp 1048-1055 (1984).

13. Marconi, F., "Shock Induced Vortices on Elliptic Cones in Supersonic Flow," AIAA Paper No. 85-0433 (1985).

14. Marconi, F., "Supersonic Inviscid, Conical Corner Flow Fields," AIAA J, Vol 18, No. 1, pp 78-84 (1980).

15. Moretti, G., "The λ-Scheme," Computers and Fields, Vol 7, pp 191-205 (1979).

16. Marconi, F., Salas, M.D. and Yaeger, L., "Development of a Computer Code for Calculating the Steady Super/Hypersonic Inviscid Flow Around Real Configurations," NASA CR-2675 (1976).

17. Rudman, S., "Multinozzle Plume Flow Fields - Structure and Numerical Calculations," AIAA Paper 77-710 (1977).

18. deNeef, T. and Moretti, G., "Shock Fitting for Everybody," Computers and Fluids, Vol 8, pp 327-334 (1980).

19. Vorropoulos, G. and Wendth, J. F., "Laser Velocimetry Study of Compressi-bility Effects on the Flow Field of a Delta Wing," AGARD CPP-342 (1983).

20. Nielsen, J.N., Kuhn, G.D., and Klopfer, G.H., "Euler Solutions of Supersonic Wing-Body Interference at High Incidence Including Vortex Effects," NEAR TR 263 (1982).

21. Siclari, M. and Visich, M., "Shock Fitting in Conical Supersonic Full Potential Flows with Entropy Effects," AIAA Paper 84-0261 (1984).

22. Rainbird, W.J., "The External Flow Field About Yawed Circular Cones," AGARD CP30 (1968).

23. Siclari, M.J., "Supersonic Nonlinear Potential Flow with Implicit Isentropic Shock Fitting," AIAA J, Vol 20, No. 7, p 924 (1982).

24. Squire, L.C., "Leading-Edge Separations and Cross-Flow Shocks on Delta Wings," *AIAA J*, Vol 23, No. 3, p 321 (1985).